Design for Humanity

The book is a comprehensive guide for anyone who wants to design with purpose, empathy, and impact. Rather than rushing to solutions, the book champions the art of framing and refining a problem, revealing how powerful design begins with asking smarter questions. It takes readers on a journey from uncovering user needs to creating and validating meaningful solutions. Built around the new GRIP (gather, refine, ideate, produce) framework, the book emphasizes designing with people, not just for them. It introduces themes like design for all, empathy, creativity and ideation, sustainability, and ethics while offering hands-on tools such as storytelling, persona building, advanced prototyping techniques, including augmented reality/virtual reality (AR/VR) testing.

From shaping lovable products to aligning designs with market needs, the book combines the artistry of empathy with the rigor of engineering.

The book is ideal for students of design, engineering, and innovation as well as educators, entrepreneurs, and early-career professionals. It covers different methods and is intended to spark a mindset shift. Readers will be equipped to build better products to shape experiences that are accessible and delightful to use. This is more than a book on design; it is a roadmap for creating solutions that matter.

Ergonomics Design and Management: Theory and Applications

Series Editor

Waldemar Karwowski

Industrial Engineering and Management Systems University of Central Florida (UCF) – Orlando, Florida

Published Titles

Knowledge Service Engineering Handbook
Jussi Kantola and Waldemar Karwowski

Neuroadaptive Systems
Theory and Applications
Magdalena Fafrowicz, Tadeusz Marek, Waldemar Karwowski, and Dylan Schmorrow

Safety Management in a Competitive Business Environment
Juraj Sinay

Ergonomics in Developing Regions
Needs and Applications
Patricia A. Scott

Human Factors of a Global Society
A System of Systems Perspective
Tadeusz Marek, Waldemar Karwowski, Marek Frankowicz, Jussi I. Kantola, and Pavel Zgaga

Working Posture Assessment
The TACOS (Time-Based Assessment Computerized Strategy) Method
Daniela Colombini and Enrico Occhipinti

ERGOCHECK for a Preliminary Mapping of Risk at Work
Tools, Guidelines, and Applications
Daniela Colombini and Enrico Occhipinti

Design for Humanity
Creating Products That Matter
Ravindra S. Goonetilleke, Ameersing Luximon, Emily Y. L. Au, Yan Luximon, and Alma Maria Jennifer Gutierrez

For more information about this series, please visit: www.crcpress.com/Ergonomics-Design--Mgmt-Theory--Applications/book-series/CRCEDMTAPPS

Design for Humanity

Creating Products That Matter

Ravindra S. Goonetilleke, Ameersing Luximon,
Emily Y. L. Au, Yan Luximon, and
Alma Maria Jennifer Gutierrez

CRC Press
Taylor & Francis Group
Boca Raton London New York

CRC Press is an imprint of the
Taylor & Francis Group, an **informa** business

Designed cover image: Getty Images

First edition published 2026
by CRC Press
2385 NW Executive Center Drive, Suite 320, Boca Raton FL 33431

and by CRC Press
4 Park Square, Milton Park, Abingdon, Oxon, OX14 4RN

CRC Press is an imprint of Taylor & Francis Group, LLC

ISBN: 9781032738925 (hbk)
ISBN: 9781032759166 (pbk)
ISBN: 9781003476177 (ebk)

DOI: 10.1201/9781003476177

Typeset in Warnock Pro
by Newgen Publishing UK

Contents

Preface

We were a bunch of design enthusiasts emailing each other, trying to birth a design book. At first, we were all gung-ho, ready to conquer the design world. But then reality hit. Do we need another design book when there are already so many out there? To make things even more challenging, we couldn't agree on just one book. It was like trying to pick a single favorite snack for all when each of us had different tastes and preferences. Rather than assembling a random assortment, we have carefully curated a seamless fusion, integrating the best elements from the latest literature to create a cohesive and enriching experience.

Before you roll your eyes and think, "Great, another book to gather dust on my bookshelf", hold your bookmarks! This is not your run-of-the-mill design manifesto. It is more of a journey of fun through the world of design. Who is it for? Well, if you are curious about design, you are in. For undergraduate and first-year graduate students specializing in design and development, it provides a structured yet flexible approach to innovation, equipping them with the tools, methodologies, and mindset needed to develop impactful solutions. Whether you are looking to refine your design thinking skills, explore market-driven strategies, or bridge the gap between concept and commercialization, this book serves as a comprehensive guide to navigating the complexities of modern design and innovation. What is inside? The secrets of design, creativity hacks, and a guide to navigating the sometimes wacky world of designing for people. All theories with no application can be dull, while all applications with no theory can make it harder for those with a logical mindset to grasp concepts easily. We have created the ideal mix. At times, we have focused more on theory; at other times, we have emphasized application to ensure an engaging learning experience. True innovation begins with empathy and understanding; this book reveals why mastering design for humanity is the first step toward creating solutions that genuinely matter. Grab a comfortable seat, if you know what that means, and enjoy the ride. The book is not just another item on your shelf (if you have one these days); it is here to be your design buddy, creative pal, and go-to for learning.

While the term "designer" may carry a certain weight, it is essential to acknowledge the diverse nature of design disciplines. Consider a scenario where a group of designers, each specializing in machines, electronics, buildings, fashion, airplanes, or industrial design, is brought together in a room and tasked with creating something groundbreaking for their business-minded boss. While their combined expertise may be impressive, the sheer diversity of their backgrounds and approaches could lead to misalignment, making it challenging to achieve a truly cohesive and market-ready innovation. Each "designer" operates within a unique framework, facing distinct challenges and employing resources specific to their field. Despite the apparent distinctions in these design branches, a unifying thread exists. At its core, design is the process of solving problems within defined constraints, using available resources to achieve specific goals. What varies across disciplines is how the problem is framed: for some, design is about creating a functional artifact to perform a task; for others, it is about achieving aesthetic excellence. How the differing needs and experiences of users are identified and addressed across disciplines is a central question explored in depth throughout this book. It presents core principles and problem-solving methodologies that are broadly applicable across design fields. The journey does not end there. In the later chapters, we introduce methodologies and tools designed to help you tackle critical challenges and contribute to making the world a better place. Finally, we explore strategies for bringing unique, innovative products to market.

While the book follows a logical progression from idea to market, you do not need to read it in order. The chapters are highly modular, allowing you to start wherever you find the most value as long as you have a general understanding of the key concepts. We hope this book empowers you to address pressing issues affecting you and society, ultimately driving meaningful change. As this is the first edition, we are eager to refine and expand upon it with your help. Your feedback, insights, and suggestions are extremely valuable.

About the Authors

Emily Y.L. Au is an Operations Director at a fabless semiconductor company and a part-time lecturer at UC Berkeley's Jacobs Institute for Design Innovation, where she has taught since 2017. From 2011 to 2015, she was an Assistant Professor of Engineering Education at the Hong Kong University of Science and Technology (HKUST). With over 15 years of experience in operations management and human-centered design, she leads enterprise-wide process optimization to improve efficiency, strengthen collaboration, and expand knowledge access. Across both industry and academia, her work focuses on user-centered design, bridging theory and practice in technology, education, and product development.

Ravindra S. Goonetilleke is a professor in the Department of Management Science and Engineering. Prior to joining Khalifa University in the United Arab Emirates (UAE), Dr. Goonetilleke was a professor and held a joint appointment in the Division of Integrative Systems and Design and the Department of Industrial Engineering and Decision Analytics at the Hong Kong University of Science and Technology. His research focuses on ergonomics and human factors engineering and their applications on product design and human–computer interaction. Dr. Goonetilleke is recognized as a leading researcher on the ergonomics of footwear. He is a fellow of the International Ergonomics Association (IEA), Human Factors and Ergonomics Society and the Hong Kong Ergonomics Society. He is a senior member of the Institute of Electrical and Electronics Engineers (IEEE).

Alma Maria Jennifer Gutierrez is the first Filipino to be awarded the International Ergonomics Association (IEA) Fellow and the South East Asian Network of Ergonomics Societies (SEANES) Educator in 2018 in Thailand. She is a Full Professor at the De La Salle University Industrial & Systems Engineering Department. She also help leadership position as former Department Chairperson and Vice Chairperson. Aside from this, she is the conference chair of the Asian Network of Ergonomics Societies (ACED-SEANES, 2020). She is the lead Editor of Convergence of Ergonomics and Design Proceedings in Advances in Intelligent Systems and Computing. She conducts consulting projects on Safety, Productivity Enhancement through Ergonomics Development (SPEED) program for the government and private sector.

Ameersing Luximon currently serves as a Professor of Practice at the Georgia Tech Shenzhen Institute, Tianjin University in Shenzhen, China, while also holding the position of Adjunct Associate Professor at the Georgia Institute of Technology in Atlanta, USA. He is the Vice President and a former President of the Hong Kong Ergonomics Society (HKES) and serves as a council member for both the International Ergonomics Association (IEA) and the Asian Council on Ergonomics and Design (ACED). Previously, Prof. Luximon was an Associate Professor at The Hong Kong Polytechnic University. He is one of the inventors of i.Dummy, a robotic mannequin, and has edited the Handbook of Footwear Design and Manufacture (1st Edition, 2013; 2nd Edition, 2021). With over a decade of experience in academia and another decade in industry, Prof. Luximon earned his PhD from the Hong Kong University of Science and Technology. His research interests include ergonomics in design, AI in design, Smart design, wearables, social robots, digital human modeling, and entrepreneurship.

Yan Luximon is a full professor and the Associate Dean (Research) in the School of Design at The Hong Kong Polytechnic University. She also serves as Chair of the School Research Committee and as Lab Leader for the Asian Ergonomics Design Lab. In addition, she is the President of the Hong Kong Ergonomics Society (HKES). She has published over 150 peer-reviewed journal articles, book chapters, patents, and international conference papers. Her research interests include 3D digital human modeling and CAD, computer graphics, AI in design and visualization, deep learning, ergonomic design, head- and face-related products, human-computer interaction, virtual reality, and cultural differences.

The Wild Terrain of Design

1

> Humanity... lies in man's capacity to question the known and imagine the unknown.
>
> —Margaret Mead

1.1 EVOLUTION OF DESIGN

Ask five people what design means, and you will likely receive five different, sometimes conflicting, answers. This variation arises because design is interpreted differently across disciplines, professions, and cultures, each framing it according to their goals, methods, and values (Buchanan, 1992; Cross, 2001). For some, design is about solving problems; for others, it is about expressing ideas, shaping experiences, or bringing visions to life (Lawson, 2006; Nelson & Stolterman, 2012). An architect sees blueprints, a chef imagines plating, and a game designer thinks in pixels and plays. Design is one of those words that wears many hats. At its core, a design is a carefully made idea or plan for creating a physical object, a process, or a system. It is rarely a random outcome. More often, it results from someone making deliberate choices to achieve a goal or solve a problem.

The term design can refer to a finished product, like a sketch or a model (a noun), and to the creative process (a verb) that led to it (Cross, 2011). However, not all design work begins with a clear plan. Design can also emerge through action and reflection, even if there was no formal plan at the start (Simon, 1996). Consider areas like arts and crafts. People often begin making with only a rough idea in mind. Along the way, they make a series of small decisions that shape the outcome. Even without a detailed blueprint, the result still reflects thoughtful choices, which is also considered Design (Dormer, 1997).

In 1959, when architects, designers, and design teachers were asked to name the 100 best-designed products, their

DOI: 10.1201/9781003476177-1

choices were as diverse as their opinions (Bentley, 2020). Some items earned their spot because they looked good, while others set new trends or captured the mood of the time. A few were chosen out of personal memories or emotional attachment. Then, there were items like the Talon zipper, which seemed to belong more to the world of invention than to the practice of deliberate design (Doblin, 1970). In his book, Doblin (1970) emphasized that the process (the verb) should be more important than the product (the noun). He maintained that human values, such as education, intelligence, creativity, and contributions to society, would eventually emerge from pride in ownership.

Yet, the evolution did not stop there. The design world grew even further, ushering in the era of human-centered design, which put emotions and user experiences at the forefront. This approach aligns with the principles of empathy and inclusivity, which seek to address a broad range of human experiences and challenges. This paradigm shift emphasized the importance of identifying the right problems and embracing an iterative process that involved prototyping, rigorous testing, and continuous refinement to meet human needs effectively (Norman, 2013).

Making a movie is very much like designing a product: in both cases, you build an experience for someone else. Just like a filmmaker imagines how the audience feels during every scene, a product designer imagines how users feel during every interaction. Filmmakers ask, if I tell the story this way, will the audience be thrilled, confused, or bored? (Bordwell & Thompson, 2010). Designers ask, "If I design it this way, will users be delighted, frustrated, or lost?" Both crafts are acts of empathy, stepping into the user's or viewer's shoes, anticipating their reactions, and shaping the experience to create connection, excitement, and satisfaction. Whether it is a movie or a product, the real star is how it makes people feel. The *what-ifs* take center stage!

1.2 PEOPLE-CENTRIC DESIGN

The origins of human performance studies can be traced back to early pioneers like the Gilbreth's and Taylor, who in the early 1900s laid the groundwork for understanding how we work and how well we perform (Gilbreth & Gilbreth, 1924; Taylor, 1911). However, it wasn't long before industries realized that achieving optimal results isn't just about improving humans or technology in isolation. The key to success lies in how well humans, machines, and processes interact as part of a cohesive *system*. In this sense, the key to success is the interaction between all parts, not simply the efficiency of each one. With the advent of ergonomics in design in the 1960s and 1970s, products underwent a transformative shift. They evolved to become not only easier, safer, and more efficient to use, but also marked a departure from mere functionality. This marked the inception of user-friendly design, progressing to user-centered, and eventually people-centered design. This paradigm shift necessitated a deep understanding of the capabilities and limitations of individuals during the design process. The focus was no longer solely on how well a product worked but on how well it worked for the people using it.

This is where human factors professionals or ergonomists step in. They focus on designing systems with human experience at the core, considering how people interact with technology and processes. Effective design requires a multidisciplinary approach, integrating knowledge from fields such as medicine, physiology, engineering, computer science, statistics, and psychology. Understanding how all these elements fit together helps ensure that systems are functional and designed to maximize performance, reduce errors, and provide a seamless user experience (UX).

A good design usually serves a purpose and fits a particular situation. It aims to meet specific goals while respecting any applicable limits or rules. Design also considers how the result will look, how it will work, and how people will interact with it (Papanek, 1985). In 2020, a contemporary survey used the same methodology as Doblin (1970), aiming to identify the best product designs. This time, the participants included educators, independent creators, corporate design teams, and digital influencers. This broadened demographic reflects the evolving definition of who is considered a designer in today's interdisciplinary and digitally connected world. The updated poll revealed a strong consensus on the top 25 products selected for attributes such as adaptability, societal and environmental impact, ease of use, commercial success, and the ability to redefine their respective categories. This shift indicates a move from valuing designs solely for their aesthetic appeal toward appreciating how effectively a product or service fulfills its intended purpose and contributes positively to users and society. This evolution aligns with current trends in product design, where functionality, sustainability, and user experience are paramount. Minimalist design dominates, emphasizing simplicity and clarity to enhance usability. The mid-twentieth-century Scandinavian design movement is rooted in simplicity and minimalism (Fiell & Fiell, 2024). The work of Dieter Rams shows Scandinavian design through its focus on clarity, purpose, and reducing form to essential elements (Lovell, 2011). Lagom, a Swedish concept meaning "just the right amount," shapes the philosophy behind a balanced approach to living and designing that avoids excess and embraces purpose (Brantmark, 2017). The heart of minimalist design is simple but powerful: remove the clutter so that what truly matters can shine through (Norman, 2013). This approach is not just about looking clean or elegant. It is grounded in how our minds work. Since people can only hold a limited amount of information at one time, overly complex designs can confuse users and get in the way of what they are trying to do (Sweller, 1988).

In a world where we are constantly bombarded with information, minimalist products and interfaces offer relief. By cutting out the noise, they make it easier for people to focus, understand, and act. This is not just good design; it is intelligent design. It leads to easier learning, more efficient tasks, and experiences that feel calm and clear rather than overwhelming (Tuch et al., 2012). Consider the wearables we use today. These devices, such as smartwatches and fitness trackers, can capture physiological metrics like heart rhythms, heart rate, and oxygen saturation and even pinpoint events through the Global Positioning System (GPS) data. While they deliver individual data points effectively, their potential goes far beyond this. Unlocking that potential means understanding what users truly want and figuring out how best to present the needed information.

In this data-driven world, visualization is paramount. With sensors capturing vast amounts of data from our surroundings, the challenge lies in making sense of everything. This is where the right visualization tools come into play (Cairo, 2012). A new wave of software, such as Tableau (Tableau.com), Microsoft Excel (www.microsoft.com/en-us/microsoft-365/excel), Datawrapper (academy.datawrapper.de), Google Charts (developers.google.com/chart), Qlik Sense (www.qlik.com), Looker (cloud.google.com/looker), Domo (domo.com), Yellowfin (yellowfinbi.com), Visme (visme.co), has emerged, transforming raw data into compelling narratives and driving informed decision-making through visually appealing graphics. These tools elevate traditional charts, bar, line, pie, and scatter plots, into engaging visual stories. Some even harness AI to create interactive dashboards that adapt to user needs. But with these innovations come the questions: How do we ensure clarity in the data deluge? How can we make visualizations intuitive and impactful? The journey of turning data into actionable insights is just beginning, inviting us to explore, innovate, and refine. Ultimately, great design is not just about what a product does or how it looks. Its value lies in how well it connects with people's needs, desires, and goals and contributes to the greater good.

1.3 HUMAN NEEDS OVER GENERATIONS

While languages, landscapes, and lifestyles may differ globally, basic human needs are universal. They are life's psychological and physiological bedrock (Diener, Inglehart, & Tay, 2013). The Silent Generation (1928–1945) emerged with values of frugality, stability, and material conservation (Gilleard, 2004). Psychologically, their consumption behavior was primarily safety and physiological security. Material goods were not status symbols but practical investments; possessions were expected to last and serve a functional role. The emotional restraint often associated with this generation translates into a low-consumption mindset, where repair over replacement was the norm. Their identity was anchored in collective duty and family roles, not personal expression through products. As a result, Abraham Maslow organized the needs into a hierarchy (Figure 1.1), starting with survival-level needs like food and safety and progressing toward more abstract aspirations like self-actualization (Maslow, 1943). But the real human experience is not that orderly. Recent cross-cultural research reveals that the sequence in which these needs are met does not reliably predict subjective well-being. People can report high levels of life satisfaction even when foundational needs are only partially fulfilled (Ryff, 1989; Tay & Diener, 2011). As we examine past and present products, the factors that have shaped their development become increasingly apparent.

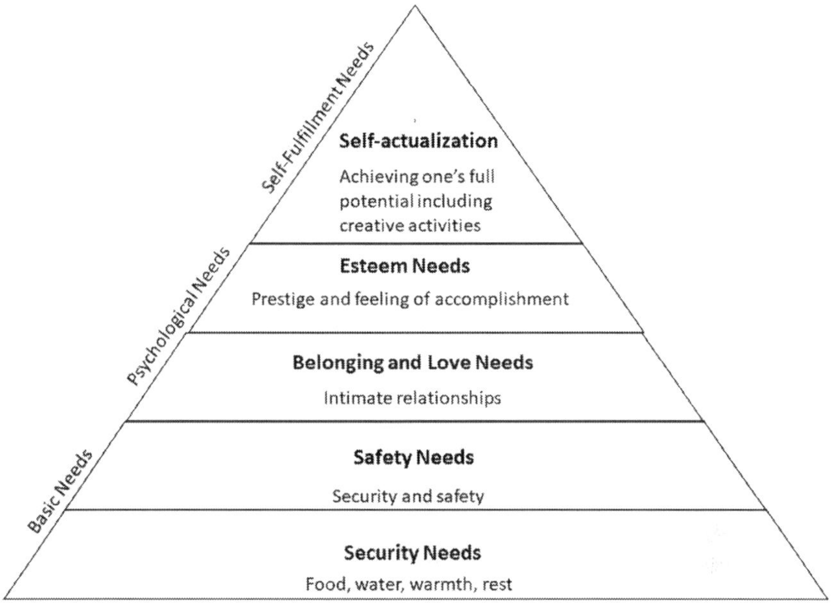

FIGURE 1.1 Maslow's hierarchy of needs.

Baby boomers (1946–1964) grew up during post-war prosperity and the rise of mass consumerism. They sought meaning through work, family, and legacy creation. Their consumption reflects materialistic values, with goods functioning as symbols of achievement, success, and social identity (Richins & Dawson, 1992). Physical products were extensions of personal success and stability and thus became intertwined with self-worth and achievement (Belk, 1988). Their acquisition of homes, vehicles, and branded goods reflected externalized identity validation and a belief in material acquisitions as life progressed.

The Generation Xers (1965–1979) matured in a time of economic volatility, rising divorce rates, and early technological disruption, fostering a psychological orientation toward self-reliance and skepticism (Mitchell, 1995; Coupland, 1991). They view products pragmatically, emphasizing function, independence, and customization (Jenkins, 2006; Williams & Page, 2011). Unlike Boomers, they favor durability over status and tend to critique over-consumption. Cognitively, they reflect an internal locus of control and critical thinking, often embracing minimalism and quality.

Millennials (1981–1996), called experience-driven consumers, prioritize autonomy, sustainability, and meaningful engagement over material accumulation (Inglehart, 1997). Their identity is fluid and constructed through experiences rather than fixed possessions (Arnett, 2000). They are more likely to rent, share, or co-own, viewing material goods as props in a larger narrative of self-expression, not the narrative itself. Their emphasis on authenticity, ethics, and aesthetic alignment reflects an advanced form of psychological branding with buying what works and "feels right."

Then came Generation Z, also termed iGen (1997–2012), who grew up in an era of instant connectivity, environmental crisis, and financial instability. Their consumer behavior leans toward minimalism, ethical consumption, and experiential value over ownership (Francis & Hoefel, 2018). This shift reflects a post-materialist orientation where digital identities and access (e.g., streaming, renting) outweigh traditional ownership (Kaur, Dhir, & Rajala, 2020). Psychologically, Gen Z demonstrates higher levels of anxiety and the need for control, often using technology and product ecosystems (e.g., Apple, gaming platforms) to establish identity boundaries and self-regulation (Twenge, 2017).

Generation Alpha (2013–2025) is the most globally connected generation in history to date (McCrindle & Fell, 2021). Their environment is strongly influenced by tech-savvy, education-conscious parents. Psychological research suggests they may develop shorter attention spans but stronger multitasking abilities and face rising levels of digital fatigue and mental health issues (Christakis et al., 2018; Bedford et al., 2016; Holloway, Green, & Livingstone, 2013). Despite being hyper-connected digitally, Generation Alpha may have fewer face-to-face interactions in early development due to screen use and remote schooling (Livingstone & Blum-Ross, 2020). This generation will likely form relationships across borders through online platforms but may face challenges in developing offline social skills, empathy, and teamwork.

The next Generation Beta (~2026–2040) will likely grow up in a fully immersive, connected world where AI assistants and augmented/virtual reality (AR/VR) become standard tools from infancy. AI-driven education with gamification could result in quick adaptation and learning across multiple formats (e.g., text, visuals, VR). However, this might also lead to reduced deep learning or overreliance on digital tutors, weakening traditional memory skills. AI-driven devices will likely play a central role in emotional development, helping children to recognize, articulate, and regulate emotions. This could result in higher emotional intelligence (EQ) than previous generations. Despite growing up with advanced emotional support tools, mental health issues may still arise due to digital dependency, with potential social isolation, anxiety, and identity crises tied to the digital world. Blending of the virtual and physical worlds will allow the Generation Beta to engage in a highly interwoven online/offline social life, making friendships and activities multidimensional. They will likely face the blurring of self-identity through digital interfaces, AI avatars, and constantly evolving virtual realities. Raised amidst the climate change crisis, eco-anxiety, and sustainability movements, Beta will likely prioritize environmentally conscious products. They will demand eco-friendly alternatives to traditional products, including recycled materials, sustainable manufacturing processes, and carbon-neutral production. Beta is predicted to favor brands that align with their ethical values, including zero-waste packaging, ethically sourced materials, and carbon-neutral shipping. Their moral framework will demand that their purchases reflect an eco-conscious mindset (Leyts, 2024).

1.4 DESIGN FOR HUMANITY

Traditionally associated with aesthetics or market-driven solutions, design is being reimagined as a human-centered and ethically grounded discipline capable of addressing systemic global challenges. It is no longer just about creating products that look good or function well. It is about designing with the future in mind, ensuring that our work meets today's needs and contributes to the health and longevity of the people and our planet (NHS, 2018). Design for humanity builds on the foundations of human-centered design (HCD), rooted in empathy, inclusivity, and sustainability. Design for humanity advocates for participatory approaches that prioritize the dignity, agency, and well-being of all people (Government Digital Service, 2025). It draws upon various theoretical frameworks, including human-centered design, participatory design, and sustainable design. It prioritizes human needs, social equity, and environmental sustainability. The aim is to provide a comprehensive understanding of how design can be harnessed as a tool for positive societal change (Brown & Wyatt, 2010; Striegler & Hjort, 2025).

Participatory design actively involves stakeholders in the design process, ensuring their voices are heard and their needs are met. Schuler & Namioka (1993) highlighted the significance of co-creation, arguing that engaging users empowers them and leads to more relevant and effective design outcomes. This collaborative approach fosters a sense of ownership and agency among participants, essential for achieving sustainable impact.

Sustainable design is about creating solutions that minimize environmental impacts while promoting social equity (Manzini, 2015). Organic design draws from the flowing, asymmetrical forms of the natural world to create spaces to foster harmony with the environment and emotional well-being. A prime example is Frank Lloyd Wright's Fallingwater House (fallingwater.org/), which shows this natural integration (Kaufmann, 1982). Manzini (2015) posits that design should address immediate needs and consider long-term implications for future generations. This perspective emphasizes the interconnectedness of social, economic, and environmental factors, advocating for holistic approaches that prioritize human well-being.

Thus, design for humanity is geared toward tackling complex and intricate problems, often called "wicked problems," to enhance the quality of people's lives. It signifies a holistic and empathetic approach, weaving together innovation, problem-solving, and emotional resonance to address humanity's multifaceted challenges. The application of design for humanity principles in healthcare has gained significant attention. An example is the design of a low-cost infant warmer for neonatal care in developing countries (TED, 2010), exemplifying the potential of design to address pressing social issues (IDEO, 2012). Projects such as the design of low-cost medical devices and user-friendly healthcare environments have demonstrated the potential of design to improve patient outcomes and enhance the overall healthcare experience (Buchanan, 2001; Bason, 2017). For instance, the "Design for Health" initiative in the UK involved collaboration between designers, healthcare professionals, and patients to create solutions that address specific health challenges (NHS, 2018).

Design for humanity can be traced back to the early twentieth century, with movements such as the Bauhaus advocating for functional and socially responsible design. The Bauhaus philosophy emphasized integrating art and technology to meet societal needs (Gropius, 1965). This foundation laid the groundwork for subsequent design movements that prioritized social impact. Together, these movements aim to craft environments that are aesthetically pleasing and practical and emotionally nurturing, sustainable, and deeply respectful of both human needs and the natural world.

In the late twentieth century, the emergence of design activism marked a significant shift in the discourse surrounding design for humanity. Designers began to engage in social issues, using their skills to address challenges such as poverty, inequality, and environmental degradation. This period saw the rise of initiatives that sought to leverage design as a catalyst for social change, leading to the establishment of organizations such as Designers Without Borders and Design for the World.

The twenty-first century has witnessed a further expansion of the design for humanity paradigm, driven by technological advancements and a growing awareness of global challenges. The United Nations' Sustainable Development Goals (SDGs) developed in 2015 have provided a framework for designers to align their work with broader societal objectives, emphasizing the role of design in achieving sustainable development (United Nations, 2025). This alignment has prompted a re-evaluation of design practices, encouraging designers to consider the ethical implications of their work.

The key principles of Design for Humanity include:

1. *Empathy and participation*: This is listening and co-creating with stakeholders, particularly underserved communities. Participatory design methods ensure that outcomes are not only context-specific but also empowering.
2. *Equity and inclusion*: Address inequalities by challenging designers to ensure voices are heard and values are encoded for those who benefit. It highlights the invisible lines of privilege and exclusion embedded in everyday systems (Costanza-Chock, 2020).
3. *Sustainability and resilience*: This is not about sticking a green label. It is about rethinking how we live, build, and embed ecological, social, and economic regeneration at every stage of the design process. We shift from the outdated cradle-to-grave or take-make-dispose mindset, where products end in landfills, to a cradle-to-cradle (C2C) or regenerative design approach that reduces waste and minimizes environmental impact. True resilience means creating designs that sustain life and adapt and thrive with a disturbing agent and adversity.

Designing for humanity is like a symphony. It thrives on cross-disciplinary harmony, inclusive collaboration, and design processes that are as decentralized as they are intelligent, drawing on AI, immersive technologies, and the creative potential of people. This includes leveraging open-source communities (e.g., GitHub [github.com], Arduino [arduino.cc]), open innovation platforms (e.g., OpenIDEO [openideo.com], InnoCentive [innocentive.com]), and crowdfunding platforms (e.g. Kickstarter [kickstarter.com] and Indiegogo [indiegogo.com]), all of which provide easy access to ideas, tools, and resources while fostering collective creativity. There is a growing need for education and training programs that equip designers with the skills and knowledge to navigate the complexities of social impact design. As highlighted by the Design Management Institute, fostering a culture of empathy and ethical awareness among design professionals is essential for advancing the field. When design literacy spreads beyond the studio, problem-solving becomes democratic, dynamic, and deeply human. That is the spirit we have tried to capture in this book. We hope its pages inspire you to design not just for people but with them for a more just, resilient, and compassionate world.

1.4.1 THINK OUTSIDE THE BOX

Steve Jobs said, "Think different," but here's the twist. You cannot do that if you are stuck inside the box. Where is the box, and can you think outside the box? Try this: connect the nine evenly spaced dots with three straight lines without lifting your pen (Figure 1.2). Most won't make it. Why? Because the box is comfy. It is time to think outside of it. This exercise is a reminder not to make assumptions, especially when creating something new.

Innovation occurs on a spectrum, from minor improvements to radical transformations. Incremental innovation enhances existing products or processes, like improving battery life or refining a user interface (Norman & Verganti, 2014). It is often lower risk and easier to integrate. Disruptive

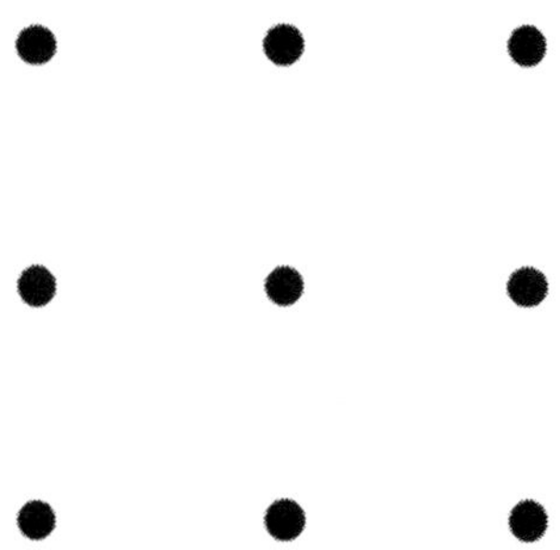

FIGURE 1.2 Connect all nine dots using three straight lines without lifting the pen.

FIGURE 1.3 Museum of the Future in Dubai, United Arab Emirates.

innovation, on the other hand, changes the game entirely, usually rendering existing solutions obsolete (Christensen, 2024). Think of how Netflix disrupted the DVD rental market or how 3D printing reshapes manufacturing. Both types of innovation are necessary: incremental changes ensure reliability and refinement, while disruptive leaps drive progress and redefine possibilities. Understanding when and how to pursue each type is crucial for any design strategist.

Our minds tend to follow familiar patterns, or "canyons," making it difficult to think beyond the usual (Rybarkiewicz, 2018). To truly think outside the box, we must break free from this patterned thinking. Look at the Dubai Future Museum of Science (Figure 1.3). What distinguishes it? The unique geometry of its exterior. We are accustomed to mostly boxy or cylindrical buildings, so anything that breaks that mold grabs attention. Radical innovation often comes from imagination, thinking beyond the typical (Samli, 2011). Designers have long created water features, but waterfalls and fountains have become predictable. Now, imagine water moving, dancing to music, making waves. That is mesmerizing. This kind of unexpected thinking breaks the norm and sparks something genuinely unique. So, what does "out-of-the-box thinking" really mean? It is about approaching problems with fresh, unconventional perspectives, moving beyond traditional solutions to ignite creativity and drive innovation. The lone genius in a quiet room is only a myth. True breakthroughs happen when diverse minds come together: integrating different cultures, challenging ideas, fostering debates, and embracing disagreements (de Paula et al., 2024). Bold ideas aren't born in silence; they emerge in the messy, vibrant space where perspectives collide and collaboration reigns.

As more of the same innovations appear, they lose their sense of appeal. Think of suspension bridges. When Thomas Telford's Menai Suspension Bridge opened in 1826, it was a breathtaking engineering marvel. People were in awe, even though the Inca civilization had used rope suspension bridges as far back as 2000 BCE. Fast forward to today, and we barely blink at the sight of one. This illustrates a critical point: when we talk about thinking "out of the box," we are not just urging you to avoid assumptions but pushing you to reimagine. People often dismiss some ideas as "impossible," but these are precisely the ones that spark real excitement if you flip your perspective and see "impossible" as "I'm possible." Some good examples are levitating structures or Frank Zapata's Flyboard Air, the world's smallest manned aircraft (www.zapata.com/). That is how innovation escapes the ordinary and ventures into uncharted territory.

1.4.2 THE MULTI-SENSORIAL EXPERIENCE

When we ask, "Do people attend concerts purely for the sound?" it is much like asking, "Do people visit a fine dining restaurant just to eat?" Of course not. Concerts, like the fine dining experience, are multi-sensorial experiences. Over the past two decades, research into concert hall design has shown that what we perceive is shaped not just by sound but by multiple senses, converging into a unified experience (Hyde, 2004). Modern performances make this undeniable. Today's concerts are no longer just auditory events but immersive spectacles. Visuals, synchronized lighting, and even kinetic elements like vibrating seats work together to transport audiences beyond the music itself. Nowhere is this evolution more striking than at the Sphere in Las Vegas, a groundbreaking venue where a 360-degree wraparound screen, dynamic soundscapes, and even haptic technology create an overwhelming multisensory 4D experience. It is not just a concert but a full-body immersion into another reality.

This blending of senses taps into something fundamental about human perception (Stein, 2012). People who attend concerts or operas aren't experiencing sound in isolation. Visual, social, and spatial factors are equally important. We are wired for multisensory integration, meaning our brain does not process sights, sounds, and tactile sensations separately but fuses them into a single coherent perception. In fact, perception isn't merely a sum of inputs; it is a product of their interaction. When sensory information overlaps, vision often dominates, a phenomenon well-illustrated by the ventriloquism effect, where what we see can alter where we believe a sound originates. Such findings highlight that "hearing" and "seeing" are deeply interconnected, especially when stimuli are presented together with comparable intensity. Thus, designing memorable experiences, whether in a concert hall, a product, or a service, demands more than perfecting individual elements. It requires orchestrating them to create powerful, unified moments that simultaneously engage multiple senses. Innovation today is no longer optimizing single dimensions; it thrives in mastering the rich interplay between the senses to create a multifaceted experience.

Considering such aspects is ***designing for people***! Designing for people means keeping users in mind while aiming to meet their needs, preferences, and limitations. For a product or system to succeed, we must also ***design with people***, which is called co-design or participatory design, actively involving users as partners and contributors throughout the design journey. When people are invited to share their lived experiences, offer feedback on prototypes, and help shape design decisions, the outcomes are more likely to be relevant, inclusive, and effective. This collaborative approach fosters empathy, reveals hidden needs, and prevents misaligned solutions that may look good on paper but fail in practice. It also builds trust and ownership, especially in systems that require user adoption and sustained engagement. In short, people-centered design is not only about understanding users; it is about empowering them as co-creators in the design process.

1.5 DESIGN FRAMEWORKS

Many frameworks have been proposed to guide the design process, each with its own vocabulary and emphasis. Examples include Inspiration–Ideation–Implementation, Dream–Design–Do, Discover–Dream–Design–Deliver, Discover–Interpret–Ideate–Experiment–Evolve, Discover–Define–Develop–Deliver, Empathy–Define–Ideate–Prototype–Test, and Plan–Do–Check–Act, among others. While the terminology may vary, the underlying logic remains consistent: understanding needs, generating ideas, and turning them into outcomes.

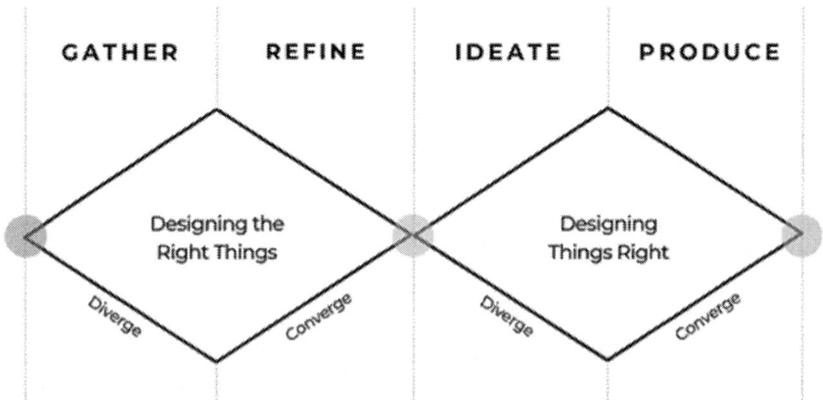

FIGURE 1.4 The GRIP framework in the double-diamond process of the Design Council (2025).

In this book, we propose the GRIP framework: gather, refine, ideate, and produce (Figure 1.4), a system that helps you take control of the design process. The four-stage model emphasizes grasping and shaping the solution from initial research to real-world impact.

It offers a structured yet flexible approach to design that emphasizes user understanding, creative problem-solving, and practical execution:

Gather: This stage focuses on exploring the context: understanding the users, their environment, their challenges, and latent needs. It involves observations, interviews, secondary research, and immersion techniques to collect meaningful insights.

Refine: Once data is gathered, it must be analyzed and interpreted. This phase transforms raw observations into well-defined problems. It includes clustering findings, identifying patterns, synthesizing user needs, and framing the right questions to be solved.

Ideate: With clear problem definitions in place, this phase encourages the generation of diverse and imaginative ideas. Brainstorming, sketching, role-play, and scenario-building help explore multiple solutions before selecting the most promising concepts. Ideas are translated into tangible prototypes, tested, and refined.

Produce: This final step brings ideas to life. This stage involves clearly specifying the design, validating that it solves the right problem, verifying that it works as intended, and finally preparing to launch it onto the market. It is where concepts are transformed into tangible outcomes that deliver real value to users.

Together, the GRIP framework helps structure the design journey, ensuring that it remains rooted in real user needs while fostering creativity and thoughtful execution.

1.6 SIGNIFICANCE OF THE BOOK

Design, as a discipline and practice, has evolved beyond traditional boundaries. It is no longer limited to physical artifacts or visual aesthetics. Design now plays a strategic role in shaping experiences, systems, and services. This book aims to equip learners and practitioners with foundational design thinking principles and practical tools that transcend disciplinary silos. It is needed now more than ever, as problems faced by society, ranging from climate change to healthcare inequities, require better solutions, and better problem framing. The growing emphasis on interdisciplinary collaboration and human-centered design accentuates this need (Brown, 2009; Buchanan, 1992).

In an era driven by rapid technological change and shifting user expectations, design education must evolve to foster not only creativity but also critical thinking, empathy, and adaptability. This book supports that evolution by guiding students and professionals through structured methods, emphasizing utility and user experience. It moves beyond the "how" of design execution to examine the "why" behind design choices (Sinek, 2009). Sinek's "Golden Circle" offers a simple but transformative framework for understanding purpose-driven design. Most organizations start with what they do (the product or service); some know how they do it (their processes), but few articulate why they do it, which is the core belief or purpose. Design that begins with "why" is more likely to resonate with users and build loyalty. Apple's success is not due to its sleek devices (the what), but its commitment to empowering creativity and challenging the status quo (the why). This book encourages designers to lead purposefully, ensuring that every decision aligns with a meaningful value proposition.

Each morning offers a choice to stay in your comfort zone or enter the unknown. Creativity doesn't bloom in comfort; it thrives on courage. To innovate, you need the spirit of a warrior: restless, curious, and unafraid to try something bold and unfamiliar, asking why. People often ask me, why do you keep asking why? That fearless mindset is not optional; it is essential. The people you associate with shape the boundaries of your imagination. Spend your days with people who work to get by, and your creativity may quietly idle. But surround yourself with curious minds, those driven to explore, question, and build a better world, and your thinking begins to spark. This book is for that second group: the brave thinkers, the restless doers, the ones who don't just think outside the box but redesign it entirely. It is essential to first cultivate a clear vision. Vision provides the anchor that aligns design goals with user needs, stakeholder values, and long-term impact. Design becomes a mechanical exercise in aesthetics or functionality without vision, disconnected from context. Design is inherently a goal-directed activity. What differentiates successful designers is their technical skill and clarity of purpose (Simon, 1996). A compelling vision is a compass, helping teams navigate ambiguity and complexity with coherence and intent.

Designers are uniquely positioned to challenge assumptions and question the status quo. Conventional wisdom often constrains innovation by framing problems too narrowly or accepting existing limitations as fixed and unquestionable. Schön (2017) describes the reflective practitioner as one who "frames and re-frames" problems, often uncovering hidden biases. This book fosters that reflexivity by encouraging readers to critique design briefs, user stereotypes, and industry norms. By doing so, learners are empowered to envision alternatives that are better, more inclusive, and sustainable.

The book provides a blend of established and contemporary methods, offering you the flexibility to apply them in ways that suit your needs. Like strength training or sprinting, progress comes not from repetition alone but from pushing beyond what feels easy. Growth happens when things get uncomfortable; that signals you are stretching into new terrain. When faced with difficulty, consider it a clear indicator that you are moving beyond your comfort zone. We aim for this book to serve as a tool that encourages you to confront those challenges, helping you push your limits and advance every time you engage with its content.

1.7 CHAPTER INTRODUCTIONS

Let us look at how this book is organized, chapter by chapter. Each chapter is designed to be modular and self-contained, allowing it to stand on its own while still maintaining clear connections to related chapters. This structure reflects the reality of design as a non-linear, iterative process. As you move through a project, you may find yourself revisiting earlier chapters to refine ideas, reassess requirements, or explore new possibilities. Navigating back and forth is not a setback; it is a hallmark of thoughtful design to achieve an optimized, people-centered outcome.

Design starts with people. To design effectively, we must begin by understanding human capabilities and limitations, a foundation explored in Chapter 2. Ergonomics, the science related to designing tools, tasks, and environments, seeks to align work demands with what people can realistically perform physically, cognitively, and perceptually (Kroemer, 2017). Although the human body is remarkably adaptable, that adaptability can be misleading. People may perform tasks in awkward or constrained postures, but sustained exposure to such conditions can result in discomfort, fatigue, injury, and long-term musculoskeletal disorders. Effective ergonomic design is rooted in human-centered thinking: it starts with the user's natural abilities and creates systems that support them rather than expecting people to conform to poorly designed tasks. This leads to safer, more comfortable, and more productive environments, products, and systems.

Ergonomics must also embrace the rich diversity within human populations. People vary greatly in age, gender, physical stature, functional abilities, and cultural background. Designing with this diversity in mind means creating products and environments that are accessible, fair, and usable by all, not just those who fall within average norms. Addressing a broader range of human capabilities helps reduce exclusion and enhances the overall quality and effectiveness of the user experience. Chapter 3 discusses how these human variations can be thoughtfully integrated into the design process, drawing on established frameworks and real-world examples that support equity and participation for all.

The key question is: how do we uncover what people truly need to design effectively? This involves identifying their challenges and addressing them through innovative solutions, whether incremental improvements or disruptive breakthroughs. Understanding these needs requires empathy-driven inquiry to reveal unspoken frustrations and unmet expectations. It is not only about asking users what they want but deeply understanding the context of their experiences, struggles, and aspirations. Through this empathic approach, designers can uncover deeper insights that may not be immediately obvious. Chapter 4 focuses on how to gain these critical insights through empathy, laying the groundwork for meaningful and impactful design that genuinely resonates with users.

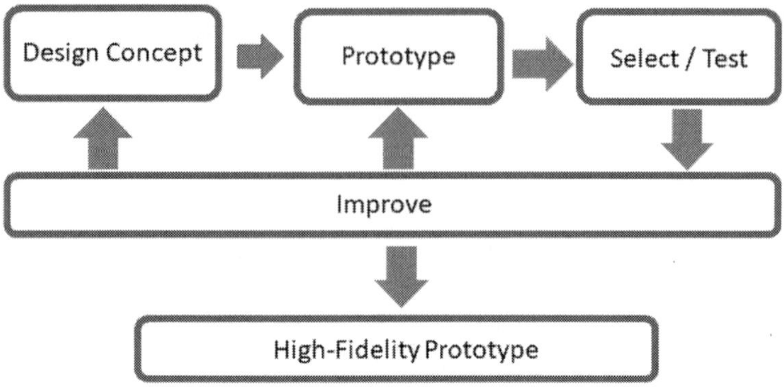

FIGURE 1.5 The iterative improvements are made using the design–prototype–test cycle.

An often-overlooked insight in design is that innovative products and services provoke new questions. The reflexive relationship between solutions and inquiry is central to creativity and innovation. The resulting product is more likely to succeed in real-world contexts if the right questions are anticipated early in the design process (Verganti, 2009). Questioning also shifts the mindset from reactive problem-solving to proactive opportunity-seeking. In fact, some of the most breakthrough innovations, from the iPhone to Airbnb, have emerged not by answering existing questions but by redefining what questions needed to be asked (Christensen, 2024). Chapter 5 encourages readers to adopt a questioning mindset that draws deeply on user insights to identify and address the right problems. Rather than relying solely on the assumptions or perspectives of the designers in the room, the emphasis is on framing the problem through the lived experiences, needs, and challenges of real users.

Chapters 6, 7, and 8 are about ideating, prototyping, and prototype selection, the iterative cycle to obtain a high-fidelity prototype (Figure 1.5). The concept of Intelligent Fast Failure (IFF) advocates for structured, rapid experimentation to test assumptions and uncover unexpected insights (Matson, 1991; Tahirsylaj, 2012). It helps fail safely and strategically. For example, IDEO's design teams often create low-fidelity prototypes early to elicit feedback before investing in refinement. However, IFF is inappropriate in high-stakes environments where lives or safety are at risk (e.g., aerospace or medical device design). In such contexts, simulations, redundancies, and extensive testing must substitute for real-world failure. When applied wisely, failure becomes a powerful trigger for creativity, inviting alternative approaches that would not emerge in risk-averse settings.

As we move further into the twenty-first century, the connection between design and sustainability has never been more apparent. Designers today face the exciting challenge of blending innovation with environmental responsibility, creating solutions that inspire and protect our planet. This means thinking about the entire lifecycle of a product, everything from sourcing raw materials to manufacturing, use, and ultimately, disposal. Sustainable design is inherently tied to the concept of circular design that creates a closed-loop system, where products are crafted for easy disassembly, repair, and recycling, ensuring they live on long after their initial use. With this in mind, we have dedicated Chapter 9 to Design for X, where X can represent anything you envision, highlighting the endless possibilities for sustainable innovation.

Once prototypes are tested and refined, detailed specifications prevent production errors and ensure the product meets people's multifaceted needs while maintaining functional and safety standards, ultimately delivering a delightful product. Consider the development of a robot-assisted walker for individuals with mobility challenges. The specification outlines the critical aspects, such as sensor placement, battery life, weight capacity, and user interface features, such as a mobile app for personalized settings. If the design specifies that the walker must detect uneven surfaces and adjust its stability, the specification ensures the sensors and algorithms are accurately tuned for different surfaces. Chapter 10 details the process of creating the specifications for successful manufacturing.

Chapter 11 on verification and validation is a critical step in the design journey to ensure that the final product performs as intended and meets user needs using a pre-production prototype. Validation asks whether we are solving the right problem that delivers real, meaningful value to users. Verification checks if we've solved the problem correctly, ensuring the solution is reliable, safe, and functions as designed. In short, validation asks, "Did we solve the right problem?" while verification asks, "Did we solve the problem right?" (Figure 1.4). Together, they form the foundation of responsible people-centered design.

Once all checks confirm that the product is ready, functionally sound, user-validated, and aligned with its intended purpose, the focus shifts to how it can be successfully brought to market. This transition from design to delivery requires strategic thinking about commercialization pathways, whether launching a new venture or innovating within an existing organization. Chapter 12 explores entrepreneurship, where individuals or teams build new business models around novel solutions, and intrapreneurship, where innovation is driven from within established companies to create value without forming a new entity.

Figure 1.6 maps how each chapter aligns with the GRIP framework, guiding you through a purposeful, adaptable design journey. The pathways introduced in this book aren't just steps; they are mindset shifts, empowering you with the tools to create products that are not only innovative and competitive but also sustainable and meaningful. And remember, great design is not linear. You will circle back, revisit, and rethink; that is not failure; that is design done right. So, whether you are just starting out or deep into prototyping, let this book be your companion in building products that truly matter. Let's dive in!

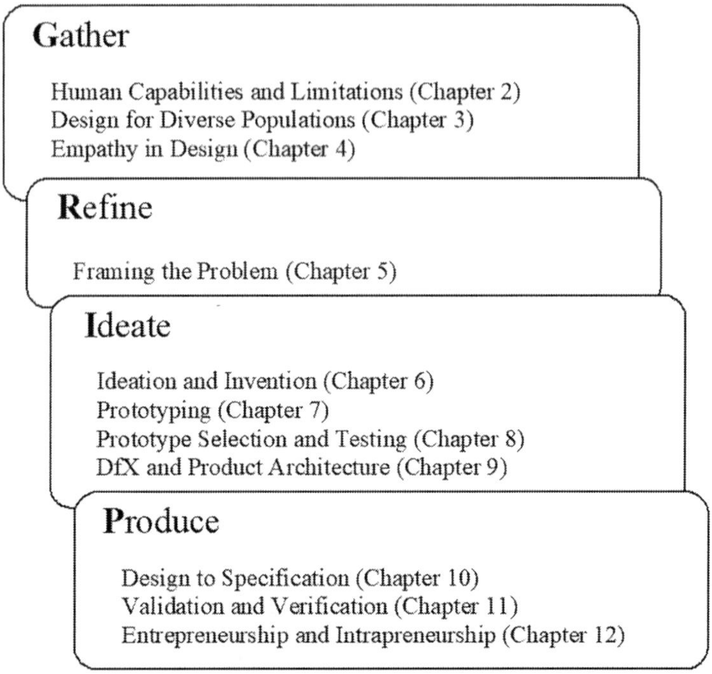

FIGURE 1.6 Overview of the book content mapped onto the GRIP framework.

REFERENCES

Arnett, J. J. (2000). Emerging adulthood: A theory of development from the late teens through the twenties. *American Psychologist*, 55(5), 469–480. https://doi.org/10.1037/0003-066X.55.5.469

Bason, C. (2017). *Leading Public Design: Discovering Human-Centred Governance.* Policy Press, Bristol University. https://policy.bristoluniversitypress.co.uk/leading-public-design

Bedford, R., Saez De Urabain, I. R., Cheung, C. H., Karmiloff-Smith, A., & Smith, T. J. (2016). Toddlers' fine motor milestone achievement is associated with early touchscreen scrolling. *Frontiers in Psychology*, 7, 1108. https://doi.org/10.3389/fpsyg.2016.01108

Belk, R. W. (1988). Possessions and the extended self. *Journal of Consumer Research*, 15(2), 139–168. https://doi.org/10.1086/209154

Bentley, D (2020, March 16). The greatest designs of modern times. *Fortune*, https://fortune.com/longform/100-best-designs/.

Bordwell, D., & Thompson, K. (2010). *Film Art: An Introduction* (10th ed.). McGraw-Hill.

Brantmark, N. (2017). *Lagom: Not Too Little, Not Too Much: The Swedish Art of Living a Balanced*, Happy Life. Harper Design.

Brown, T. (2009). *Change by Design: How Design Thinking Creates New Alternatives for Business and Society*. HarperBusiness.

Brown, T., & Wyatt, J. (2010). Design thinking for social innovation. *Stanford Social Innovation Review*, 8(1), 30–35.

Buchanan, R. (1992). Wicked problems in design thinking. *Design Issues*, 8(2), 5–21. https://doi.org/10.2307/1511637

Buchanan, R. (2001). Design research and the new learning. *Design Issues*, 17(4), 3–23.

Cairo, A. (2012). *Functional Art, The: An Introduction to Information Graphics and Visualization*. New Riders.

Christakis, D. A., Ramirez, J. S. B., & Ferguson, S. M. (2018). How early media exposure may affect cognitive function: A review of results from observations in humans and experiments in mice. *Proceedings of the National Academy of Sciences*, 115(40), 9851–9853. https://doi.org/10.1073/pnas.1812717115

Christensen, C. M. (2024). *The Innovator's Dilemma, with a New Foreword: When New Technologies Cause Great Firms to Fail.* Harvard Business School Press.

Costanza-Chock, S. (2020). *Design Justice: Community-Led Practices to Build the Worlds We Need.* MIT Press. https://doi.org/10.7551/mitpress/12255.001.0001

Cross, N. (2001). Designerly ways of knowing: Design discipline versus design science. *Design Issues*, 17(3), 49–55. https://doi.org/10.1162/074793601750357196

Cross, N. (2011). *Design Thinking: Understanding How Designers Think and Work.* Berg Publishers.

Coupland, D. (1991). *Generation X: Tales for an Accelerated Culture.* St. Martin's Press.

Design Council (2025). History of the double-diamond. https://www.designcouncil.org.uk/our-resources/the-double-diamond/history-of-the-double-diamond/

de Paula, L. D., Branco, A., & Glaveanu, V. (2024). Thinking outside of the (western) box: cultural psychology perspectives on creativity in education. *Journal of Creative Behavior*. https://doi.org/10.1002/jocb.655

Diener, E., Inglehart, R., & Tay, L. (2013). Theory and validity of life satisfaction scales. *Social Indicators Research*, 112(3), 497–527. https://doi.org/10.1007/s11205-012-0076-y

Doblin, J. (1970). *One Hundred Great Product Designs.* Van Nostrand Reinhold.

Dormer, P. (1997). *The Culture of Craft: Status and Future.* Manchester University Press.

Fiell, C., & Fiell, P. (2024). *Scandinavian Design*: 40th Anniversary Edition. Taschen.

Francis, T., & Hoefel, F. (2018). *True Gen: Generation Z and Its Implications for Companies.* McKinsey.

Gilleard, C. (2004). Cohorts and consumption: A sociological analysis of generational change. *Sociological Review*, 52(1), 1–20. https://doi.org/10.1111/j.1467-954X.2004.00439.x

Gilbreth, F. B., & Gilbreth, L. M. (1924). *Applied Motion Study: A Collection of Papers on the Efficiency of Work.* Van Nostrand.

Government Digital Service (2025). *Government Design Principles.* www.gov.uk/guidance/government-design-principles.

Gropius, W. (1965). *The New Architecture and the Bauhaus.* MIT Press.

Holloway, D., Green, L., & Livingstone, S. (2013). Zero to eight: Young children and their internet use. EU Kids Online. https://eprints.lse.ac.uk/52630/1/Zero_to_eight.pdf

Hyde, J. R. (2004). The concert hall experience: how visual stimuli can affect what we hear. 147th ASA Meeting, New York, NY. https://acoustics.org/pressroom/httpdocs/147th/Hyde.htm

IDEO (2012). *Design for Social Impact: How to Guide.* IDEO.org. https://new-ideo-com.s3.amazonaws.com/assets/files/pdfs/news/IDEO_RF_Guide.pdf

Inglehart, R. (1997). *Modernization and Postmodernization: Cultural, Economic, and Political Change in 43 Societies.* Princeton University Press.

Jenkins, H. (2006). *Convergence Culture: Where Old and New Media Collide.* New York University Press.

Kaufmann, E., Jr. (1982). *Frank Lloyd Wright: Writings and Buildings.* Meridian Books.

Kroemer, K. H. E. (2017). *Fitting the Human: Introduction to Ergonomics / Human Factors Engineering,* CRC Press.

Kaur, P., Dhir, A., & Rajala, R. (2020). Consumer well-being in the digital age: A comprehensive review of antecedents, consequences, and directions for future research. *Journal of Business Research,* 118, 365–376. https://doi.org/10.1016/j.jbusres.2020.06.055

Lawson, B. (2006). *How Designers Think: The Design Process Demystified* (4th ed.). Architectural Press.

Leyts, M. (2024, November 2). Generation beta (gen beta): understanding beta, millennials, gen zers, and boomers. https://www.generationzalpha.com/blog/gen-beta.

Livingstone, S., & Blum-Ross, A. (2020). *Parenting for a Digital Future: How Hopes and Fears about Technology Shape Children's Lives.* Oxford Internet Institute.

Lovell, S. (2011). *Dieter Rams: As Little Design as Possible.* Phaidon Press.

Manzini, E. (2015). *Design When Everybody Designs.* MIT Press.

Maslow, A. H. (1943). A theory of human motivation. *Psychological Review,* 50(4), 370–396. https://doi.org/10.1037/h0054346

Matson, J. V. (1991). *The Art of Innovation: Using Intelligent Fast Failure.* Pennsylvania State University Publishing.

McCrindle, M., & Fell, A. (2021). *Understanding Generation Alpha.* McCrindle Research. https://mccrindle.com.au/insights/blog/understanding-generation-alpha/

Mitchell, T. R. (1995). Motivation: new directions for theory, research, and practice. *Academy of Management Review,* 20(3), 679–682. https://doi.org/10.2307/257251

Nelson, H. G., & Stolterman, E. (2012). *The Design Way: Intentional Change in an Unpredictable World* (2nd ed.). MIT Press.

Norman, D. A. (2013). *The Design of Everyday Things* (Revised and expanded edition). Basic Books.

Norman, D. A., & Verganti, R. (2014). Incremental and radical innovation: design research vs. technology and meaning change. *Design Issues,* 30(1), 78–96. https://doi.org/10.1162/DESI_a_00250

NHS (2018). Design and build digital services for the NHS. National Health Service, UK. https://service-manual.nhs.uk/

Papanek, V. (1985). *Design for the Real World: Human Ecology and Social Change* (2nd ed.). Thames & Hudson.

Richins, M. L., & Dawson, S. (1992). A consumer values orientation for materialism and its measurement. *Journal of Consumer Research,* 19(3), 303–316. https://doi.org/10.1086/209304

Rybarkiewicz, D. (2018). Out of the box thinking. In: *Cognition, Meaning and Action,* Eds: P. Łukowski, A. Gemel and B. Żukowski. Jagiellonian University Press, mv pp.153–168.

Ryff, C. D. (1989). Happiness is everything, or is it? Explorations on the meaning of psychological well-being. *Journal of Personality and Social Psychology,* 57(6), 1069–1081. https://doi.org/10.1037/0022-3514.57.6.1069

Samli, A. C. (2011). *Thinking Outside of the Box. In: From Imagination to Innovation.* Springer. https://doi.org/10.1007/978-1-4614-0854-3_5
Schön, D. A. (2017). *The Reflective Practitioner: How Professionals Think in Action.* Routledge.
Schuler, D., & Namioka, A. (1993). *Participatory Design: Principles and Practices.* Lawrence Erlbaum Associates.
Simon, H. A. (1996). *The Sciences of the Artificial* (3rd ed.). MIT Press.
Sinek, S. (2009). *Start With Why: How Great Leaders Inspire Everyone to Take Action.* Portfolio Publishers.
Stein, B. E. (2012). *The New Handbook of Multisensory Processing.* MIT Press.
Striegler, S. G. & Julie Hjort, J. (2025). *Navigating Societal Change through Design: Leading Missions for a Prosperous Future.* Policy Press, Bristol University
Sweller, J. (1988). Cognitive load during problem solving: Effects on learning. *Cognitive Science,* 12(2), 257–285. https://doi.org/10.1207/s15516709cog1202_4.
Tahirsylaj, A. S. (2012). Stimulating creativity and innovation through intelligent fast failure. *Thinking Skills and Creativity,* 7(3), 265–270. https:// doi:10.1016/j.tsc.2012.05.005.
Tay, L., & Diener, E. (2011). Needs and subjective well-being around the world. *Journal of Personality and Social Psychology,* 101(2), 354–365. https://doi.org/10.1037/a0023779
Taylor, F. W. (1911). *The Principles of Scientific Management.* Harper & Brothers.
TED. (2010, January 28). A warm embrace that saves lives| Jane Chen [Video]. *YouTube* . www.youtube.com/watch?v=IwidCkCmWg4.
Tuch, A. N., Roth, S. P., Hornbæk, K., Opwis, K., & Bargas-Avila, J. A. (2012). Is beautiful really usable? Toward understanding the relation between usability, aesthetics, and affect in HCI. *Computers in Human Behavior,* 28(5), 1596–1607. https://doi.org/10.1016/j.chb.2012.03.024.
Twenge, J. M. (2017). *iGen: Why Today's Super-Connected Kids Are Growing Up Less Rebellious, More Tolerant, Less Happy—and Completely Unprepared for Adulthood.* Atria Books.
United Nations (2025). The SDGs in action. https://www.undp.org/sustainable-development-goals. Last accessed 28 Aug 2025.
Verganti, R. (2009). *Design Driven Innovation: Changing the Rules of Competition by Radically Innovating What Things Mean.* Harvard Business Press.
Williams, K. C., & Page, R. A. (2011). Marketing to the generations. *Journal of Behavioral Studies in Business,* 3, 1–17.

2 Exploring Human Limits and Capabilities

> The human body has limitations; the human spirit is boundless.
>
> —Dean Karnazes

2.1 INTRODUCTION TO HUMAN LIMITS AND CAPABILITIES

To create products that delight users, designers need to understand how our bodies and minds operate. This chapter explores the basics, laying the groundwork for designing products that meet users' needs on every level.

Designs should aim for a "neutral" posture, providing customized support for the spine, limbs, and joints during diverse activities to reduce strain on the musculoskeletal system. Mastering the intricacies of human body dimensions and diversities is key to crafting products that snugly fit and cater to various body shapes and sizes. Regarding interactive features like controls and buttons, accessibility is paramount, ensuring they are within easy reach and sight to prevent strain or discomfort. Incorporating adjustable features further enhances versatility, accommodating diverse postures and personal preferences. Design should emphasize comfort and safety, shielding users from musculoskeletal disorders (MSDs) and injuries by providing sufficient support, cushioning, and safeguards against repetitive injury risks. Minimizing repetitive movements can prevent fatigue and overuse injuries, while design solutions should aim to reduce unnecessary reaching, bending, and twisting motions.

Beyond physical ergonomics, effective design should also consider cognitive ease, ensuring that interfaces and interactive elements support intuitive use and minimize mental workload. Error-free interface design and the "less is more" principle help reduce cognitive strain, enabling users to process information efficiently without unnecessary complexity.

DOI: 10.1201/9781003476177-2

Design should consider sensory and cognitive limits, ensuring that spaces are free from excessive noise or visual distractions that could lead to mental fatigue or decreased focus. By optimizing environmental factors such as lighting, temperature, and sound and considering the cognitive load required to perform tasks, designs can create ergonomic workspaces that promote physical and cognitive well-being and overall productivity and satisfaction.

2.2 PHYSICAL LIMITS

People exhibit a range of capabilities and limitations in flexibility, strength, endurance, and speed, all determined by how much the muscles can exert and how long and how fast the muscles can work within the musculoskeletal system. Muscles contract and apply force through tendons to bones, enabling movement and the ability to carry loads (Saladin, 2011). Muscles are most effective when in an extended state, gradually losing the force they can exert as they contract.

2.2.1 RANGE OF MOTION

The bones of the body are connected at points referred to as joints. Each joint in the human body allows the limbs attached to it to move in different directions. The amount of movement is called the range of motion. The range of motion determines flexibility. The range of motion is greatest when the body is in a neutral posture. Figures 2.1(a) and (b) show the person's range of motion when standing upright, and Figures 2.1(c) and (d) show the arm's range of motion when fully bent forward. This is something you can try with someone. You will

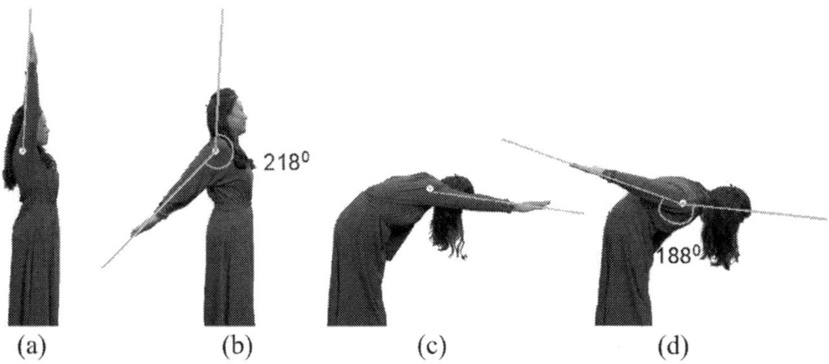

FIGURE 2.1 An example of human's range of motion: (a) arm fully extended upward with protractor set to zero; (b) arm fully back with protractor reading 218°; (c) fully bent forward with arm extended and set to zero; (d) fully bent forward with arm fully back at 188°.

see that the arm's range of motion is reduced in the bent position comparing to standing upright. The change could be as much as 30 degrees. This illustrates the importance of maintaining neutral postures. Try the "TikTok broomstick challenge" (2024) if this is not good enough. That will help you appreciate the stark difference between neutral posture and poor postures. Products must be designed to maintain neutral postures so that the person has a complete range of motion. One way to determine neutral posture is to use a biofeedback muscle monitor such as the Pocket Ergometer (Trademark, Biomechanics Corporation of America), where the loudness of an audio tone correlates with muscle activity. More advanced versions of such biofeedback devices are emerging in rehabilitation (Giggins et al., 2013).

The spine, or vertebral column, plays a crucial role in supporting the body and enabling various movements, such as bending and twisting. It is divided into five regions: cervical (or neck), thoracic (or upper back), lumbar (or lower back), sacrum, and coccyx. Comprising 33 vertebrae (seven cervical, 12 thoracic, five lumbar, five sacral, and four coccygeal) stacked from the neck to the base of the spine, these bones are separated by fibrocartilaginous discs in the lumbar, thoracic, and cervical regions. The discs cushion the vertebrae, absorbing shocks and allowing slight mobility between the bones. The lumbar area in the lower spine is particularly designed for flexibility. It permits movements like bending and twisting, essential for activities such as dancing or gymnastics. However, the lumbar vertebrae are not equipped to bear weight and move simultaneously despite their flexibility. This paradoxical function makes the lumbar spine vulnerable to injuries when subjected to excessive strain or improper movement patterns.

The five lumbar vertebrae move together as a cohesive bone column, facilitating forward and backward bending movements (Figure 2.2(a)). Positioned atop the pelvic platform, these vertebrae are interconnected with the pelvis through ligaments and muscles. The pelvis serves as a crucial link between the spine, hips, and legs, exerting control over the position of the lumbar vertebrae directly above it. When the pelvis maintains a level and balanced position, it promotes the alignment of the vertebrae, contributing to the spine's optimal strength and health. This balanced alignment minimizes injury vulnerability, ensuring the spine functions at its strongest and most resilient state. A complex interplay of abdominal, low back extensor, hip extensor, and hip flexor muscles maintains pelvic balance. These muscle groups work together to stabilize the pelvis. Any imbalance among them can lead to pelvic tilt, where the pelvis tilts forward anteriorly or backward posteriorly. When the pelvis rotates anteriorly, the vertebrae above it can become misaligned (Figure 2.2(b)). This misalignment weakens the spine and increases vulnerability to injury, potentially causing compression of the disc space and discomfort in the facet joints. Conversely, when the spine is properly aligned with a balanced pelvis, it alleviates pressure on the discs, unloads the facet joints, and optimizes space for the nerve roots. For individuals experiencing backaches, maintaining a balanced pelvic position while resting in bed can help alleviate discomfort and support spinal health. Products

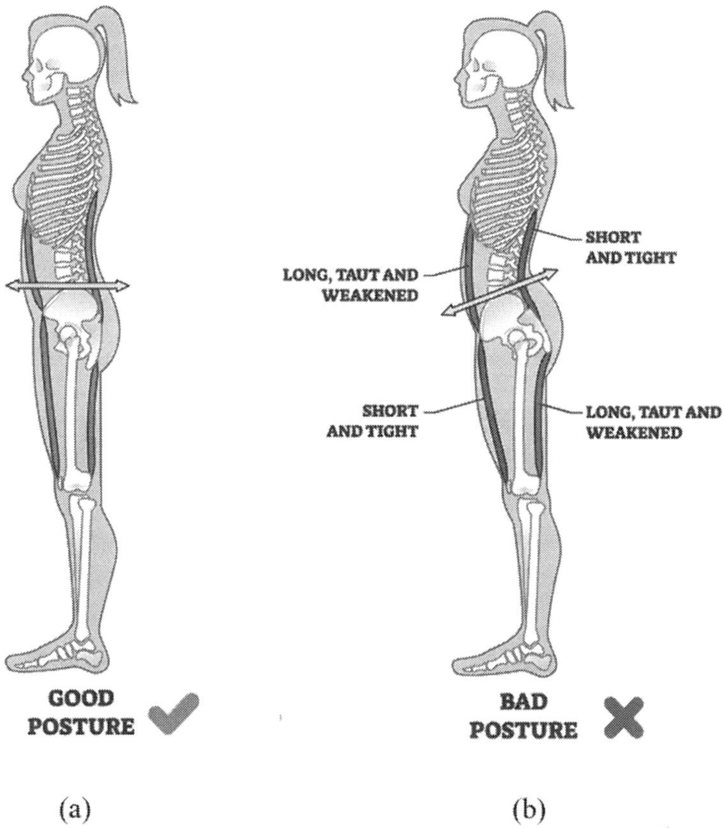

FIGURE 2.2 Spine and pelvis in (a) correct orientation and (b) with anterior pelvic tilt.

like seats should consider the spine's natural curvature and aim to support the pelvis in a balanced position.

2.2.2 STRENGTH

Ever struggled with a coffee pot that felt like it was slipping through your fingers? Or maybe wrestled with a stubborn sauce, wine, or jam bottle that just wouldn't cooperate? These little battles are all about ergonomics—the science of making everyday stuff comfortable and easy to use.

As your range of motion changes with your posture, so does your grip strength (Watanabe et al., 2005). People typically use one of two grips: the power grip or the precision grip (Figure 2.3). In a power grip, the whole hand is used to grasp the object, like gripping a racket or a pistol, providing maximum control

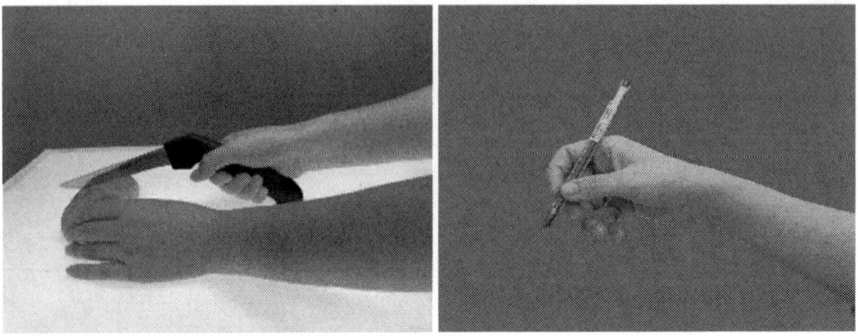

(a) (b)

FIGURE 2.3 Two types of grips: (a) power grip. Note the angle of the handle with respect to the forearm allowing a straight wrist. (b) Precision grip.

and strength. On the other hand, a precision grip involves using the index or middle finger, or both, together with the thumb, similar to how you might hold chopsticks. Each grip offers different advantages and is suited to various tasks, affecting how we interact with everyday objects. The middle finger and thumb are powerhouses, perfect for handling heavy loads. On the flip side, the index, ring, and little fingers aren't cut out for sustained heavy loads. However, the index finger stands out for its speed, making it the nimblest and quickest of all. When it comes to accuracy and performance, the size of the tool matters (Goonetilleke et al., 2009). Smaller pens or brushes excel in precision, allowing for meticulous detail work. On the other hand, larger tools are better for speed, enabling broader strokes and quicker coverage. Whether you're fine-tuning intricate designs or aiming for rapid, expansive strokes, choosing the right tool size can make all the difference in your workflow. Frequent finger movements can cause fatigue and stiffness, so minimizing finger strain is necessary to minimize the potential for injury. For example, opting for a bar/strip switch over a trigger switch is better because continuous use of triggers can lead to a condition known as trigger finger (Figure 2.4).

The grip strength, as measured using a Jamar dynamometer, varies with age, hand size, body type, and other parameters. The variation of grip strength with age for the left and right hands is shown in Figure 2.5. These data are from over 1,000 participants in the Wisconsin area (Mathiowetz et al., 1985; Mathiowetz et al., 1986) and have been curve-fitted by us. The fitted curve shows that the grip strength peaks around 35 years in males and 32 years in females. The difference between left and right hands for males is around 7–8%, while that in females is around 14–16%. Cultural differences should be considered, and the curves presented are only for a reference to illustrate the differences that exist among the young and old.

The change in power grip force at various wrist angles is well documented (Figure 2.6) (Adams & Berlin, 2017). Depending on the direction in which the

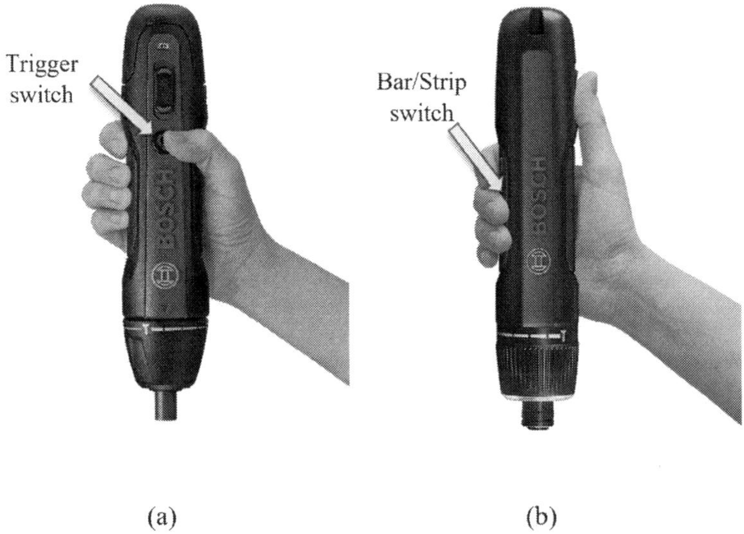

FIGURE 2.4 Two types of switches: (a) Bosch Go 2 trigger switch; (b) Bosch Go 3 bar/strip switch to relieve finger strain.

hand is bent, the grip strength will reduce from 100% of the straight hand posture to as low as 45%. This is mainly because the muscle acts to bend the wrist, and how much it is bent will limit what the muscle can provide to apply force. Thus, the higher the deviation, the lower the applied force. Can people apply the same force for any grip size? No, they cannot. Figure 2.7 shows the grip force changes with grip size. It is an "inverted U" shape; hence, it can be accurately modeled using a polynomial of order 2. There is an optimum grip size for any handle so that we can exert the maximum force, and that handle size tends to be the most comfortable (Goonetilleke et al., 1997). For racket sports, power is important, and hence grip size should be one that gives the maximum force. But how do people grip an object? This is where designers need to be careful. As we use a power grip, with the hand in a straight or neutral posture, the needed angle will be around 110° from the forearm (Figure 2.3(a)). Based on the aforementioned factors, a practical design guideline is that when using a power grip, the exerted forces should be kept below 10 percent of the user's maximum capacity.

When designing equipment like a tennis racket, getting the right grip size is just one part of the equation. Enhancing the dynamic characteristics to make controlling the ball easier and dampen harmful vibrations generated by each stroke is also important. Players typically aim to hit the ball "somewhere in the middle" of the racket, known as the "sweet spot." This is the area of the strings where the player feels minimal vibration upon impact, also called the "nodal" region. The damping characteristics of a tennis racket, the attachment of damping

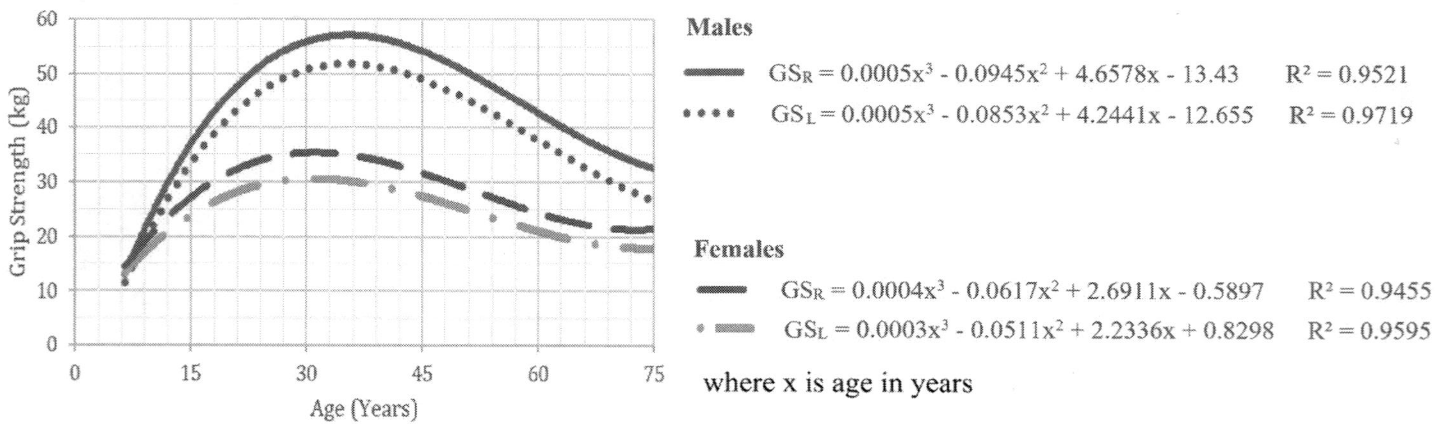

FIGURE 2.5 Variation of grip strength with age for left and right hands of males and females. Data from Mathiowetz et al. (1985, 1986). Each data point is from 20 to 50 participants.

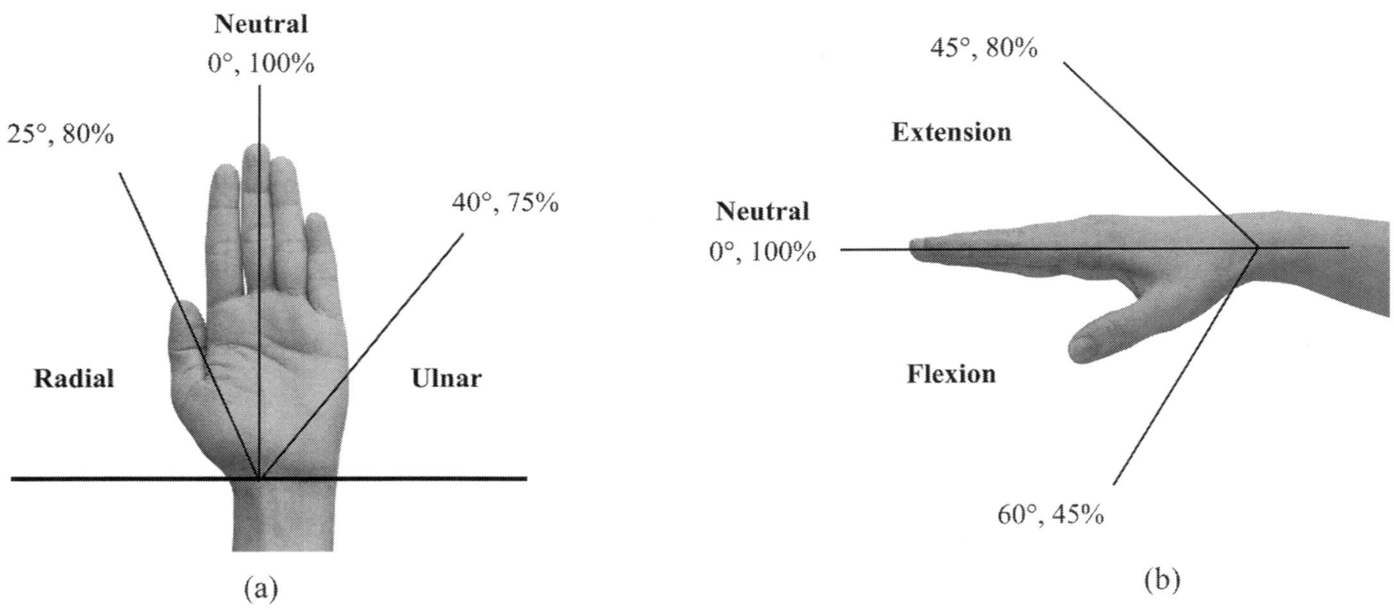

FIGURE 2.6 Effect of grip strength relative to a straight wrist posture or a neutral posture at (a) radial/ulnar angle and (b) flexion/extension angle. (Adapted from Adams & Berlin, 2017.)

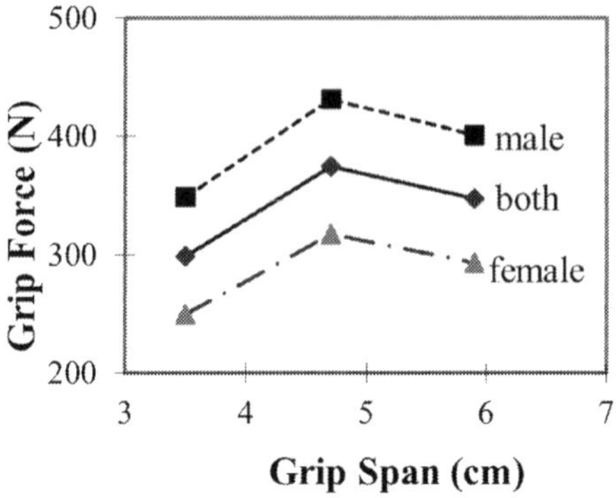

FIGURE 2.7 Effect of grip span on grip force.

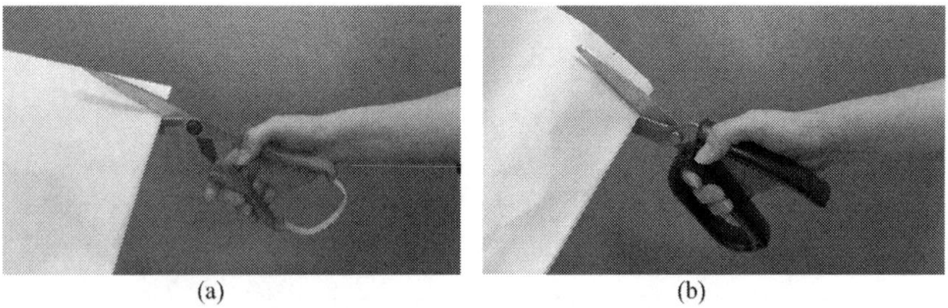

FIGURE 2.8 Two designs of spring-loaded scissors that open automatically once squeezed to minimize effort: (a) the link between the two arms acts as the spring; (b) spring is inside the two arms and is not visible.

materials to the grip, and changes in stiffness all influence the size of the sweet spot. This means the sweet spot can vary significantly from racket to racket, depending on the materials used in its construction. Additionally, the shape of the racket head plays a crucial role. By altering the head's shape, manufacturers can enlarge or shift the power and nodal regions, thereby expanding the sweet spot. The first oval-shaped heads, introduced in 1976, immediately enlarged the impacting area without changing the head's length. Since then, manufacturers have experimented with various sizes and shapes beyond the traditional oval to further enhance the racket's performance.

Tools like scissors or metal snips typically require users to apply pressure against a metal or plastic handle to open the jaws, which can lead to calluses and soft tissue trauma on the back and sides of the hand. An automatic spring-opening mechanism in tools such as scissors and pliers allows operators to utilize stronger hand-closing muscles rather than weaker hand-opening muscles (Figure 2.8). These designs of scissors also overcome the common problem of left-handedness as most scissors are designed for right-handed people.

2.2.3 ENDURANCE

Muscles work by using the nutrients in the blood supply to create energy through a process called metabolism. This can be aerobic, when sufficient oxygen is present, or anaerobic when oxygen to the muscle from the blood is limited. The ability to exert for a long time is called endurance. When a muscle is contracted, the blood flow is limited; thus, the incoming blood to the muscle is constrained, and the muscle cannot take out the by-products of the metabolism, causing fatigue. One of those unwanted by-products is lactic acid. Thus, the blood circulation governed by the heart plays an important role. When there is insufficient oxygen, the anaerobic process will create more and more lactic acid that accumulates in the muscle, and the work that the muscle can do will gradually diminish. Thus, endurance has limits. Rohmert (1960) describes our ability to hold a given level relative to the maximum that muscle or muscle group can exert.

$T_H = -90 + (126/p) - (36/p^2) + (6/p^3)$	Rohmert's formulation for holding time
T_H= Holding time (seconds), $p = f/f_{max}$, f = force exerted and f_{max} = maximum force	

When p =1 (i.e., the force exerted is the maximum force one can exert), the holding time is 6 seconds (s). Our experiments have shown that 6 s is an overestimate and more in the range of 1–3 s (Figure 2.9). When someone attempts to hold the maximum force, the force exerted will drop very rapidly, as shown in Figure 2.9. For two participants, the force exerted is only around 50 percent of their maximum after 30 s. This illustrates the need for regular movement, avoiding extended periods of static postures.

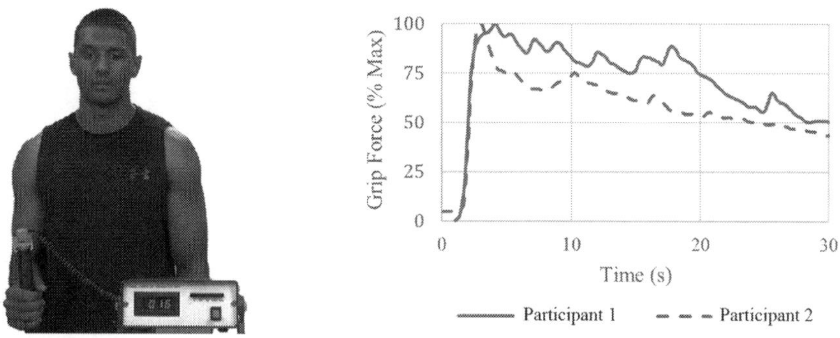

(a) (b)

FIGURE 2.9 Grip strength: (a) grip strength endurance measurement; (b) the change in percentage maximum force during max force holding for two participants, 1 and 2.

The law of motion economy emphasizes that hand movements should be symmetrical and simultaneous to prevent "balance fatigue." When both hands are engaged, it creates a sense of balance and rhythm, boosting productivity. If one hand is idle while the other works, the body strains to maintain balance, leading to more fatigue than when both hands are active. Naturally, our hands move in symmetrical patterns, which is evident when trying to draw a circle with one hand while drawing a square with the other—an inherently challenging task.

2.2.4 MOVEMENT SPEED

Most human limbs pivot around joints, making continuous curved motions easier than straight-line ones, especially those with sudden, sharp changes in direction. Abrupt direction changes force the body part, like the hand, to decelerate, change direction, and accelerate. As a result, curved motions are faster per unit of distance. Additionally, the less mass you need to move, the faster the movement. It is easier to move just the fingers than to move the fingers and wrist or the entire lower arm.

Voluntary movements fall into two categories: controlled (slow or feedback-based) and ballistic (fast or pre-programmed). Throwing a ball is a ballistic movement, whereas placing an object on a shelf or writing will be a controlled movement. Controlled movements involve the contraction of opposing muscle groups. When these muscle sets engage unevenly, movement occurs. Conversely, balanced muscle forces maintain limb stability in a fixed position. For instance, writing with a finger-thumb grip is an example of a controlled movement, where precise muscle coordination dictates motion. On the other hand, ballistic movements are more like sudden bursts of energy, where muscles act swiftly and forcefully in one direction, such as throwing a ball or swinging a bat. These movements rely less on precision and more on speed and power, resembling the rapid and decisive movements of an athlete in action. The time (*T*) taken for a ballistic movement is proportional to the square root of the distance (*D*) moved (Gan & Hoffmann, 1988).

$T = a + k\sqrt{D}$ where *a* and *k* are constants.

Controlled movements are governed by Fitts' Law (1954), where *W* is the width of the target and $\log_2 \frac{2D}{W}$ is referred to as the index of difficulty (*ID*).

$$T = b + c^{*}\log_2 \frac{2D}{W}$$ where *b* and *c* are constants.

2.2.5 CONSEQUENCES OF POOR PHYSICAL DESIGNS

Bernardino Ramazzini, way back in 1717, was one of the first to recognize the health hazards posed by unnatural body postures. Today, these issues are known

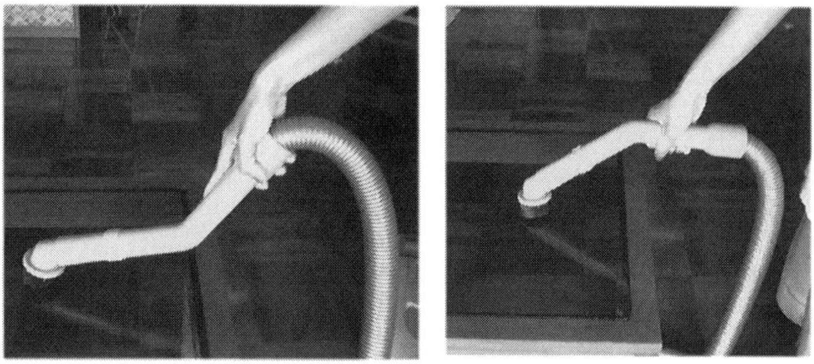

(a) (b)

FIGURE 2.10 A vacuum cleaner: (a) tube installed incorrectly and hand flexed with ulnar deviation; (b) same tube installed correctly and hand in neutral posture.

as cumulative trauma disorders (CTD), occupational overuse injuries (OOS), repetitive strain injuries (RSI), and work-related musculoskeletal disorders (WMSD). These injuries do not happen overnight; they develop gradually over weeks, months, or even years due to repeated stress on specific body parts. The main culprits? High forces, repetitive motions, awkward postures, vibration, lack of rest, and individual factors all contribute to these debilitating conditions. This highlights the importance of following the fundamental principles mentioned at the start of the chapter.

Lack of awareness can lead to misuse of equipment, placing unnecessary stress on the human body. For instance, Figure 2.10(a) illustrates a vacuum cleaner with tubes installed incorrectly, forcing the user into awkward postures while cleaning. To prevent such misuse, designs should be foolproof, ensuring that equipment can only be assembled and used correctly (Figure 2.10(b)). Toyota started mistake-proofing through the concept of poka-yoke (Shingo, 1986).

When designing, we need to account for the different loading patterns on each body part so that we minimize the transmitted forces and the deformations in the greater part of the body. Should the stiffness of a shoe be the same as that of a chair? A chair will undergo static compression while a shoe has to withstand the impact forces as the foot hits the ground and dissipate those to minimize the transmitted forces to the rest of the body.

2.3 SENSORY AND PERCEPTION LIMITS AND CAPABILITIES

Humans cannot get the information without sensors. Do you know what are the common human senses? We mainly perceive information from the world through five senses: vision, hearing, touch, taste, and smell. The capabilities of human sensory system and its limitations are an important consideration (Noyes et al., 2004), in order to make sure that users can receive, process, and understand the information when using products, interfaces, and workspaces.

2.3.1 VISION

Our eyes help us perceive the world, especially when illuminated by light. Just like sound, light possesses distinct physical properties: intensity and wavelength. The wavelength of light is a defining characteristic of waves within the electromagnetic spectrum. Human vision is limited to a specific range of wavelengths, known as visible light, which spans from 380 to 700 nm (Figure 2.11). Light intensity, also known as illuminance, is the measure of how much light falls on a surface expressed in luminous flux (lumens) per square meter (lux or lx) (International Bureau of Weights and Measures, 2019). By measuring illuminance, we can assess whether there is sufficient light for visual tasks. For example, the recommended light level in a typical classroom is approximately 300–750 lux (International Organization for Standardization, 2025). Luminance, on the other hand, refers to the reflected light from a surface, indicating how much light the human eye perceives when looking at a surface from a particular angle. Simply stated, luminance represents the brightness of a surface. Luminance across a space plays a significant role in visual comfort (International Organization for Standardization, 2025).

Light enters our eyes directly or via reflections from various surfaces, allowing us to see. The technical term for what we normally refer to as color is hue. For simplicity, we will use the term color instead of hue.

When you look into someone's eyes, the iris (black, blue, brown, etc.) controls light intake, while the pupil is the dark central opening where light enters. The pupil contracts to 2 mm in bright light and dilates over 8 mm in dim light (Spector, 1990). Pupil size isn't just about light. It also changes with our emotions and focus. When we are highly interested in something, our pupils enlarge—a fact savvy traders use to gauge a customer's interest in a product. Shrewd customers often wear dark glasses to hide this tell-tale sign during negotiations. In low-light conditions, pupils dilate to allow more light into the eye, enhancing vision. So, candlelight, with its soft, dim glow, dilates the pupils, making lovers appear more attractive, attentive, and engaged (Liao et al., 2021).

The next time you look into someone's eyes, remember: there's a lot more going on than meets the eye!

The retina contains two photoreceptors: rods for low light and cones for bright light. Nocturnal animals (e.g., owls) have only rods, diurnal animals (e.g., pigeons) have only cones, and those active both day and night (e.g., humans, monkeys) have both. Vision in bright light is called photopic; in dim light, it is scotopic. Rods don't detect color, so the world appears in black, white, and gray in dim environments. For example, when entering a dark room or stepping outside at night, colors fade as rods take over from cones, which are less active

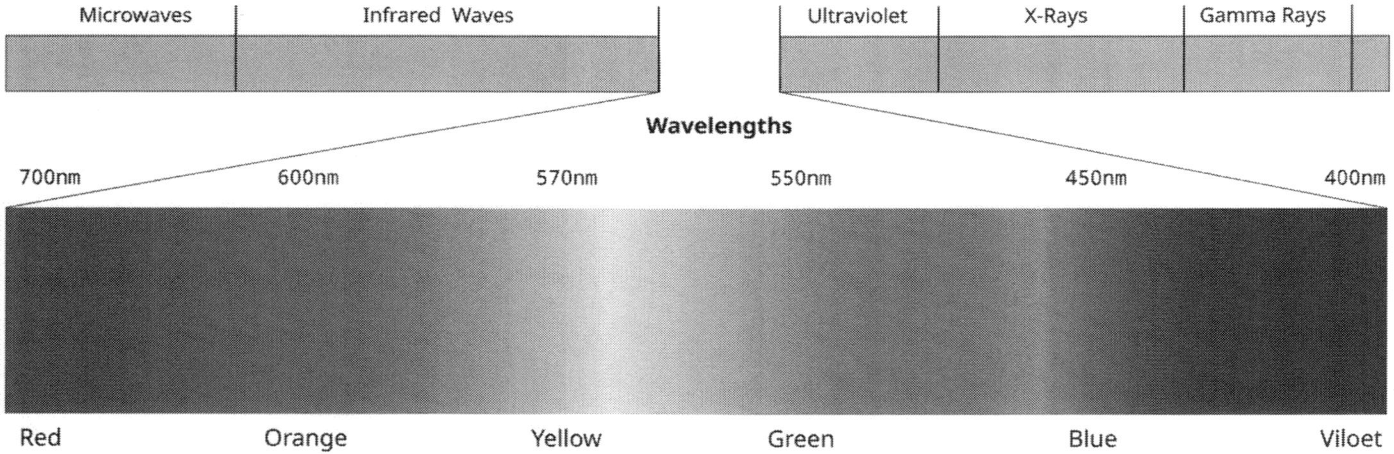

FIGURE 2.11 The electromagnetic spectrum together with that of visible light. (From www.dreamstime.com.)

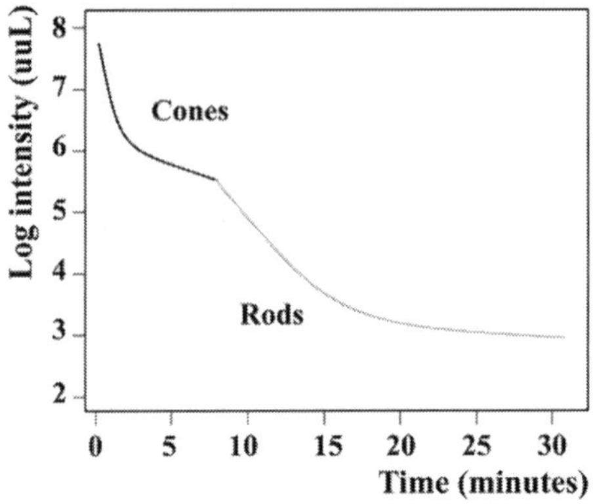

FIGURE 2.12 Dark adaptation curve. (Adapted from Pirenne, 1962.)

in low light. Rods are excellent at detecting light intensity but can't differentiate colors. Transitioning from light to dark takes about 20–30 minutes for full dark adaptation, while light adaptation occurs much faster, in less than a minute (Figure 2.12). This adjustment process allows our eyes to adapt to changing lighting conditions.

The fovea, a small depression at the center of the retina, is densely packed with cones, the photoreceptors responsible for sharp, detailed, and color vision. Moving outward, cone density decreases while rods, which support peripheral and night vision, become more prevalent. The optic nerve transmits signals from both photoreceptors to the brain, where they are processed into images—sometimes leading to visual illusions. If an image falls solely on the optic nerve, it becomes invisible, though binocular vision usually compensates. However, individuals with vision in one eye may experience this blind spot more frequently. The cones are of three types: long-wavelength sensitive (L-cones), middle-wavelength sensitive (M-cones), and short-wavelength sensitive (S-cones). As shown in Figure 2.13, each type of cone is most sensitive (where the curve peaks) to a specific color—red, green, or blue—but can also detect a range of colors as seen from their distribution over the range of wavelengths. This overlap is what allows our brain to perceive millions of colors. For instance, yellow is seen when both green and red cones are stimulated, but not the blue cones. When all three types of cones are equally stimulated, we perceive the color white. Interestingly, our rods, which are more active in low light, can also contribute to our perception of white. This intricate system of overlapping signals and brain integration creates the rich tapestry of colors we experience every day.

Blue light, a vibrant part of the visible spectrum, spans from 380 to 500 nm, the shortest wavelength and highest energy among all visible colors (Wong & Bahmani, 2022). Blue light, a high-energy part of the visible spectrum, is naturally abundant in sunlight but also emitted by artificial sources like fluorescent lights and LED screens of TVs, computers, smartphones, and tablets. This "energetic"

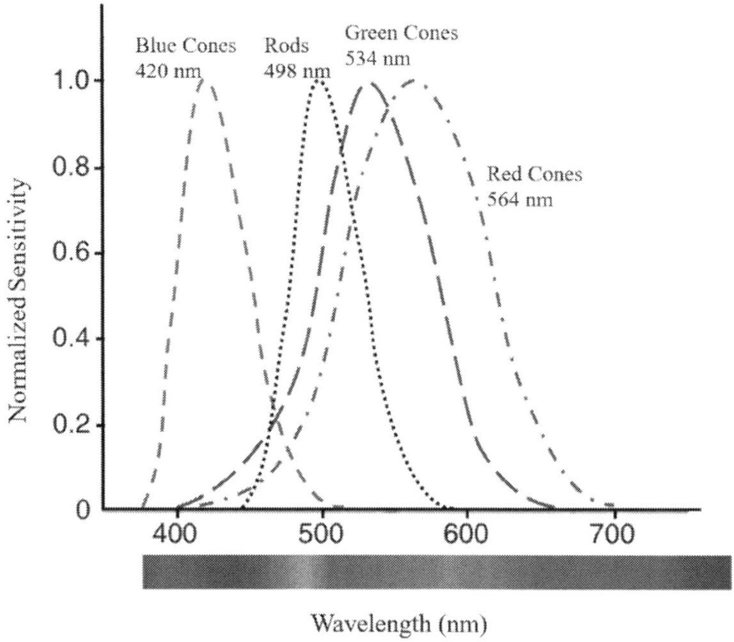

FIGURE 2.13 The three types of cones, rods, and their sensitivities at different wavelengths. (Adapted from Ramamurthy & Lakshminarayanan, 2015.) The visible spectrum ranges in wavelength from 400–700 nm (nm). As the wavelength increases, the frequency (Hertz (Hz) and photon energy (electronvolt (eV)) of light decrease.

blue light plays multiple roles in our daily lives. It enhances alertness, boosts memory, and uplifts mood while also regulating our body's natural sleep–wake cycle, known as the circadian rhythm (Gaggioni et al., 2014). However, exposure to blue light at night can disrupt this cycle by suppressing melatonin production, affecting our ability to sleep.

Although screen-based blue light exposure is small compared to natural sunlight, concerns linger over its long-term effects. Prolonged exposure can pose potential risks, especially from excessive screen time and close viewing distances. Our eyes, inefficient at filtering out blue light, allow almost all of it to penetrate deep into the retina, the light-sensitive cells crucial for visual processing. Over time, this chronic exposure may lead to retinal cell damage, contributing to conditions like age-related macular degeneration, cataracts, and other eye issues. Understanding these impacts encourages us to strike a balance in our digital habits, protecting our eye health while enjoying the benefits of modern technology.

In sensory perception, the concept of vibration in touch and pressure finds its parallel in vision as flicker. At low frequencies, each light flash is distinct and perceived individually. As the frequency increases beyond a certain threshold, the flashes blend into a continuous stream of light. This critical frequency, called the critical flicker frequency (CFF), marks where flicker transitions to perceived steady illumination. The critical frequency is proportional to the logarithm of the illumination intensity (Ferry–Porter Law). CFF helps to predict visual discomfort.

The highest CFF of the eye typically ranges between 45 and 55 Hz. As a result, fluorescent tube lights operating at 100 or 120 Hz do not usually exhibit noticeable flicker, even though they can create a stroboscopic effect on moving reflective objects. However, as fluorescent tubes age or malfunction, a slow, noticeable flicker may develop at the edges of the tubes. This flicker is often caused by a minor 50 Hz oscillation superimposed on the main 100 Hz frequency, likely originating from uneven electrode emissions. This subtle pulsation of alternating current can be perceptible and potentially lead to visual discomfort. To mitigate these effects, the lighting industry has developed phase-shifting equipment that stabilizes light output, reducing the flicker from fluorescent lights. Many modern light units now utilize multiple phase-shifted tubes. In addition to flicker, direct or reflected glare can also impede vision by making objects difficult to see. This glare occurs when light sources shine directly into the eyes or create unwanted reflections.

2.3.1.1 VISUAL ACUITY

If objects are too small, our eyes may not be able to see the details. The ability to see fine details of objects or clarity of vision is called visual acuity. Visual acuity commonly refers to distance acuity (e.g., US Feet "20/20 vision" equals to decimal "1.0"). To enable users to read and recognize the information easily and correctly, we need to ensure that text, icons, and other visual elements are large enough (Figure 2.14). To allow users to see the object, the minimal visual angle, α should be 28 seconds of arc (Curcio et al., 1990). Visual acuity also has a significant relationship with background luminance (Riggs, 1965). In addition, designers must consider other factors such as line thickness, line spacing, size,

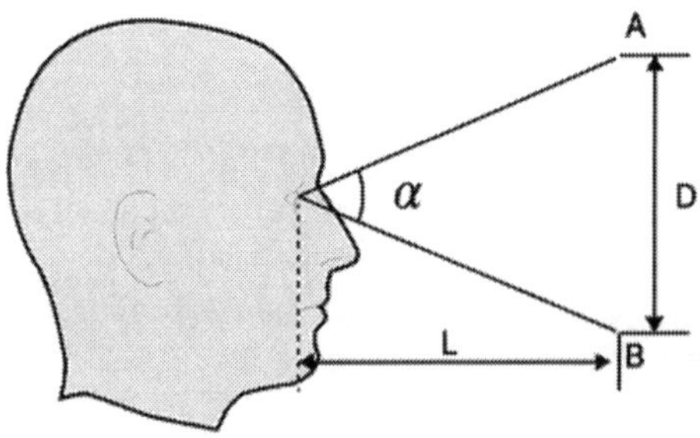

$$\alpha = 2\,arctan\frac{D}{2L}$$

FIGURE 2.14 Minimal visual angle.

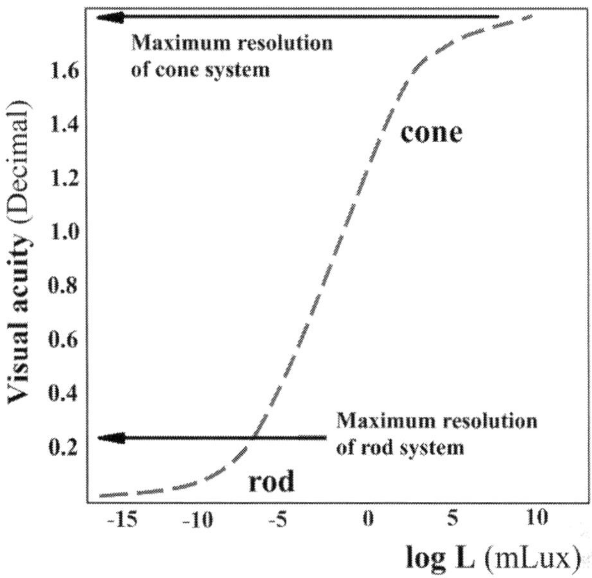

FIGURE 2.15 Relationship between background luminance and visual acuity. (Adapted from Riggs, 1965.)

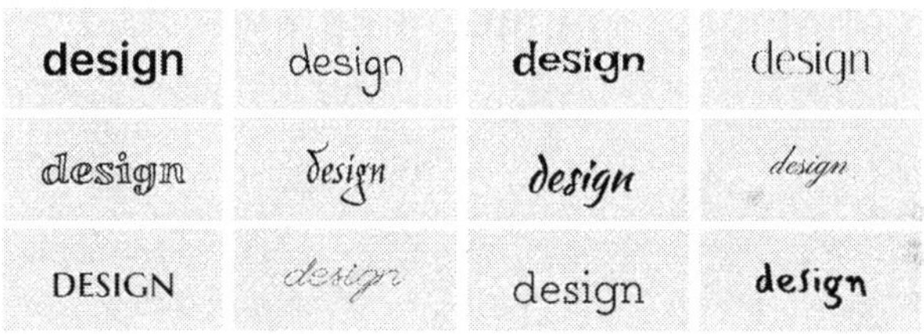

FIGURE 2.16 Visual acuity influenced by font styles.

and font styles when designing graphical symbols and text or signs (Figure 2.16) (Legge & Bigelow, 2011). Pena (2016) has proposed a method for designing and evaluating various typographies using visual acuity.

2.3.1.2 COLOR VISION AND CONTRAST

Ever struggled to read gold text on a red bottle? Certain color combinations can be a nightmare for readability, even for those with perfect vision. These tricky combinations can make text appear blurry or completely illegible, highlighting the importance of choosing contrasting colors for optimal clarity. The appropriateness of color contrast when using text or icons against a background can be determined with a metric known as the contrast ratio. This ratio ranges from 1:1

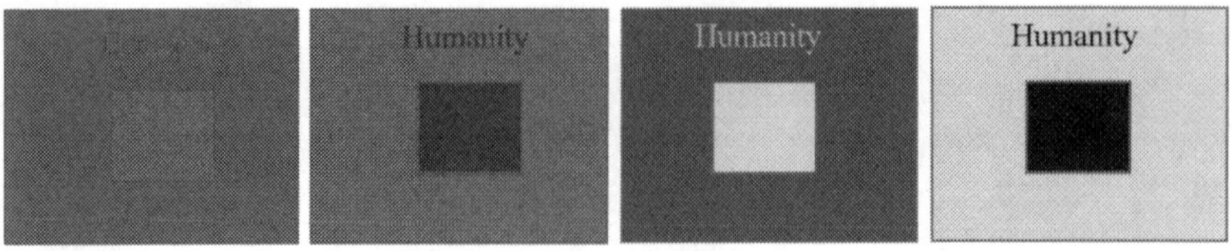

Foreground: (255,4,75) Background: (0,135,32) CR = 1.2	Foreground: (0,0,255) Background: (255,0,0) CR = 2.15	Foreground: (68,255,115) Background: (21,95,156) CR = 5.02	Foreground: (7,19,41) Background: (0,247,231) CR = 13.63

FIGURE 2.17 Contrast ratios of different foregrounds and backgrounds with text and icons showing examples of difficult to easiest to read.

(no contrast) to 21:1 (the highest possible contrast with black and white). While high contrast is generally beneficial, an extreme contrast is not always ideal. A moderate contrast between these extremes is a good choice (Figure 2.17). The contrast ratio is determined using luminance, which reflects how our eyes perceive color. Green light significantly contributes to luminance, accounting for most of the light perceived by humans, while blue light contributes the least and red in-between the two.

CONTRAST RATIO

If the RGB value of a color is known, then the relative luminance of that color can be calculated using the procedure given below.

Rs = R/255 Gs = G/255 Bs = B/255

if $Rs \leq 0.03928$ then $R_N = Rs/12.92$ else $R_N = ((Rs+0.055)/1.055)^{2.4}$
if $Gs \leq 0.03928$ then $G_N = Gs/12.92$ else $G_N = ((Gs+0.055)/1.055)^{2.4}$
if $Bs \leq 0.03928$ then $B_N = Bs/12.92$ else $B_N = ((Bs+0.055)/1.055)^{2.4}$

Relative luminance $L = 0.2126 * R_N + 0.7152 * G_N + 0.0722 * B_N$
The maximum value of L is 1 (white), and the minimum value of L is 0 (black).

Contrast Ratio = (L1 + 0.05) / (L2 + 0.05)
where L1 is the larger value (brighter) and L2 is the smaller of the two (darker).

Examples:

1. Using L1=1 (white) and L2=0 (black) gives a contrast ratio = 1.05/0.05 = 21.
2. Bright yellow (RGB = 254,231,21) on dark charcoal (RGB=16, 24, 32) will give L1=0.7827 and L2 = 0.0086 and a contrast ratio = 14.21, which is a visually striking, high-contrasting color combination.

The visual presentation of text and images of text should have a contrast ratio of at least 4.5:1. Exceptions include large-scale text and images, where a contrast ratio of at least 3:1 is required (Figure 2.17). This 3:1 ratio is the minimum level recommended for standard text and vision (International Organization for Standardization, 1992). Many sites offer tools for conveniently performing color contrast checks (Adobe Color, 2024; Siegemedia, 2024; Snook.ca, 2024). Another useful source is the World Wide Web Consortium (W3C, 2024), which provides standards and guidelines based on accessibility, internationalization, privacy, and security principles. This guideline defines two levels of contrast ratio: AA (minimum contrast) and AAA (enhanced contrast). Level AA requires a contrast ratio of at least 4.5:1 for normal text and 3:1 for large text (at least 18-point text size) or bold text (Figure 2.18). This level is the minimum in some standards as well for text and vision (International Organization for Standardization, 1992). Level AAA demands a higher standard, with a contrast ratio of at least 7:1 for normal text and 4.5:1 for large or bold text.

In Figure 2.19, the black-on-orange combination exceeds the AAA level for contrast, ensuring high accessibility for individuals with visual limitations such as color blindness. Conversely, the white-on-orange combination fails to meet even the AA level. Despite this, many people find white-on-orange clearer in terms of readability, showcasing the subtle relationship between contrast guidelines and perceived readability. But this claim may not be true when there is excessive light, making text blurry due to excessive reflections.

The system just described has its limitations. Black text on a white background is not perceived to be the same as white text on a black background, even though the contrast ratio is the same. There is no consensus on which is better. White text on a black background may be a better option for low-light conditions.

2.3.1.3 USABLE FIELD OF VIEW

Usable field of view refers to the ability to use the whole visual field to perceive detail in the area being looked at and the surrounding area. Ergonomic design considers the user's fields of view (Figure 2.20) (Tilley, 2002). We arrange key visual information and controls within the normal viewing angles to minimize the need for excessive eye and head movements (Clarkson, 2008). For example, when we position visual displays, controls, and work surfaces, we should put them at optimal heights and angles (primary field of view) to minimize neck strain as a result of deviations from neutral posture and eye strain from prolonged use due to muscle fatigue. We can position secondary information, such as monitoring displays, within the visual limit (peripheral field of view).

The usable field of view tends to narrow with age, with peripheral vision gradually diminishing, sometimes resulting in "tunnel vision" (Figure 2.21). This reduction can significantly impact mobility and spatial awareness, making it more difficult for individuals to navigate their surroundings safely and confidently.

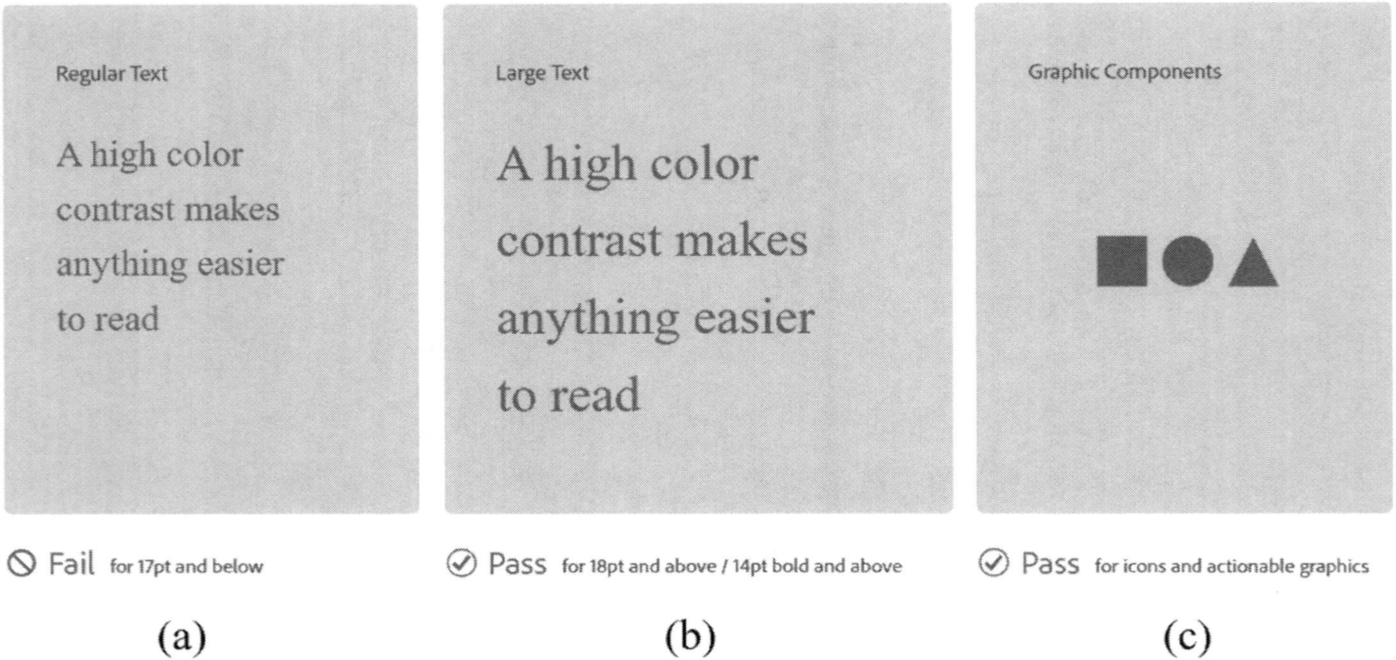

FIGURE 2.18 The sample colors from Adobe Color (2024) show how a contrast ratio of 3.01:1 is (a) unsuitable for smaller text, (b) suitable for larger text, and (c) suitable for icons.

(a) CR = 7.3 (b) CR = 2.86

FIGURE 2.19 Black and white text on an orange background (R=228, G=127, B=0) values and their respective contrast ratios.

2.3.1.4 SUPPLEMENTARY FACTORS

Other factors such as lighting, glare, daylight, and individual health conditions can lead to eye strain, headaches, and fatigue (Dhayal & Jha, 2023). Figure 2.22 demonstrates the effect of glare on an ATM machine screen. Users will not be able to read easily. Proper lighting levels and glare reduction are required to optimize visual comfort and performance (Liu et al., 2024). For instance, an illuminance level above 300 lux is generally acceptable for indoor spaces where energy savings are important. However, increasing illuminance to 500 lux provides a more comfortable lighting environment while an increase to 900 lux may be necessary for high-performance tasks. Visual comfort tends to plateau around 500 lux (Liu et al., 2024).

In today's digital age, prolonged computer use often leads to dry eyes. Adjusting the monitor height to lower levels encourages a downward gaze, which naturally increases eyelid coverage over the eyes. This simple adjustment prompts more frequent unconscious blinking and stimulates the production of more lubrication, helping to alleviate dry eye discomfort associated with screen time.

2.3.2 SOUND, NOISE, AND HEARING

Sound travels as a pressure wave through mediums like air, water, or even steel, but it cannot move through a vacuum (Norton & Karczub, 2012). Think about plucking a guitar string—when it vibrates, it compresses the air around it. These vibrations create waves of compressions and rarefactions, similar to the ripples on a pond when you toss in a rock. The ripples spread out, moving across the water without actually carrying the water along.

The speed of sound varies with the medium. For instance, at 15°C, sound travels about 340 meters per second through the air, but it zips through water around four times faster (1480 m/s) and shoots through steel or glass at another four times the speed (~ 5800 m/s). The formula $C = 20.06\sqrt{T}$ m/s, where T is the temperature in Kelvin (273 + °C), helps us calculate the speed at different temperatures. Interestingly, the speed increases by about 60 cm/s for each degree rise in temperature. When a jet or another object zooms faster than the speed of sound, it creates an exciting phenomenon: the sound waves can't "get out of the way" in time. Imagine it like this—there's an eerie silence for a split second,

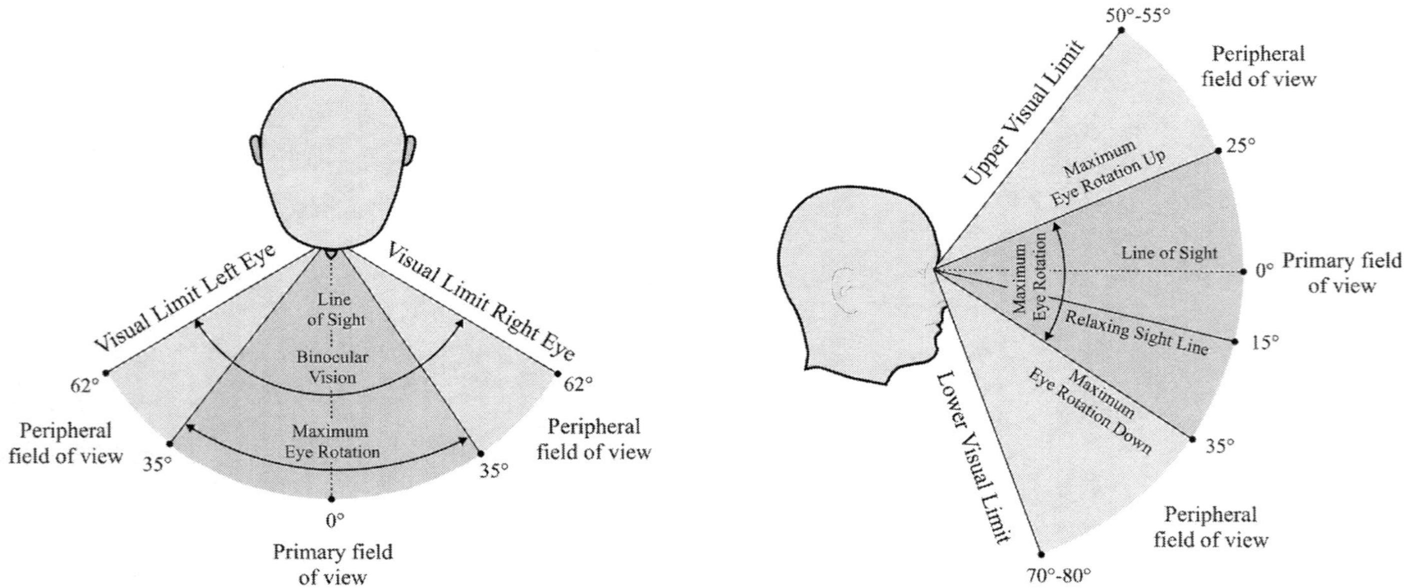

FIGURE 2.20 Usable field of view (a) horizontal (b) vertical. (Adapted from Tilley, 2002.)

(a) (b)

FIGURE 2.21 Demonstration of tunnel vision: (a) full image; (b) tunnel vision view.

FIGURE 2.22 Glare on an ATM machine screen.

and then, BOOM! (NASA, 2017). You hear an explosive shock wave known as a sonic boom. It is the sound catching up all at once, a thrilling reminder of just how fast that jet is moving.

Noise, which is essentially unwanted sound, can be quite a nuisance. Defined as "audible acoustic energy that adversely affects the physiological or psychological well-being of people" (Kryter, 1985), noise can lead to noise-induced hearing loss (NIHL), a common occupational hazard (Berger, 1986), with potentially permanent effects.

2.3.2.1 CHARACTERISTICS OF SOUND: FREQUENCY, COMPLEXITY, AND AMPLITUDE

Sound has some intriguing characteristics: frequency (measured as cycles per second or Hertz), complexity, and amplitude. Human hearing spans from 20 Hz to 20 kHz, but individual abilities can vary significantly. Any frequency below 20 Hz is termed infrasound. Speech typically falls within 100–5,000 Hz, most concentrated between 300 and 700 Hz. On the other hand, dog whistles operate at a much higher range of 23,000 to 54,000 Hz. Musical instruments also vary widely in their frequency ranges: a pipe organ can go as low as 20 Hz, while a violin can reach up to 15,000 Hz. These frequency variations shape our perception of different instruments, making each one sound distinct and unique. It is these frequency differences that allow us to distinguish between the mellow strum of a guitar and the resonant clang of a cymbal to the symphony of sounds we enjoy.

Our ability to hear high frequencies diminishes with age (Figure 2.23). By the time we reach 50 years, high-frequency perception starts to decline. By age

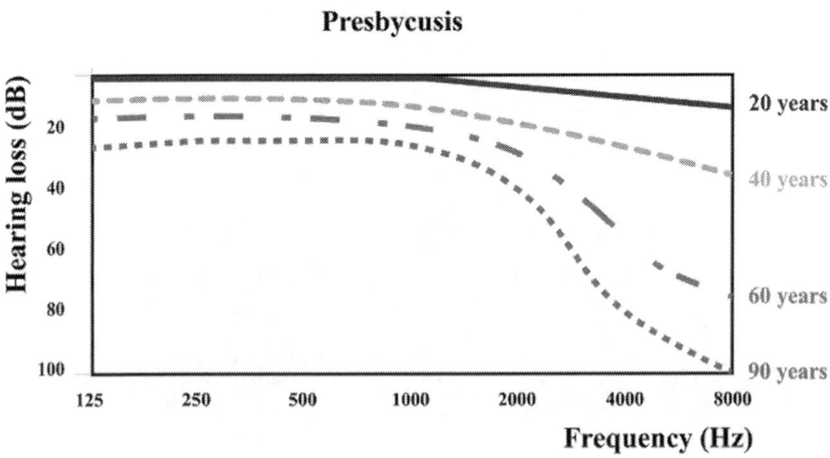

FIGURE 2.23 Age-related hearing loss. (Adapted from Wasano et al., 2021.)

70, hearing loss can affect frequencies as low as 3000 Hz (Kim et al., 2020). That means these people may not hear the phone ringing, the doorbell, birds chirping, or even the voices of their grandkids. Hearing loss in the 2,000–8,000 Hz range can make everyday conversations a frustrating game of guesswork. This type of hearing loss primarily affects the ability to understand consonant sounds, which are important for speech clarity (Rishiq et al., 2020; Woods et al., 2012). Imagine trying to follow a conversation where every third word sounds like gibberish—that's what it feels like. It is not just about volume; it is about clarity. This loss makes it especially hard to hear the higher-pitched voices of children and women. People often say, "I can hear, but I can't understand," because while they might catch the vowels, the crucial consonants slip away, thus guessing what was said. This subtle yet significant gap in hearing can turn simple interactions into bewildering puzzles.

When sounds include multiple frequencies, these various components interact in often complex ways. Noise sources like road traffic or factory machinery typically feature a broad spectrum of frequencies, not just a few dominant ones. To understand this mix, analysis techniques are used to break down and identify the spectrum of components. Essentially, when a sound wave is complex, containing more than one frequency component, it shapes what we perceive as *timbre*. This *complexity* of sound gives it its unique character, making the hum of a city street or the clatter of an industrial setting distinct and identifiable.

The amplitude of sound is characterized by sound intensity (I), which can be measured in terms of pressure (p), power, or energy. Think of it this way: sound intensity is the power carried by sound waves per unit area in a direction perpendicular to that area, and it is measured in watts per square meter (W/m^2). While engineers rely on physical measurements like amplitude and frequency to quantify sound, these measurements do not capture our subjective perception. This is because the human ear does not have a linear frequency response; it filters some frequencies more than others. For instance, a sound at 20 Hz will seem much quieter to us than a sound at 10 kHz, even if both have the same physical intensity. This discrepancy highlights the fascinating complexity of how we experience sound.

Let's look deeper into the world of sound, where frequency and amplitude create our auditory sensation of *pitch* and *loudness* (Fletcher & Munson, 1933). Pitch is determined by the frequency of the sound wave. Loudness, however, is how powerful or intense that sound feels to your ears. It is measured in phons, a unit tied to the decibel level of a standard 1,000 Hz tone (Figure 2.24). Now, consider this: our ears are incredibly sensitive, capable of detecting sounds as quiet as 20 μPa and as loud as 200 Pa of pressure. That is a range of a billion times! We use the decibel (dB) scale, named after Alexander Graham Bell, to manage this vast range. The decibel scale helps us simplify and compare these huge differences in sound pressure. Two variables differ by one decibel (1 dB) if they are in the ratio $10^{1/10}$ (~1.260) or by 3 dB if they are in the ratio $10^{3/10}$ (~2). This clever system compresses everything into a more manageable scale, which,

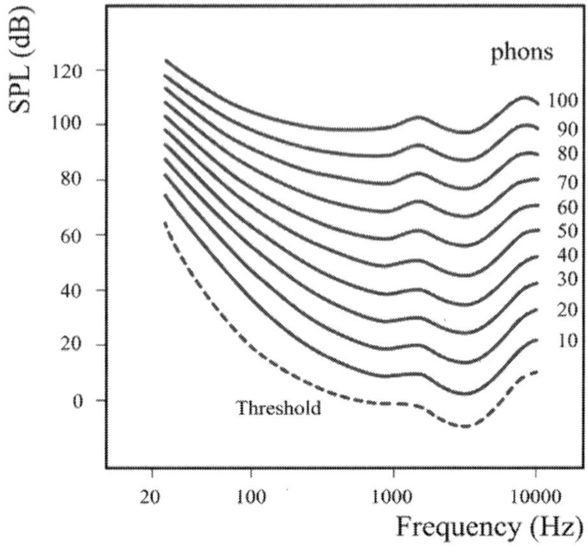

FIGURE 2.24 Loudness contours given in phons from International Organization for Standardization (2003). (Adapted from Florentine et al., 2011.)

when related to human hearing, would range from 0 to 160 dB. Such a scale makes it easier to understand and measure the sounds we hear every day.

Sound Pressure Level (SPL) (dB) = $10 \log_{10} (I/I_0) = 10 \log_{10} (p^2/p_0^2) = 20 \log_{10}(p/p_0)$

I_0 = reference level intensity at the threshold of hearing (10^{-12} Wm^{-2})

Because intensity is proportional to the square of the pressure, we can substitute p^2 for I.

p_0 = reference level amplitude at threshold of hearing ($2*10^{-5}$ Nm^{-2})

Think of your ears as a finely tuned instrument, but not every note sounds the same. Researchers have created equal loudness curves to capture this difference, which shows how we perceive different frequencies at various sound pressure levels (Figure 2.24). These curves relate the physical attributes of sound pressure level and frequency with the subjective perceptions of loudness. These equal-loudness curves have been revised over the years (International Organization for Standardization, 1987; International Organization for Standardization, 2003; Suzuki & Takeshima, 2004; Suzuki, et al., 2024). At lower frequencies, like 30 Hz, a sound needs to be significantly louder to be heard as clearly as a mid-frequency tone. Imagine a curve that dips and bows. At low loudness levels, the bow in the curve is very high, implying that low frequencies, such as a sound at 30 Hz with a sound pressure level (SPL) of around 70 dB, will be heard at the same loudness level as an SPL of 10 dB at 1,000 Hz (10 phons curve). The bottom-most curve called the minimum audible field (MAF) represents the quietest sounds we can hear.

Our ears are most sensitive to frequencies between 1,000 and 8,000 Hz, as seen from the curves (Figure 2.24), requiring the least intensity to be heard. This is why fire alarms are engineered at around 3,000 Hz, with sound levels blasting between 75 to 90 dB, ensuring they grab our attention effectively.

Another fascinating aspect of equal loudness curves is how the "bow" in the curve changes with increasing loudness. At around 100 phons (the curve right at the top), the bow is relatively small, which explains why music at concerts sounds so pleasing—different frequencies are heard at about the same level because of the high loudness, and very little equalizing is required. These equal loudness curves have significant implications for designing audio products like equalizers. An equalizer is a device that lets you adjust certain frequency ranges to improve or change the sound quality. Because our hearing sensitivity varies across frequencies, equalizers can amplify low and high-frequency sounds to ensure all frequencies are heard at the same loudness level or to cater to our preferences such as more treble or more bass. This means that an equalizer setting isn't one-size-fits-all—it varies with the type of music and from person to person. That is why some people prefer certain brands of headphones, especially those with a bass-heavy profile. For instance, pop music, which focuses on vocals and mid-range sounds, benefits from boosting mid-range frequencies and slightly enhancing the low-mid and high-mid frequencies. This fine-tuning helps create a balanced and immersive listening experience, making every note and beat resonate just right.

2.3.2.2 MEASURING SOUND

Because our hearing sensitivity varies with frequency, simply measuring sound pressure level does not tell us how loud a sound will seem. To account for this, sound levels are often measured on a decibel scale adjusted to match the ear's characteristics. The A, B, and C scales match our hearing response to sounds of low, moderate, and high intensity. The most common is the dBA scale, which best correlates with how we perceive loudness and noise annoyance. The A-scale, designed to match the 40-phon equal loudness contour, approximates the human ear's response characteristics closely (Salomons & Janssen, 2011). The B-scale corresponds to the 70-phon contour, while the C-scale aligns with the 90-phon contour and weights all frequencies almost equally, providing a flat rating. The D-weighted scale is tailored to amplify high frequencies, making it ideal for assessing high-frequency noise, like aircraft noise, more accurately.

2.3.2.3 EFFECTS OF EXTREME NOISE

Exposure to intense noise can cause temporary hearing loss, lasting from a few hours to several days. However, prolonged exposure to intense noise can lead to irreversible hearing loss as the ear's delicate, sound-sensitive cells gradually deteriorate. While temporary hearing loss may occur after short-term exposure, repeated or continuous noise, particularly at high frequencies or from sudden loud sounds, can result in permanent damage. This noise-induced deafness results from the gradual degeneration of the ear's sound-sensitive cells. The louder and more prolonged the noise, the greater the damage, with high-frequency noise being particularly harmful. Intermittent high-intensity noise and even a single loud explosion can also cause permanent hearing damage. To combat this, many countries have set permissible noise exposure limits to minimize noise-induced hearing loss. Standards, such as American National Standards Institute (2016) and Canadian Standards Association (2018), vary by country and provide guidelines for determining noise exposure and estimating the risk of hearing loss.

Hearing loss, background noise, and poor acoustics can make it challenging for users to clearly understand speech, especially in group settings or noisy environments. When designing communication systems, public address systems, and user interfaces, speech intelligibility should be optimized. In special situations, it is necessary to provide clear and distinctive audio cues with alarms, alerts, and confirmations.

2.3.3 TOUCH

When we wear clothes, shoes, or neckties, sleep on a bed, sit on a chair, use a headset, handle tools during manual work, or rely on armrests or elbow supports to reduce muscle fatigue, the forces from these external sources exerted on our bodies are quite varied. However, they all act on limited contact areas. The contact pressure that acts at the human interface has received considerable attention over the years because it can cause discomfort, improve comfort, or even lead to injury and pain. Minimizing pain and discomfort is crucial for many products that people use daily. Unfortunately, the strategy to enhance sensation or diminish discomfort varies. Floyd and Roberts (1958) suggested focusing the load on bony areas, while others proposed distributing the force uniformly over the contact area to reduce discomfort (Sanders & McCormick, 1987; Sprigle et al., 1990; Weichenrieder & Haldenwanger, 1986). Imagine sitting on a very soft sofa versus a hard one or wearing a very soft shoe versus a stiff one. The very soft surfaces may feel comfortable initially, but they often lack the support needed to keep the rest of the body stable. Over time, this can lead to greater overall discomfort as the body "slips" away into awkward postures. Designing to effectively manage loads and the various loading rates is not a trivial task.

FIGURE 2.25 A touch-and-feel playbook.

Designers face a significant challenge in accounting for the variation in pressure sensitivity across different body parts when designing products (Hirao et al., 2022; Weerasinghe et al., 2017). For instance, the plantar surface of the foot can be divided into five sensitivity clusters: arch, forefoot, lateral midfoot, outer heel, and inner heel (Weerasinghe et al., 2017). This means that the stiffness of a shoe's midsole should vary across these regions to ensure optimal comfort and support. Similarly, Hirao et al. (2022) found that the buttocks and thighs have different pressure sensitivities, indicating that acceptable pressure levels may vary depending on the contacting body part. Product design must carefully consider these differences to enhance user comfort and prevent discomfort or injury. Designs should be based on discomfort thresholds, which are typically 40–50% of the pain thresholds (Rodrigo et al., 2013; Weerasinghe et al., 2017).

Besides pressure, touch also refers to the sensory perception and evaluation of texture, temperature, and vibration. Users can interact with and receive tactile feedback from products, surfaces, and interfaces (Figure 2.25). Effective haptic feedback can enhance control, engagement, and user experience. Tactile stimuli can affect physiological responses, influencing factors like arousal, stress levels, and cognitive performance. These sensory inputs can also lead to noticeable behavioral changes, highlighting the connection between touch and human reactions (Peck & Wiggins, 2006).

Incorporating tactile cues for spatial and navigational information can enhance users' ability to orient themselves and navigate their surroundings more effectively. This approach is especially beneficial in designing tactile way-finding systems, particularly for individuals with visual impairments. Moreover, tactile information can be seamlessly integrated with other sensory inputs, such as vision and sound, to enrich cognitive processing, decision-making, and the overall user experience.

Touch experiences are generally confined to specific physical products or surfaces, lacking portability or transferability across different contexts. Additionally, touch perception and sensitivity can vary widely between individuals, influenced by factors such as age, gender, skin condition, culture, and personal experiences (Samain-Aupic et al., 2024; Vafeiadou et., 2022). In addition to individual factors, touch experiences are also influenced by environmental factors such as temperature, humidity, and surface textures, all of which can significantly affect the user's subjective experience (Steven & Hooper, 1982; Liao et al., 2018). Furthermore, designers must consider safety and regulatory guidelines when integrating tactile elements, particularly in relation to material selection, surface finishes, and physical interactions, ensuring that the tactile experience is both functional and safe for users. These variations highlight the complexity of designing tactile interactions that are universally effective and meaningful for diverse user groups.

2.3.4 TASTE AND SMELL

Taste refers to the user's ability to detect and distinguish various flavor profiles, such as sweet, sour, salty, bitter, and umami (Briand & Salles, 2016). These taste experiences are often influenced by factors such as age, food or drink temperature, consumption time, and the sequence in which different tastes are experienced (Weiffenbach, 1984; Talavera et al., 2007). The perception and enjoyment of taste are shaped by external factors like the physical environment, social setting, and cultural background, all of which can significantly affect how flavors are perceived and enjoyed (Bschaden et al., 2020).

Smell refers to the sensory perception and evaluation of olfactory stimuli (Köster, 2002). Olfactory perception and sensitivity can vary significantly among individuals due to factors such as age, genetics, medical conditions, and personal experiences (Weiffenbach, 1984). Olfactory experiences are highly dependent on temporal factors, such as the duration of exposure, the concentration of the scent, and environmental conditions, such as temperature and airflow.

Taste and smell may not always take center stage, but they have a powerful influence on human emotions (Köster, 2002). These senses often work together, with taste and smell combining to create a rich, multidimensional flavor experience (Schiffman & Graham, 2000). Beyond taste and smell, other senses such as touch can be integrated into information design to create more immersive and effective experiences. By combining multiple sensory inputs, designers can enhance the overall impact and compensate for deficiencies in any one sense—especially important in emergency situations where clear, multisensory communication is essential.

2.4 COGNITIVE LIMITS AND CAPABILITIES

Cognition limits refer to the inherent boundaries or constraints of the human brain's capacity for processing information and performing cognitive tasks (Nicolay & Walker, 2005). In other words, it's the point at which our brains become overwhelmed and can no longer function at an optimal level. Cognition process includes the few key steps in our brain: perception; attention; memory; information processing; decision-making; and action. This cognition process happens rapidly and continuously as we engage with the world. It's a remarkably complex feat that our brains accomplish with incredible speed and efficiency, most of the time without us even being aware of it. Understanding the cognition process helps us appreciate the amazing capabilities of the human mind. It also highlights areas where we may have cognitive limitations that need to be accounted for in design, learning, and performance.

2.4.1 MEMORY

Memory is the process in which information is encoded, stored, and retrieved. There are three types of memory: sensory memory, short-term memory or working memory, and long-term memory (Winkler & Cowan, 2005; Cowan, 2008).

Memory has limited capacity. Sensory memory can take large amounts in few seconds but forget them after that. The human short-term or working memory has a relatively limited capacity, typically being able to hold around seven (±two) items of information at a time (Miller, 1956). Information stored in human memory is subject to forgetting and decay over time, particularly in the case of short-term memory (Figure 2.26). The solid line demonstrates the memorization level from sensory memory with repeated rehearsal over time can reach to long-term memory. The dotted line represents decay of memory without rehearsal. Long-term memory requires learning and higher repetition rate, enabling it to store large amounts of information for a lifetime (Figure 2.26). Short-term memory can be sensitive to interruption such as a spatial shift. People often forget what they need to do or what they are thinking about once they enter another room, known as the doorway effect (Figure 2.27). New information can interfere with the retention and recall of existing memories, leading to confusion or the inability to remember specific details.

Memory is influenced by factors such as attention, encoding strategies, emotions, stress and anxiety, gender, age, personality, medical conditions, and other individual factors (Blasiman & Was, 2018). It is also highly context-dependent, meaning information retrieval is affected by the environment,

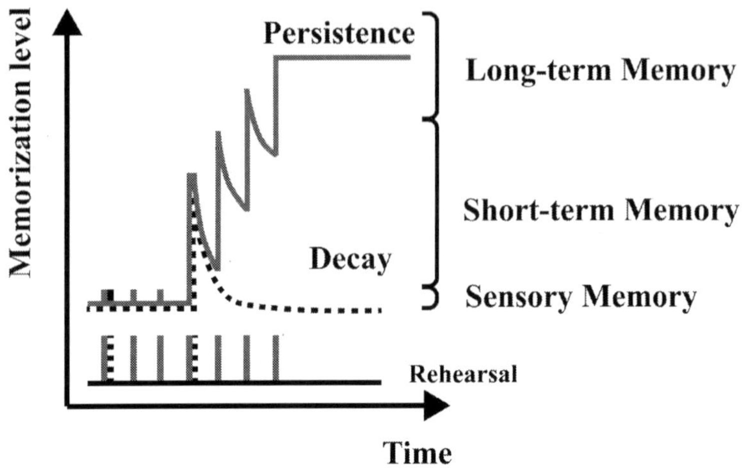

FIGURE 2.26 How repetition rate causes different memory level. (Adapted from Ohno et al., 2011.)

sensory input, and other external factors (Zhang, 2025). Designers must minimize the cognitive load on users, particularly when introducing a new product. A person's ability to learn and use a product is closely tied to memory – specifically recalling past experiences or relying on habits (Labrecque et al., 2017). When new information does not align with prior knowledge or routines, the learning process slows down, leading to more errors and frustration. Keeping the user's memory load helps in creating a smooth, intuitive experience.

2.4.2 ATTENTION

Attention involves selectively focusing on certain information while filtering out distractions (Cowan et al., 2024). Humans have a limited capacity for attention. They can either concentrate on a single task or stimulus (focused attention) or split their attention between multiple tasks or stimuli, but only to a limited extent (divided attention) (Wickens et al., 2022). When the tasks involve complex cognitive processing, humans will have difficulty maintaining attention on multiple tasks or stimuli simultaneously. Therefore, they may selectively attend to certain aspects of their environment or design, sometimes ignoring or overlooking other elements (Luximon & Goonetilleke, 2012). Attention abilities can vary significantly among individuals due to factors such as age, cognitive abilities, cultural backgrounds, and personal preferences (Gausby, 2015; Fortenbaugh et al., 2015; Goonetilleke & Luximon, 2010; Simon et al., 2023). Highly salient or novel stimuli can initially capture users' attention. However, prolonged exposure can lead to attention fatigue and a decline in engagement.

FIGURE 2.27 The doorway effect.

Designers must prioritize the most critical information and interactions to help users focus their attention effectively. By using visual cues, smart layouts, and design principles, they can guide attention where it is needed most, reducing the strain of divided attention. Clear visual hierarchies and intuitive task flows are essential for minimizing distractions and ensuring seamless user experiences.

2.4.3 DECISION-MAKING

Decision-making involves the cognitive processes that shape how users make choices and selections. It is influenced by various factors, including cognitive capacity, the volume and organization of information, feedback, time pressure, context, biases, habits, and emotional states (Ahituv et al., 1998; Keane & Thorp, 2016; Hertwiq & Todd, 2003; Lebiere & Anderson, 2011). These elements can significantly impact the quality and speed of decisions, making it essential for designers to consider how these factors interact to create smoother, more informed user experiences. Humans lack the cognitive capacity to make perfectly rational decisions, as they are limited by their knowledge and information-processing abilities. When bombarded with large amounts of information in a short time, confusion, frustration, and poor decisions follow. Users also tend to rely on habits and routines, making them resistant to change or new decision-making processes. For example, drivers or travelers may instinctively look at the wrong side of the road in countries with different driving rules (Ye et al., 2021). Beyond these factors, the environment, social influences, and emotions all control decision-making. How information is presented and framed can significantly impact users' perceptions and choices, leading to different outcomes. Clear, timely, and meaningful feedback on the consequences of decisions not only helps users make better choices but also improves their decision-making skills over time.

2.4.4 LEARNING

Learning is all about how users acquire, retain, and use new knowledge, skills, and behaviors when interacting with environments through the senses (Lachman, 1997).

Motivation, engagement, and prior knowledge can greatly impact in how well and how willing they are to learn (Maslow, 1943; Schiefele, 1991; Alonso et al., 2020). That is why designs need to be fun, rewarding, and spark curiosity, giving users a sense of accomplishment as they go. By introducing new concepts gradually and offering the right support and guidance, users can learn more easily. The best designs also let users apply their newfound skills across different tasks and situations.

Feedback plays a critical role in learning (Luft, 2014). Clear, timely, and meaningful feedback on users' actions and performance can greatly accelerate their learning and skill development. Designers should strive to make their creations inclusive and adaptable, accommodating the diverse learning needs and preferences of all users. This approach not only enhances the learning experience but also ensures that no one is left behind.

2.4.5 MENTAL WORKLOAD

As work shifts from physical labor to cognitive tasks, mental workload is becoming increasingly relevant. Mental workload refers to how much effort your brain exerts at any given moment. It's like your brain's workout level—how hard it is working to get things done. When you are juggling multiple tasks or ideas, your mental workload increases. Imagine trying to write a paper while listening to music and responding to text messages; that is a recipe for high mental workload! Constantly switching your attention between these tasks can be exhausting and lead to decreased focus and productivity. Managing your mental workload is key to staying sharp and effective.

2.4.5.1 FACTORS INFLUENCING MENTAL WORKLOAD

Several key factors influence mental workload. First, task characteristics play a significant role. Task complexity, which reflects the difficulty and cognitive demands of a task, directly impacts mental workload; more complex tasks generally require greater cognitive effort (Galy et al., 2012). Additionally, multitasking can dramatically increase mental workload, especially for monochrons—those who prefer focusing on one task at a time (Zhang et al., 2005). Time constraints also add pressure; tight deadlines can heighten mental workload and stress (Galy et al., 2012).

On the individual level, factors such as cognitive abilities, experience, and strategies vary widely, affecting the mental workload for the same tasks (Zhang & Goonetilleke, 2005; Galy et al., 2012). Moreover, health conditions like lack of sleep, fatigue, and high stress levels can further increase mental workload, stressing the importance of considering both task demands and individual capabilities.

2.4.5.2 MENTAL WORKLOAD MEASUREMENTS

Researchers have crafted a variety of innovative techniques to measure and quantify mental workload. From cutting-edge brain imaging to subjective surveys and performance metrics, these methods unveil the hidden efforts of our minds. By tapping into physiological responses or tracking cognitive tasks, scientists can gain valuable insights into how we manage mental demands, providing a clearer picture of our brain's busy work behind the scenes.

- Subjective rating scales such as NASA Task Load Index (NASA-TLX), the Subjective Workload Assessment Technique (SWAT), and the Workload Profile (WP), allows users to self-report their experienced mental effort (Luximon & Goonetilleke, 2001; Zhang & Luximon, 2005; Rubio et al., 2004). NASA-TLX has six dimensions (mental demand, physical demand, temporal demand, performance, effort, and frustration level). SWAT has three dimensions (time load, mental effort load, and psychological stress load). WP uses multiple resource theory to evaluate the mental workload on four dimensions (Stage of processing, code of processing, input and output).
- Physiological measures like heart rate variability, pupil dilation, and brain activity (Charles & Nixon, 2019) can be used to evaluate the mental workload. Functional magnetic resonance imaging or

functional MRI (fMRI) and Electroencephalography (EEG) both can demonstrate where is and level of the activated brain regions.
- Other performance-based metrics like task completion time, error rates, and secondary task interference (Kaber & Riley, 1999) also can indicate the mental workload level.

The design of something can make all the difference in how hard your brain has to work to use it. A well-thought-out design simplifies tasks, guiding you effortlessly through the process. On the flip side, poor design can turn even simple tasks into a mental workout. The goal of great design? To ease the mental load so users can focus on what they want to achieve—not on figuring out how to get there.

2.4.6 CONSEQUENCES OF POOR COGNITIVE DESIGNS

We have all encountered designs that look fine, but are frustrating to use or prone to errors. Don Norman has provided various good examples in his renowned book *The Design of Everyday Things* (Norman, 2013). Users might struggle to see or hear important information, misunderstand instructions, or simply forget key details—leading to mistakes. A common example is the stove design in Figure 2.28 (a). It features four burners, each with its own control, but the confusing layout causes users to frequently turn on the wrong burner. This design flaw stems from poor spatial mapping, forcing users to remember switch positions or check the printed instruction again and again instead of relying on intuitive and natural associations. Figure 2.28(b) shows an example of design using natural mapping principle and it obviously does not require any additional mental workload. Designers must understand how human biases and mental workload affect decision-making and prioritize layouts that support easy information processing and reduce the chance for errors (de la Fuente et al., 2014).

(a) (b)

FIGURE 2.28 An example of stove design: (a) arbitrary burner arrangement; (b) natural mapping burner arrangement. (From www.poorvika.com.)

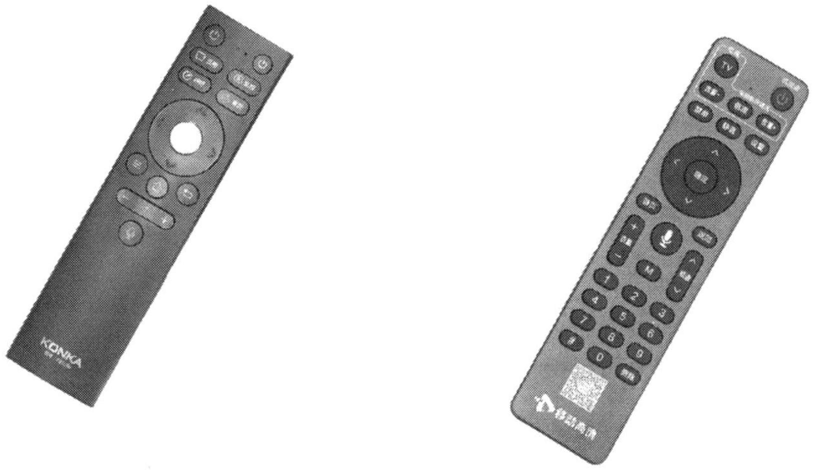

(a) (b)

FIGURE 2.29 Examples of remote controllers: (a) simple design without detailed labels; (b) complex design with more labels.

For the products with more complex content, designers should try to make it easier for users to use and make good decisions. They should present information in a clear and organized way, and give users tools to help them decide, like visual guides or step-by-step processes. Also, the design should make users easily recognize rather than recall. Figure 2.29 shows examples of remote controllers. Remote controls without detailed labels on the buttons (Figure 2.29(a)) are sometimes very difficult for elderly users to remember how to use them comparing to the ones with clear labels (Figure 2.29(b)). The overall user experience and situation should also be considered, to create an environment that supports good decision-making. Designers should strive for transparency, so users understand the options and what they mean.

2.5 PHYSIOLOGICAL LIMITS AND CAPABILITIES

When one does physical work, the body's energy demands soar, much like a car engine revving up on a steep hill. To meet these increased demands, the muscles call for more oxygen and need to dispose of accumulated waste more efficiently. Imagine a bustling city where traffic suddenly doubles—more buses are dispatched to shuttle people around, traffic signals change faster, and street sweepers work overtime to clear the debris. Similarly, your vascular activity ramps up, with more blood being pumped to the muscles in need. As your

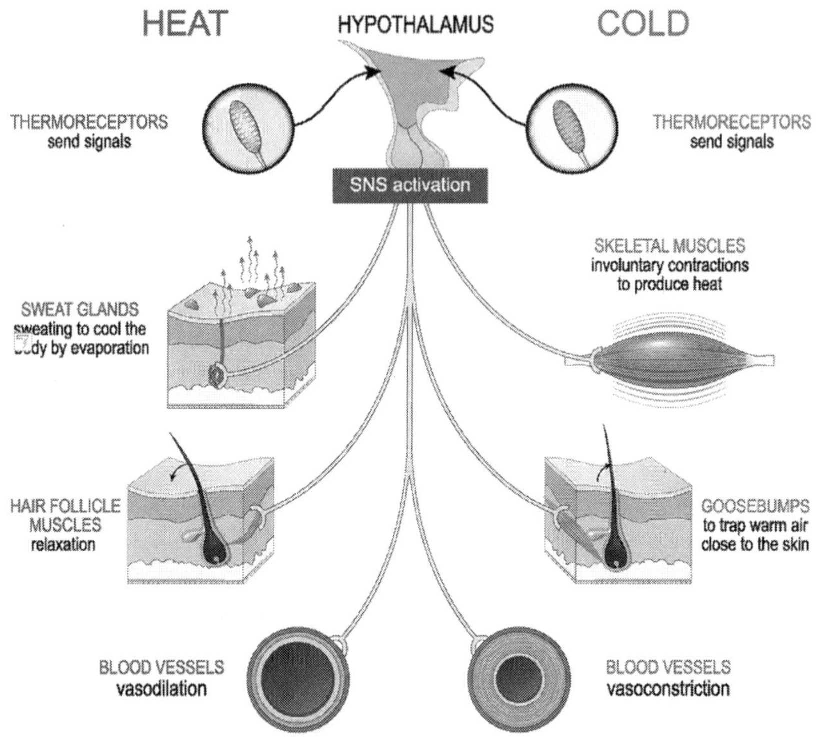

FIGURE 2.30 Mechanisms with which the body copes with heat and cold. (From www.shutterstock.com/.)

muscles crave more oxygen, your breathing becomes faster and deeper. This heightened respiratory activity ensures that more oxygenated blood reaches the lungs within a given time. Concurrently, the heart rate spikes, enabling more blood to be processed and circulated per minute. With the surge in muscular activity, heat production also rises, turning your body into a furnace. To prevent overheating, the body activates its cooling system—sweating (Figure 2.30). Sweat evaporates from your skin, carrying away heat and helping to maintain a stable internal temperature. Thus, oxygen consumption and heart rate are the primary indicators of physical strain, reflecting how hard your body is working to keep you going. Just as the RPM gauge of a car and temperature meter tell you how hard the engine is working and when it might overheat, these physiological markers reveal the intensity of the effort your body is enduring.

2.5.1 OXYGEN CONSUMPTION AND HEART RATE

There is a linear relationship between oxygen consumption and heart rate, making heart rate a convenient proxy for estimating how much oxygen is used and, subsequently, how much energy is expended. Think of it as a simple yet effective speedometer for your body's engine. However, this neat relationship can get thrown off in certain scenarios. For instance, imagine you are hammering nails into a ceiling. Your arms are overhead, and suddenly, your heart rate skyrockets. This isn't because you're running a marathon but because working with your arms raised sends your heart rate through the roof. Similarly, working in a hot environment will increase the heart rate as the body tries to cool itself down. Emotional stress can also push the heart rate higher than expected.

Increasing the work rate generally leads to a linear increase in oxygen uptake—up to a point. Beyond this threshold, oxygen uptake plateaus, no matter how much harder you push yourself. It is like your car hitting its maximum speed; you can press the pedal to the metal, but you will not go any faster. Moreover, humans cannot sustain high-intensity activities indefinitely. Just as a car needs a break to avoid overheating, your body requires rest to recuperate. The working heart rate (the increase above your resting heart rate) can be used to calculate rest allowances for both static and dynamic work, ensuring you don't burn out. While the theoretical maximum heart rate (HRmax) can vary significantly, it is generally related to age. It is calculated from the formula: HRmax = 220 – age. It does have its limitations (Robergs & Landwehr, 2002) and a different formulation has been proposed: HRmax = 211 – 0.64 × age (Nes et al., 2013). The maximum heart rate is an important parameter when designing fitness trackers, smartwatches, heart rate monitors, mobile fitness apps, and cardiovascular equipment.

Monitoring physiological changes within the human body has become remarkably easy with the advent of the Internet of Things (IoT) and other miniature electronics. Wearable technology is on the rise, offering significant advancements in health monitoring (Kazanskiy et al., 2024; Lu et al., 2020; Petek et al., 2023; Wu & Luo, 2019).

2.5.2 THERMAL COMFORT

Did you know our bodies are like well-tuned thermostats? Normal body temperature is around 36.8°C (37 ± 0.5°C or 98.6 ± 1.8°F), but everyone runs a little hot or cold (Geneva et al., 2019). Imagine you are at the gym, sweating it out. Your body generates heat, and to stay comfortable, it has to get rid of that extra warmth. This balance is crucial for what we call the "comfort zone" (Sanders & McCormick, 1993).

Think of your body as having two layers: the core, which includes vital organs, and the shell, which is about 25 mm of skin. Our internal "thermostat" cares about the core temperature, usually about 4°C higher than our skin temperature. Each individual is different and that is why you might feel hot while someone else feels cold and vice versa, even in the same room. It is also why gyms and swimming pools are too warm for some and too cold for others—there is no one-size-fits-all when it comes to heat balance!

When our bodies generate heat from metabolism, it has to go somewhere, or our core temperature will increase. We keep cool through:

1. *Evaporation* (E): Sweat is like our natural air conditioning. When we perspire, sweat absorbs heat from our skin and turns into water vapor, which air currents then carry away (Figure 2.30).
2. *Convection* (C): This is a heat exchange between our skin and the surrounding air. When there is a temperature difference, heat moves from our skin to the air or vice versa.
3. *Radiation* (R): Our bodies radiate heat to cooler surfaces in our environment. If your skin is warmer than the surroundings, you will lose heat, helping to keep you at a comfortable temperature.

Next time you feel a cool breeze or break a sweat, know that your body is hard at work keeping your core temperature in check! When our core temperature dips below 33.5°C (92.3°F) or soars above 40.5°C (106.7°F), the human body is in a crisis. These extreme temperatures can swiftly impair normal functions, damage cells, and ultimately be fatal (Cramer et al., 2022). In June 2024, tragedy struck during the Hajj pilgrimage to Mecca as scorching temperatures soared above 50°C (122°F), claiming the lives of at least 1,300 pilgrims (AlArabia News, 2024). As we approach 40.5°C, confusion sets in, and if it continues, we lose consciousness. Excessive hyperthermia (>40.5°C), where the body overheats beyond its ability to cool down, even with sweating, leads to heat stroke, a life-threatening condition (Chesire, 2016). On 16 June 2024 alone, over 2,764 cases of heat-related illnesses were reported in Mecca, Saudi Arabia, highlighting the severe impact of extreme heat on participants (Khaleejtimes, 2024).

Sweating is our natural cooling mechanism, but it might not be enough in environments with high temperatures and humidity. When the body's cooling systems are overwhelmed, heat accumulates faster than it can be released, leading to heat stress. This can result in heat-related illnesses like heat exhaustion, heat stroke, or heat cramps. Indices such as thermal work limit (TWL) and wet-bulb globe temperature (WBGT) are used to monitor and manage heat stress. International Organization for Standardization (2017) offers guidelines to assess the level of heat stress workers face, helping to determine if their environment poses a risk.

Thermal comfort is challenging as it hinges on a complex interplay of environmental and personal factors that influence how individuals perceive comfort (Tabor et al., 2020). While physical elements such as temperature, air movement, humidity, radiant heat, and personal factors like physical activity and clothing choice play critical roles, the true measure of thermal comfort isn't just in these variables alone. It is often gauged by the number of people expressing discomfort in their thermal environment.

Thermal comfort refers to the set of conditions where the majority feel comfortable. International Organization for Standardization (2005) outlines methods to predict the overall thermal sensation and degree of discomfort (thermal dissatisfaction) of individuals exposed to typical thermal conditions.

These methods employ PMV (predicted mean vote) and PPD (predicted percentage of dissatisfied). They aim to define environmental conditions that are generally conducive to comfort, as well as those that may cause local discomfort.

Achieving optimal thermal comfort is not just about comfort. It is a critical factor in safety, productivity, and overall well-being. During intense physical activity, heat becomes a critical challenge. As the body generates excess heat, it activates its primary cooling mechanisms, such as sweating, to maintain a stable internal temperature. However, the body's natural defenses can quickly become overwhelmed in extreme conditions, whether in high heat or oppressive humidity. The cardiovascular system acts as the first line of defense, with blood vessels in the skin dilating to release heat. This raises skin temperature and facilitates heat loss through convection and radiation, provided the surrounding air is cooler. If that fails, the heart rate rises in an attempt to cool the body, pumping more blood to dissipate heat. Yet, under prolonged or intense heat exposure, core body temperature can rise to dangerous levels, leading to heat-related illnesses such as heat rash, heat cramps, heat exhaustion, or, worst cases, fatal heat stroke. To prevent such scenarios, heat stress indices like the WBGT are used to monitor the risks. At the same time, using advanced cooling technologies, personal protective equipment (PPE), and scheduled rest breaks are ways to mitigate danger (Tumram, 2020).

2.5.2.1 PERSONAL PROTECTIVE EQUIPMENT

Employers can take several steps to protect workers from heat stress beyond environmental acclimatization, rest breaks, and proper hydration. One important measure is ensuring workers have the right personal protective equipment (PPE) (Golbabaei et al., 2022). For those in construction, manufacturing, or outdoor activities, protective clothing is essential to prevent heat stress (Chan et al., 2016). Here are key factors in selecting the ideal protective clothing:

- *Weight*: Bulky, non-breathable protective clothing significantly increases the risk of heat-related illness. To combat this, many companies are now focusing on lighter, breathable products to help workers stay cool.
- *Breathability*: Heat trapped against the skin can cause temperatures to rise, contributing to heat stress. Breathable clothing allows hot air to escape, keeping the wearer cooler and minimizing the rise in skin temperature. "Spacers" between the body and clothing allow heat to be dissipated through convection.
- *Moisture Wicking*: Excessive perspiration can lead to skin irritation and heat rash. Moisture-wicking apparel, made from a blend of hydrophilic (water-loving) and hydrophobic (water-hating) fibers, keeps workers dry. Hydrophilic fibers absorb moisture, pulling it away from the skin, while hydrophobic fibers push moisture away and spread it

out so it can dry quickly. This fast drying helps mitigate heat stress through evaporative cooling.

- *Design*: Looser-fitting garments improve breathability and moisture release. Features like venting and gussets enhance airflow, keeping workers more comfortable in hot conditions.

Heat-resistant clothing is revolutionizing the way we combat heat stress. These garments are crafted to endure high temperatures while providing maximum comfort, thanks to advanced materials that offer superior heat insulation and breathability. By forming a barrier against external heat, heat-resistant clothing significantly lowers the risk of heat stress. Modern cooling garments are equipped with cutting-edge technologies to keep the body cool in hot environments (Engeroff et al., 2023; Kayacan & Kurbak, 2010; Ren et al., 2022; Zhou et al., 2023; Zhang et al., 2024). These garments often feature the following:

- *Built-in cooling panels*: Strategically placed panels that provide a cooling effect.
- *Evaporative cooling fabric*: Materials that cool the body through the evaporation of sweat.
- *Phase-change materials*: Substances that absorb and release heat to regulate body temperature.

By incorporating these advanced technologies, cooling garments help regulate body temperature, ensuring you stay comfortable and protected even in extreme heat.

Mitigating heat stress not only enhances worker health and safety but also boosts productivity. When workers are free from the distraction of physical discomfort, they can concentrate better on their tasks. While cooling vests are designed to combat heat stress, an ineffective one can become more of a burden than a benefit. Adding an extra layer of clothing increases both weight and thermal insulation, creating a barrier that hinders the evaporation of sweat. This can actually exacerbate heat stress instead of alleviating it. Other drawbacks of ineffective cooling vests include (Chan et al., 2016):

- *Short cooling time*: Some vests provide only temporary relief, requiring frequent replacements or recharges.
- *Inflexibility*: Rigid designs can restrict movement and cause discomfort during physical activities.
- *Excessive weight*: Heavier vests add to physical strain, making it harder for workers to perform their tasks efficiently.
- *Lack of industry-specific design*: Generic designs may not meet the unique needs of different industries, reducing their effectiveness.

On the flip side, in icy environments where temperatures plunge below freezing, maintaining body heat becomes a critical challenge. According to International

Organization for Standardization (2001), once the air thermometer reads below 0°C (32°F), the body's natural heat production cannot keep up with the rapid loss to the cold. This imbalance pushes internal temperatures dangerously downward, leading to hypothermia, a state no one can sustain for long (Rathjen et al., 2019). The wind-chill index is a useful measure in cold environments, as workers lose fine motor skills, then wrist-finger coordination, and eventually gross motor control in freezing conditions. Cold-weather gear is essential in these situations. Its mission is to shield against biting winds, staunch the escape of precious body heat, and adapt to the body's movements seamlessly, all while ensuring thermo physiological comfort in the harshest of conditions. In essence, these specialized garments aren't just clothing; they are tools that empower individuals to defy the cold, ensuring safety and performance even when the mercury plummets (Watson et al., 2019). Whether in sweltering heat or freezing cold, proper thermal protection is vital, safeguarding health and maintaining productivity.

2.6 EMOTION

Emotion is a complex psychological state involving subjective experiences, physiological changes, and behavioral responses (Mauss & Robinson, 2009). It represents a person's affective reaction to internal or external stimuli. Common emotional responses include happiness, fear, anger, sadness, disgust, and surprise (Ekman, 1992). Positive emotions reflect pleasurable engagement, while negative emotions signal feelings of distress. Emotions influence how we perceive and interact with the world, shaping decisions, actions, and social interactions. Design can influence our emotions and in turn, emotions can affect the user behavior and experience (Norman, 2004).

Different stimuli can invoke a range of emotions. Elements like color, font, images, and sound can generate either positive or negative feelings depending on the context. For example, color psychology suggests that black often invokes sadness, fear, and hate, while red can trigger feelings of love and danger (Jonauskaite et al., 2020). Colors can be culture-specific too (Figure 2.31). For example, Chinese use red but Westerners use green for good luck. Beyond what we call color which is hue, saturation and brightness also have a powerful impact on emotional response, subtly shaping how we feel and react to what we see (Valdez & Mehrabian, 1994).

Emotion and cognition are closely intertwined, both influencing human behavior (Pessoa, 2008). When stimuli are perceived with emotional significance, they trigger strong sensory activation (Pessoa et al., 2002). Positive emotions have been shown to enhance cognitive control, improving focus and decision-making (Wang et al., 2017). Conversely, negative emotions can impair cognitive performance, leading to reduced efficiency and errors (Uher et al., 2014). Figure 2.32 illustrates how uppercase letter can increase users' environmental concerns by leveraging fear arousal and emotional congruence comparing to lowercase letter (Song & Luximon, 2019).

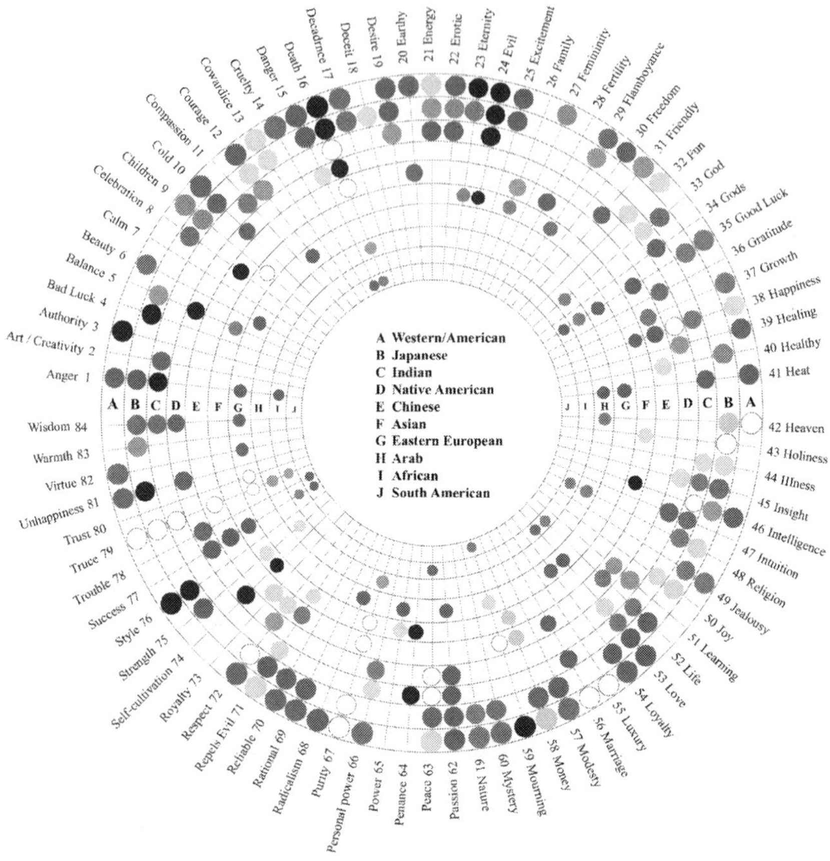

FIGURE 2.31 Color and culture diversity. (Redraw based on https://informationisbeautiful.net/visualizations/colours-in-cultures/.)

FIGURE 2.32 The influence of letter case on subjective feeling. (Image from Song & Luximon, 2019.)

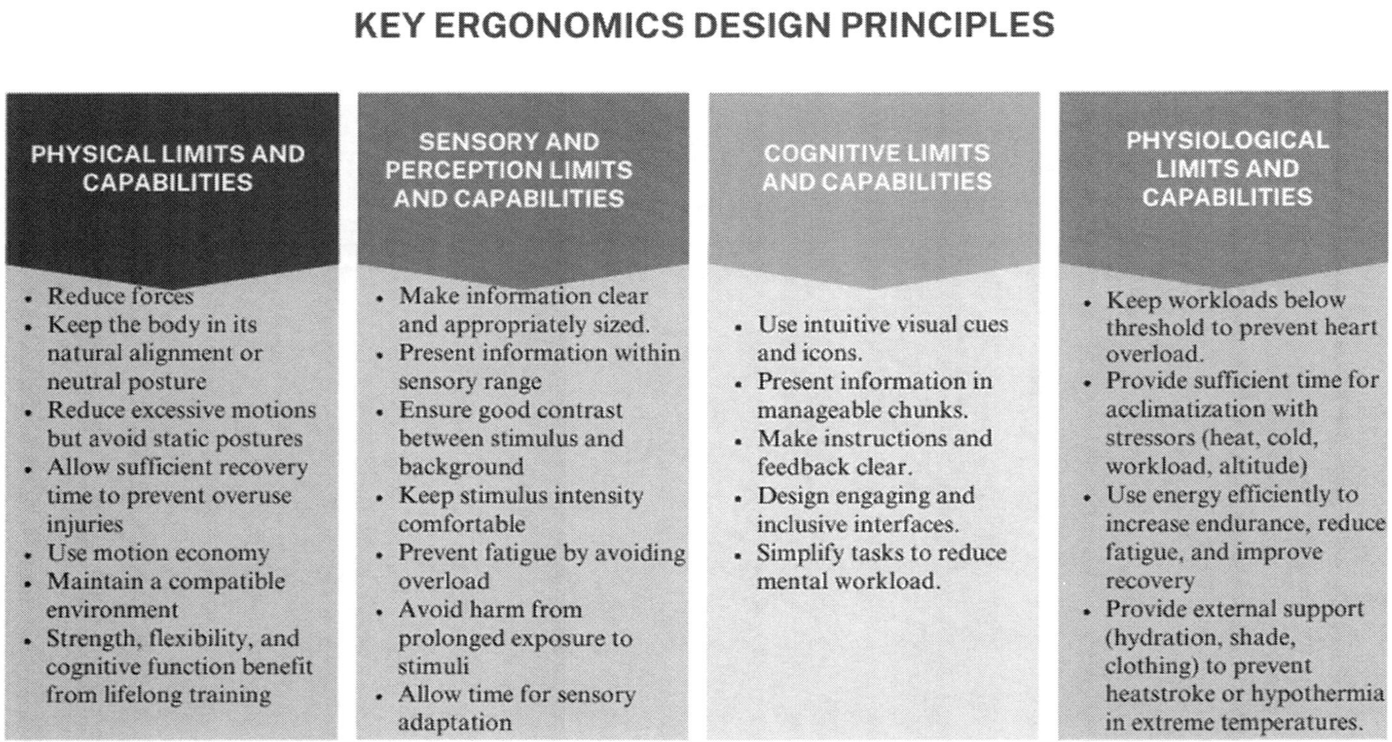

FIGURE 2.33 The key ergonomics design principles.

Designers should strive to evoke positive emotional responses in users, such as delight, joy, and surprise through thoughtful product or interface design. Elements like aesthetics, tactile feel, and overall user experience play a key role in shaping these emotions. Negative emotions such as fear and anxiety can hinder users' cognitive abilities, particularly decision-making. Therefore, designers should carefully consider the emotional factors to achieve their objectives.

2.7 CONCLUSION

By incorporating an understanding of human capabilities and limits into the design process, designers can create products and experiences that are user-friendly, efficient, safe, and enjoyable for a wide range of users. Design should prioritize the needs, abilities, and limitations of the target users. Figure 2.33 lists some key ergonomics design principles. Understanding their physical, sensory, cognitive, physiological limits and capabilities, and emotional status helps ensure that the design is accessible and inclusive. Failing to account for the stressors can overload users, leading to frustration, errors, and even harm. Great design aligns with human capabilities, creating systems that not only work well but also support well-being, efficiency, and positive experiences. It is the foundation for human-centered design.

> Once we accept our limits, we can surpass them.
>
> —Albert Einstein

REFERENCES

Adams, C. & Berlin, C. (2017). *Production Ergonomics: Designing Work Systems to Support Optimal Human Performance.* Ubiquity Press. doi: https://doi.org/10.5334/bbe

Adobe Color (2024). Contrast Checker. https://color.adobe.com/create/color-contrast-analyzer. Last accessed 1 July 2024.

Ahituv, N., Igbaria, M., & Sella, A. V. (1998). The effects of time pressure and completeness of information on decision making. *Journal of Management Information Systems*, 15(2), 153–172.

Alonso, A., van der Meij, J., Tse, D., & Genzel, L. (2020). Naïve to expert: Considering the role of previous knowledge in memory. *Brain and Neuroscience Advances*, 4, 2398212820948686.

AlArabia News (2024). Saudi health minister says 1,301 people died during Hajj. https://english.alarabiya.net/News/saudi-arabia/2024/06/23/saudi-health-minister-says-1-301-died-during-hajj. Last Accessed 25 June 2024.

American National Standards Institute. (2016). Acoustic-Estimation of Noise-induced Hearing Loss—Part 1: Method for Calculating Expected Noise-induced Permanent Threshold Shift (a modified nationally adopted international standard). (ANSI/ASA S3.44-2016/Part 1 / ISO 1999:2013 (MOD)).

Berger, E. H. (1986). Methods of measuring the attenuation of hearing protection devices. *Journal of the Acoustical Society of America*, 79, 1655–1687. doi: 10.1121/1.393228

Blasiman, R. N., & Was, C. A. (2018). Why is working memory performance unstable? A review of 21 factors. *Europe's Journal of Psychology*, 14(1), 188.

Briand, L., & Salles, C. (2016). Taste perception and integration. In Patrick Etiévant, Elisabeth Guichard, Christian Salles, Andrée Voilley (eds), Flavor. 101–119. Woodhead Publishing.

Bschaden, A., Dörsam, A.F., Cvetko, K., Kalamala, T., & Stroebele-Benschop, N. (2020). The impact of lighting and table linen as ambient factors on meal intake and taste perception. *Food Quality and Preference*, 79, 103797.

Canadian Standards Association. (2018). Measurement of noise exposure (CSA Z107.56-18:2018).

Chan, A. P. C., Yi, W., & Wong, F. K. W. (2016). Evaluating the effectiveness and practicality of a cooling vest across four industries in Hong Kong. *Facilities*, 34(9/10), 511–534. doi: 10.1108/F-12-2014-0104

Charles, R. L., & Nixon, J. (2019). Measuring mental workload using physiological measures: A systematic review. *Applied Ergonomics*, 74, 221–232.

Clarkson, J. (2008). Human capability and product design. Chapter 6. In Schifferstein, H. N. J. and Hekkert, P. (eds), Product Experience. Elsevier.

Cheshire, W. P. (2016). Thermoregulatory disorders and illness related to heat and cold stress, *Autonomic Neuroscience*, 196, 91–104, doi: 10.1016/j.autneu.2016.01.001

Cowan, N. (2008). What are the differences between long-term, short-term, and working memory? *Progress in Brain Research*, 169, 323–338.

Cowan, N., Bao, C., Bishop-Chrzanowski, B. M., Costa, A. N., Greene, N. R., Guitard, D., ... & Ünal, Z. E. (2024). The relation between attention and memory. *Annual Review of Psychology*, 75(1), 183–214.

Cramer, M. N., Gagnon, D., Laitano, O., & Crandall, C. G. (2022). Human temperature regulation under heat stress in health, disease, and injury. *Physiological Reviews*, 102(4), 1907–1989. doi: 10.1152/physrev.00047.2021

Curcio, C. A., Sloan, K. R., Kalina, R. E., & Hendrickson, A. E. (1990). Human photoreceptor topography. *Journal of Comparative Neurology*, 292, 497–523.

de la Fuente, J., Gustafson, S., Twomey, C., & Bix, L. (2014). An affordance-based methodology for package design. *Packaging Technology and Science*, 28, 157–171. doi: org/10.1002/pts.2087

Dhayal, P., & Jha, B. (2023). Indoor visual comfort: A review of factors and assessments. *Journal of the International Society for the Study of Vernacular Settlements*, 10 (11). doi: 10.61275/ISVSej-2023-10-11-03

Ekman, P. (1992). Facial expressions of emotion: New findings, new questions. *Psychological Science*, 3 (1), 34–38.

Engeroff, L., Niederer, D., Groneberg, D., Vogt, L., & Engeroff, T. (2023). Do cool shirts make a difference? The effects of upper body garments on health, fluid balance and performance during exercise in the heat. *BMC Sports Science, Medicine and Rehabilitation*, 15, 154. doi: 10.1186/s13102-023-00768-3

Fitts, P. M. (1954). The information capacity of the human motor system in controlling the amplitude of movement. *Journal of Experimental Psychology*, 47(6), 381–391.

Fletcher, H., & Munson, W. A. (1933). Loudness, its definition, measurement and calculation. *.Journal of the Acoustical Society of America*, 5, 82–108.

Florentine, M., Popper, A. N., & Fay, R. R. (2011). *Loudness*. Springer. doi: 10.1007/978-1-4419-6712-1

Fortenbaugh, F. C., DeGutis, J., Germine, L., Wilmer, J. B., Grosso, M., Russo, K., & Esterman, M. (2015). Sustained attention across the life span in a sample of 10,000: Dissociating ability and strategy. *Psychological Science*, 26 (9), 1497–1510. doi:10.1177/0956797615594896

Floyd, W. F., & Roberts, D. F. (1958). Anatomical and physiological principles in chair and table design. *Ergonomics*, *2*(1), 1-16.

Gaggioni, G., Maquet, P., Schmidt, C., Dijk, D.-J., & Vandewalle, G. (2014). Neuroimaging, cognition, light and circadian rhythms. *Frontiers in Systems Neuroscience*, 8, 126. doi: 10.3389/fnsys.2014.00126

Galy, E., Cariou, M., & Mélan, C. (2012). What is the relationship between mental workload factors and cognitive load types? *International Journal of Psychophysiology*, 83(3), 269–275.

Gan, K. C., & Hoffmann, E. R. (1988). Geometrical conditions for ballistic and visually controlled movements. *Ergonomics*, 31(5), 829–839.

Gausby, A. (2015). Microsoft attention spans research report. Microsoft Corporation. https://www.scribd.com/document/265348695/Microsoft-Attention-Spans-Research-Report

Geneva, I. I., Cuzzo, B., Fazili, T., & Javaid, W. (2019). Normal body temperature: A systematic review. *Open Forum Infectious Diseases*, 6(4), ofz032. doi: 10.1093/ofid/ofz032

Giggins, O.M., Persson, U.M., Caulfield, B. (2013). Biofeedback in rehabilitation. *Journal of Neuroengineering Rehabilitation*, 10, 60. doi: 10.1186/1743-0003-10-60

Golbabaei, F., Heydari, A., Moradi, G., Dehghan, H., Moradi, A., & Habibi, P. (2022). The effect of cooling vests on physiological and perceptual responses: A systematic review. *International Journal of Occupational Safety and Ergonomics*, 28(1), 223–255. doi: 10.1080/10803548.2020.1741251

Goonetilleke, R. S., Hoffmann, E. R., & Luximon, A. (2009). Effects of pen design on drawing and writing performance. *Applied Ergonomics*, 40(2), 292–301. doi: 10.1016/j.apergo.2008.04.015

Goonetilleke, R. S., Hamad, B. O., & So, R. H. Y. (1997). Grip span and arm position effects on grip strength. Proceedings of the ASEAN 97 Conference, Kuala Lumpur, Malaysia, pp. 99–104.

Goonetilleke, R. S., & Luximon, Y. (2010). The relationship between monochronicity, polychronicity and individual characteristics. *Behaviour & Information Technology*, 29(2), 187–198.

Hertwig, R., & Todd, P. M. (2003). More is not always better: The benefits of cognitive limits (Chapter 11, p213-231). Thinking: Psychological Perspectives on Reasoning, Judgment and Decision Making (Eds David Hardman and Laura Macchi). John Wiley & Sons.

Hirao, A., Naito, S., & Yamazaki, N. (2022). Pressure sensitivity of buttock and thigh as a key factor for understanding of sitting comfort. *Applied Sciences*, 12, 7363. doi: 10.3390/app12157363

International Bureau of Weights and Measures (2019-05-20), The International System of Units (SI) (9th ed.), ISBN 978-92-822-2272-0.

International Organization for Standardization. (1987). Acoustics—Normal Equal-Loudness Level Contours (ISO 226:1987).

International Organization for Standardization. (1992). Ergonomic requirements for display terminals (VDTs)—Part 3: Visual display requirements (ISO 9241-3:1992).

International Organization for Standardization. (2001). Ergonomics of the thermal environment—Medical supervision of individuals exposed to extreme hot or cold environments (ISO 12894:2001).

International Organization for Standardization. (2003). Acoustics—Normal Equal-Loudness Level Contours (ISO 226:2003).

International Organization for Standardization. (2005). Ergonomics of the thermal environment—Analytical determination and interpretation of thermal comfort using calculation of the PMV and PPD indices and local thermal comfort criteria. (ISO 7730:2005).

International Organization for Standardization. (2017). Ergonomics of the thermal environment—Assessment of heat stress using the WBGT (wet bulb globe temperature) index (ISO 7243:2017).

International Organization for Standardization. (2025). Light and lighting-Lighting of work places. Part 1. Indoor. (ISO/CIE 8995-1.:2025).

Jonauskaite, D., Abu-Akel, A., Dael, N., Oberfeld, D., Abdel-Khalek, A. M., Al-Rasheed, A. S., Antonietti, J.-P., Bogushevskaya, V., Chamseddine, A., Chkonia, E., Corona, V., Fonseca-Pedrero, E., Griber, Y. A., Grimshaw, G., Hasan, A. A., Havelka, J., Hirnstein, M., Karlsson, B. S. A., Laurent, E., ... & Mohr, C. (2020). Universal patterns in color-emotion associations are further shaped by linguistic and geographic proximity. *Psychological Science* 31(10): 1245–1260.

Kaber, D. B., & Riley, J. M. (1999). Adaptive automation of a dynamic control task based on secondary task workload measurement. *International Journal of Cognitive Ergonomics,* 3(3), 169–187.

Kayacan, Ö, & Kurbak, A. (2010). Effect of garment design on liquid cooling garments. *Textile Research Journal,* 80(14), 1442–1455. doi: 10.1177/0040517509358800

Kazanskiy, N. L., Khonina, S. N., & Butt, M. A. (2024). A review on flexible wearables – recent developments in non-invasive continuous health monitoring. *Sensors and Actuators A: Physical,* 366, 114993. doi:10.1016/j.sna.2023.114993

Keane, M. P., & Thorp, S. (2016). Complex decision making: the roles of cognitive limitations, cognitive decline, and aging. In *Handbook of the Economics of Population Aging* (Vol. 1, pp. 661–709). North-Holland.

Khaleejtimes (2024). Saudi Arabia: Over 2,700 cases of heat exhaustion, sunstroke recorded during Haj. www.khaleejtimes.com/world/gulf/haj-2024-over-2700-cases-of-heat-exhaustion-and-sunstroke-recorded-in-saudi Last accessed 27 August 2025.

Kim, S., Park J. M., Han, J. S., Seo, J. H., Han, K. D., Joo, Y. H., & Park, K. H. (2020). Age-related hearing loss in the Korea National Health and Nutrition Examination Survey. *PLoS One,* 15(12), e0243001. doi: 10.1371/journal.pone.0243001

Köster, E. P. (2002). The specific characteristics of the sense of smell. In *Olfaction, Taste and Cognition,* chapter 3, pp. 27–43. Cambridge University Press.

Kryter, K. (1985). *The Effects of Noise on Man* (2nd edition). Academic Press.

Labrecque, J. S., Wood, W., Neal, D. T., & Harrington, N. (2017). Habit slips: When consumers unintentionally resist new products. *Journal of the Academy of Marketing Science,* 45, 119–133.

Lachman, S. J. (1997). Learning is a process: Toward an improved definition of learning. *Journal of Psychology*, 131(5), 477–480.

Lebiere, C., & Anderson, J. R. (2011). Cognitive constraints on decision making under uncertainty. *Frontiers in Psychology*, 2, 305.

Legge, G. E., & Bigelow, C. A. (2011). Does print size matter for reading? A review of findings from vision science and typography. *Journal of Vision*, 11(5), 8–8.

Liao, H., Kashino, M., & Shimojo, S. (2021). Attractiveness in the eyes: A possibility of positive loop between transient pupil constriction and facial attraction. *Journal of Cognitive Neuroscience*, 33(2), 315–340. 10.1162/jocn_a_01649

Liao, X., Li, Y., Hu, J., Ding, X., Zhang, X., Ying, B., ... & Wu, X. (2018). Effects of contact method and acclimation on temperature and humidity in touch perception. *Textile Research Journal*, 88(14), 1605–1615.

Liu, C., Li, J., Wang, W., Zang, Q., Wang, X., & Gao, W. (2024). A review of subjective evaluation, physiological indicators and cognitive performance in indoor light environment: the role of illuminance and correlated color temperature. *Journal of Building Engineering*, 110224.

Lu, L., Zhang, J., Xie, Y., Gao, F., Xu, S., Wu, X., & Ye, Z. (2020). Wearable health devices in health care: narrative systematic review. *JMIR mHealth and uHealth*, 8(11), e18907. doi: 10.2196/18907

Luft, C. D. B. (2014). Learning from feedback: The neural mechanisms of feedback processing facilitating better performance. *Behavioural Brain Research*, 261, 356–368.

Luximon, A., & Goonetilleke, R. S. (2001). Simplified subjective workload assessment technique. *Ergonomics*, 44(3), 229–243.

Luximon, Y., & Goonetilleke, R. S. (2012). Time use behavior in single and time-sharing tasks. *International Journal of Human–Computer Studies*, 70(5), 332–345.

Maslow, A. H. (1943). A theory of human motivation. *Psychological Review*, 50(4), 370–396. doi: 10.1037/h0054346

Mathiowetz, V., Kashman, N., Volland, G., Weber, K., Dowe, M., & Rogers, S. (1985). Grip and Pinch Strength: Normative data for adults. *Archives of Physical Medicine and Rehabilitation*, 66(2), 69–74.

Mathiowetz, V., Wiemer, D. M., & Federman, S. M. (1986). Grip and pinch strength: norms for 6- to 19-year-olds. *The American Journal of Occupational Therapy*, 40(10), 705–11. doi: 10.5014/ajot.40.10.705

Mauss, I. B., & Robinson, M. D. (2009). Measures of emotion: a review. *Cognition and Emotion*, 23(2), 209–237.

Miller, G. A. (1956). The magic number seven, plus or minus two: some limits on our capacity for processing information. *Psychological Review*, 63, 81–93.

NASA (2017). NASA Armstrong Fact Sheet: Sonic Booms. https://web.archive.org/web/20200511082325/ www.nasa.gov/centers/armstrong/news/FactSheets/FS-016-DFRC.html. Accessed 23 June 2024.

Nes, B. M., Janszky, I., Wisløff, U., Støylen, A., & Karlsen, T. (2013). Age-predicted maximal heart rate in healthy subjects: The HUNT Fitness Study. *Scandinavian Journal of Medicine & Science in Sports*, 23(6), 697–704.

Nicolay, C. W., Walker, A. L. (2005). Grip strength and endurance: Influences of anthropometric variation, hand dominance, and gender. *International Journal of Industrial Ergonomics*, 35(7), 605–618. doi: 10.1016/j.ergon.2005.01.007

Norman, D. A. (2004). *Emotional Design: Why We Love (or Hate) Everyday Things*. Basic Books.

Norman, D. A. (2013). *The Design of Everyday Things*. Revised and Expanded Edition. Basic Books.

Norton, M. P., & Karczub, D. G. (2012). *Fundamentals of Noise and Vibration Analysis for Engineers*. Cambridge University Press. doi: 10.1017/CBO9781139163927

Noyes, J., Garland, K., & Bruneau, D. (2004). Humans: Skills, Capabilities and Limitations. In: Sandorn, C. and Harvey, R.S. (Eds.), *Human Factors for Engineers*, The Institution of Engineering and Technology, London, United Kingdom, pp. 35–56.

Ohno, T., Hasegawa, T., Tsuruoka, T., Terabe, K., Gimzewski, J. K., & Aono. M. (2011). Short-term plasticity and long-term potentiation mimicked in single inorganic synapses. *Nature Materials*, 10, 591–595. doi: 10.1038/nmat3054

Peck, J., & Wiggins, J. (2006). It just feels good: Customers' affective response to touch and its influence on persuasion. *Journal of Marketing*, 70(4), 56–69. doi: 10.1509/jmkg.70.4.056

Peña, E. (2016). *About Visual Acuity and Type Design: A Protocol*. The Winnower.

Pessoa, L. (2008). On the relationship between emotion and cognition. *Nature Reviews Neuroscience*, 9, 148–158. doi: 10.1038/nrn2317

Pessoa, L., Kastner, S., & Ungerleider, L. G. (2002). Attentional control of the processing of neutral and emotional stimuli. *Cognitive Brain Research*, 15, 31–45.

Petek, B., Al-Alusi, M., Moulson, N., ...& Baggish, A. L. (2023). Consumer wearable health and fitness technology in cardiovascular medicine: JACC state-of-the-art review. *Journal of the American College of Cardiology*, 82(3), 245–264. doi: 10.1016/j.jacc.2023.04.054

Pirenne M. H. (1962). Dark adaptation and night vision. Chapter 5. In: Davson, H. (ed), *The Eye*, vol 2. Academic Press.

Ramamurthy, M., & Lakshminarayanan, V. (2015). Human vision and perception. In: Karlicek, R., Sun, CC., Zissis, G., Ma, R. (eds) Handbook of Advanced Lighting Technology. Springer, Cham. https://doi.org/10.1007/978-3-319-00295-8_46-1.

Rathjen, N. A., Shahbodaghi, S. D., & Brown, J. A. (2019). Hypothermia and cold weather injuries. *American Family Physician*, 100(11), 680–686.

Ren, S., Han, M., & Fang, J. (2022). Personal cooling garments: A review. *Polymers*, 14, 5522. doi: 10.3390/polym14245522

Riggs L. A. (1965). Visual acuity. Chapter 11. In: Graham, C. H. (ed), *Vision and Visual Perception*. New York: Wiley.

Rishiq, D., Harkrider, A., Springer, C., & Hedrick, M. (2020). Effects of aging on the subcortical encoding of stop consonants. *American Journal of Audiology*, 29(3), 391–403. doi:10.1044/2020_AJA-19-00044

Robergs, R. A., & Landwehr, R. (2002). The surprising history of the" HRmax= 220-age" equation. *Journal of Exercise Physiology Online*, 5(2), 1–10.

Rodrigo, A. S., Goonetilleke, R. S., & Xiong, S. (2013). Load distribution to minimise pressure-related pain on foot: A model. *Ergonomics*, 56(7), 1180–1193.

Rohmert, W. (1960). Ermittlung von Erholungspausen fur statische arbeit des menschen. *Internationale Zeitschrift für angewandte Physiologie einschließlich Arbeitsphysiologie*, 18, 123–164.

Rubio, S., Díaz, E., Martín, J., & Puente, J. M. (2004). Evaluation of subjective mental workload: A comparison of SWAT, NASA-TLX, and workload profile methods. *Applied Psychology*, 53(1), 61–86.

Sanders M. S., McCormick E. J. (1987). *Human factors in engineering design*. New York: McGraw-Hill.

Sanders, M. S., & McCormick, E. J. (1993). *Human Factors in Engineering and Design* (7th ed.). McGraw-Hill.

Saladin, K. S. (2011). *Human Anatomy* (3rd ed.). McGraw-Hill.

Salomons, E. M., & Janssen, S. A. (2011). Practical ranges of loudness levels of various types of environmental noise, including traffic noise, aircraft noise, and industrial noise. *International Journal of Environmental Research and Public Health*, 8(6), 1847–1864. doi: 10.3390/ijerph8061847

Samain-Aupic, L., Dione, M., Ribot-Ciscar, E., Ackerley, R., & Aimonetti, J. M. (2024). Relations between tactile sensitivity of the finger, arm, and cheek skin over the lifespan showing decline only on the finger. *Frontiers in Aging Neuroscience*, 16, 1387136.

Schiefele, U. (1991). Interest, learning, and motivation. *Educational Psychologist*, 26(3–4), 299–323.

Schiffman, S., & Graham, B. (2000). Taste and smell perception affect appetite and immunity in the elderly. *European Journal of Clinical Nutrition*, 54 (Suppl 3), S54–S63. doi: 10.1038/sj.ejcn.1601026

Shingo, S. (1986). *Zero Quality Control: Source Inspection and the Poka-Yoke System*. CRC Press.

Siegemedia (2024). Contrast Ratio. www.siegemedia.com/contrast-ratio#%23f50a0a-on-%23818c31. Last accessed 1 July 2024.

Simon, A. J., Gallen, C. L., Ziegler, D. A., Mishra, J., Marco, E. J., Anguera, J. A., & Gazzaley, A. (2023). Quantifying attention span across the lifespan. *Frontiers in Cognition*, 2. doi: 10.3389/fcogn.2023.1207428

Snook.ca (2024). Colour Contrast Check. https://snook.ca/technical/colour_contrast/colour.html#fg=FF0000,bg=0000FF. Last accessed 1 July 2024.

Song, Y., & Luximon, Y. (2019). Design for sustainability: The effect of lettering case on environmental concern from a green advertising perspective. *Sustainability*, 11(5), 1333.

Spectorm R.H. (1990). The pupils. In: Walker, H.K., Hall, W.D., Hurst, J.W., editors. *Clinical Methods: The History, Physical, and Laboratory Examinations* (3rd edition). Boston: Butterworths. Chapter 58. Available from: www.ncbi.nlm.nih.gov/books/NBK381/

Sprigle S., Chung K. C., Brubaker C. E. (1990). Reduction of sitting pressures with custom contoured cushions. *Journal of Rehabilitation Research and Development*, 27, 135–140.

Stevens, J. C., & Hooper, J. E. (1982). How skin and object temperature influence touch sensation. *Perception & Psychophysics*, 32, 282–285.

Suzuki, Y., & Takeshima, H. (2004). Equal-loudness-level contours for pure tones. *Journal of the Acoustical Society of America*, 116, 918–933.

Suzuki, Y., Takeshima, H., & Kurakata, K. (2024). Revision of ISO 226 "Normal Equal-Loudness-Level Contours" from 2003 to 2023 edition: The background and results. *Acoustical Science and Technology*, 45(1), 1–8. doi: 10.1250/ast.e23.66

Tabor, J., Chatterjee, K., & Ghosh, T. K. (2020). Smart textile-based personal thermal comfort systems: Current status and potential solutions. *Advanced Materials Technologies*, 5(5), 1901155. doi: 10.1002/admt.201901155

Talavera, K., Ninomiya, Y., Winkel, C., Voets, T., & Nilius, B. (2007). Influence of temperature on taste perception. *Cellular and Molecular Life Sciences*, 64, 377. doi: 10.1007/s00018-006-6384-0

Tik Tok Broom Challenge (2024). www.youtube.com/watch?v=ODS-yTvX1TA. Last accessed 5 May 2024.

Tilley, A. R. (2002). *The Measure of Man and Woman: Human Factors in Design*. Wiley.

Tumram, N. K. (2020). Personal protective equipment and personal cooling garments to reduce heat-related stress and injuries. *Medico-Legal Journal*, 88(1_suppl), 43–46. doi: 10.1177/0025817220935882

Uher, R., Brooks, S. J., Bartholdy, S., Tchanturia, K., & Campbell, I. C. (2014). Increasing cognitive load reduces interference from masked appetitive and aversive but not neutral stimuli. *PLoS One*, *9*(4), e94417.

Valdez, P., & Mehrabian, A. (1994). Effects of color on emotions. *Journal of Experimental Psychology: General*, 123(4), 394–409. doi: 10.1037/0096-3445.123.4.394

Vafeiadou, A., Bowling, N. C., Hammond, C., & Banissy, M. J. (2022). Assessing individual differences in attitudes towards touch in treatment settings: Introducing the touch & health scale. *Health Psychology Open*, 9(2), 20551029221137008.

Wang, Y., Chen, J., & Yue, Z. (2017). Positive emotion facilitates cognitive flexibility: an fMRI study. *Frontiers in Psychology*, 8. doi: 10.3389/fpsyg.2017.01832

Wasano, K., Kagac, K., Ogawa, K. (2021). Patterns of hearing changes in women and men from denarians to nonagenarians. *The Lancet Regional Health*, 9, 100131.

Watanabe, T., Owashi, K., Kanauchi, Y., Mura, N., Takahara, M., & Ogino, T. (2005). The short-term reliability of grip strength measurement and the effects of posture and grip span. *Journal of Hand Surgery*, 30(3), 603–609.

Watson, C., Troynikov, O., & Lingard, H. (2019). Design considerations for low-level risk personal protective clothing: A review. *Industrial Health*, 57(3), 306–325. doi:10.2486/indhealth.2018-0040

Weerasinghe, T. W., Goonetilleke, R. S., & Reischl, U. (2017). Pressure thresholds and stiffness on the plantar surface of the human foot. *Ergonomics*, 60(7), 985–996.

Weichenrieder A. & Haldenwanger H. (1986). *The best function for the seat of a passenger car* (SAE Tech. Paper 850484). Warrendale, PA: Society of Automotive Engineers.

Weiffenbach, J. M. (1984). taste and smell perception in aging. *Gerodontology*, 3(2), 137–146.

Wickens, C. D., Helton, W. S., Hollands, J. G., & Banbury, S. (2022). *Engineering Psychology and Human Performance*. London: Routledge.

Winkler, I., & Cowan, N. (2005). From sensory to long-term memory: evidence from auditory memory reactivation studies. *Experimental Psychology*, 52(1), 3–20.

Wong, N. A., & Bahmani, H. (2022). A review of the current state of research on artificial blue light safety as it applies to digital devices. *Heliyon*, 8(8), e10282. doi: 10.1016/j.heliyon.2022.e10282

Woods, D. L, Doss, Z., Herron, T. J., & Yund, E. W. (2012). Age-related changes in consonant and sentence processing. *Journal of Rehabilitation Research & Development*, 49(8), 1277–91. doi: 10.1682/jrrd.2011.08.0150

Wu, M., & Luo, J. (2019). Wearable technology applications in healthcare: A literature review. *Online Journal of Nursing Informatics (OJNI)*, 23(3). www.himss.org/resources/wearable-technology-applications-healthcare-literature-review

W3C (2024). Web Content Accessibility Guidelines (WCAG) 2.2. www.w3.org/TR/WCAG22/#abstract. Last accessed 1 July, 2024.

Ye, Y., Wong, S. C., Meng, F., & Xu, P. (2021). Right-looking habit and maladaptation of pedestrians in areas with unfamiliar driving rules. *Accident Analysis & Prevention*, 150, 105921.

Zhang, J. (2025). Various external and internal influential factors on short-term memory. In Addressing Global Challenges-Exploring Socio-Cultural Dynamics and Sustainable Solutions in a Changing World (pp. 936–941). London: Routledge.

Zhang, X., Wang, F., Guo, H., Sun, F., Li, X., Zhang, C., Yu, C., & Qin, X. (2024). Advanced cooling textiles: mechanisms, applications, and perspectives. *Advanced Science*, 11(10), e2305228. doi: 10.1002/advs.202305228

Zhang, Y., & Goonetilleke, R. S. (2005). The cultural effects of time usage and process complexity on control performance. In The 11th International Conference on Human–Computer Interaction (22–27 July 2005), Las Vegas, NV.

Zhang, Y., Goonetilleke, R. S., Plocher, T., & Liang, S. F. M. (2005). Time-related behaviour in multitasking situations. *International Journal of Human–Computer Studies*, 62(4), 425–455.

Zhang, Y., & Luximon, A. (2005). Subjective mental workload measures. *Ergonomia*, 27(3), 199–206.

Zhou, Y., Lou, L., & Fan, J. (2023). Quantitative comparison of personal cooling garments in performance and design: A review. *Processes*, 11, 2976. doi: 10.3390/pr11102976

Design for Diverse Populations

3

> Where we choose to put our attention changes our brain, which in time can change how we see and interact with the world.
>
> —Margaret Mead

3.1 INTRODUCTION

Imagine the number of people who may have limited or no vision, or who are deaf or mute. How can they effectively use a product or service designed primarily for individuals with full sensory capabilities? The voting process in a country is a prime example. How would a person with limited or no vision cast their vote without the help of another person? Section 508 of the US Rehabilitation Act (1998) outlines functional requirements for technologies to ensure accessibility for individuals with disabilities, including those with visual, auditory, tactile, and cognitive disabilities (www.section508.gov/manage/laws-and-policies/section-508-law/).

The barrier-free movement of the 1950s marked the start of a significant shift in public policies and design practices in the United States. Advocated by disabled veterans and supporters of people with disabilities, this movement identified physical barriers in the environment as major challenges for individuals with mobility impairments, leading to the introduction of the first accessibility standards by the American Standards Association in 1961.

3.1.1 INCLUSION, DIVERSITY, EQUITY, AND ACCESSIBILITY (IDEA)

In recent years, inclusion, diversity, equity, and accessibility (IDEA) have gained significant prominence (Rynarzewska et al., 2024; Zallio & Clarkson, 2021). They are grouped

DOI: 10.1201/9781003476177-3

together because they are deeply interconnected, with their full impact only realized when they work in unison.

Inclusion involves the extent to which organizations embrace all individuals, ensuring everyone feels valued and respected and able to fully participate, regardless of their background, identity, or abilities. It emphasizes the active engagement of diverse perspectives and the intentional effort to ensure that everyone has equal opportunities to contribute and thrive. Inclusion is about more than just physical presence; it is about meaningful involvement and ensuring that every individual can contribute to and benefit from their environment, whether in a workplace, educational setting, or community. Chapter 4 has more information on how this can be done.

Diversity centers on representing different attributes within a group, including factors such as age, gender, ethnicity, and beyond.

Equity, distinct from equality, refers to fair treatment considering individuals' unique circumstances and adjusting support and resources accordingly. Unlike equality, which assumes everyone should be treated the same through identical design solutions, equity acknowledges and addresses meeting the specific needs of users to ensure a level playing field. An example is a product that provides the same level of usability for both individuals with 20/20 vision and elderly users who may have visual impairments. The three social equity dimensions that should be catered to are (Perikangas et al., 2023):

(a) Procedural fairness by designing practices and culture, which leaves no one out;
(b) Access to products and services through tailored services for the disadvantaged;
(c) Quality of services through effective communication and collaboration.

Equitable products crafted to be unbiased and accessible to everyone, regardless of race, ethnicity, culture, economic status, or social background, can broaden a company's market and boost its profitability (Dudek et al., 2022). By embracing a holistic approach to product development, companies can side-step biases and dismantle access barriers, unlocking the full potential of their offerings and reaching a wider audience. To ensure equity is built in, ask the following questions:

(a) Is a comprehensive stakeholder map used to capture diverse perspectives?
(b) Does the user research include all relevant audiences?
(c) Is diverse feedback integrated into the iterations objectively?
(d) Are the data and insights unbiased?

Accessibility refers to designing products, devices, services, or environments in a way that ensures they are usable by people with disabilities. It is the means of enabling everyone to participate in society as independently as possible. Those most often concerned about accessibility are people with compromised abilities because of the barriers they continue to encounter. Thus, many countries have passed laws with the objective of making society more accessible (Accessible Canada Act, 2019; Disability Discrimination Act, 1992; European Accessibility Act, 2019; Fitzpatrick, 1991; Waddell, 2006).

3.1.2 DISABILITY CONSIDERATIONS

According to the World Health Organization (2023), 1.3 billion people, or 16 percent of the world's population, have significant disability. That is an increase of 1 percent from 2011 and an increase of 6 percent from 1970. Disabilities can range across sensory, auditory, cognitive, learning, physical, speech, visual, and mental health conditions. Gender, ethnicity, cultural and social identities govern the way people with limited abilities experience the world. The International Classification of Functioning, Disability, and Health (ICF) is a framework that defines disability broadly as impairments, activity limitations, and participation restrictions (World Health Organization, 2001). It evaluates disability based on the following criteria (WHO, 2013):

(a) Body function—physiological aspects of body system such as vision or force;
(b) Body structure—the part of the anatomy that supports the function, such as the eye or muscles;
(c) Activities—actions and tasks executed by individuals;
(d) Participation—problems people may experience in life situations;
(e) Environmental factors—physical, social, and attitudinal conditions that shape how individuals live and engage in daily life.

Designers and design standards have implemented various strategies to support impaired populations. For instance, pathway tiles with raised strip patterns are used to aid individuals who are blind (Courtney & Chow, 2000). With advancements in camera technology, smartphones, and radio frequency identification (RFID), along with the use of position-locator devices (PLD) that leverage global positioning system (GPS) and geographic information system (GIS) technologies, many effective electronic orientation aids (EOAs) are emerging to assist blind and visually impaired individuals with navigation (Elmannai & Elleithy, 2017; Kuriakose et al., 2020; Mashiata et al., 2022; Okolo et al., 2024).

Haptic devices are increasingly being developed to support individuals with hearing and vision impairments (Flores Ramones & del-Rio-Guerra, 2023). Common methods for assisting those with hearing impairments include sign language, captioning devices, vibrating alerts, and visual alerting systems. Additionally, brain–computer interfaces (BCIs) hold significant potential for enhancing communication for people with hearing impairments (Brumberg et al., 2018), offering innovative solutions for improved interaction and accessibility.

Even without accounting for impairments, designing for diverse populations necessitates addressing variations across physical, cognitive, social, and functional domains. This is a challenging endeavor. Differences in gender and age alone demand tailored functionalities to meet the unique needs of each group effectively. The functionality and performance of a product for a child may not be appropriate for an adult and vice versa. A good example are parks and playgrounds.

3.1.3 PLACE FOR ALL AGES

Parks are usually designed with kids in mind—tiny slides for the little ones, bigger playgrounds for the older ones, and some for the elderly. Integration of these has been slow, even though we are beginning to see their emergence (Figure 3.1). Catering to the different sizes and strengths of people has been one of the primary setbacks.

Imagine the joy of a wheelchair user watching the younger generation play, feeling the nostalgia of days gone by. But why stop at watching? We need more parks that everyone can enjoy, regardless of age! It is a win-win: kids get to play,

FIGURE 3.1 "Playground" with exercise facilities for the young and the elderly.

and the elderly stay connected, all while learning to appreciate each other's needs. It is not just about space; it is about designing an environment that brings generations together in a safe and exciting way. It helps bridge the gap between generations and fosters a deeper understanding of each other's needs.

3.1.4 THE MYTH OF THE "AVERAGE" USER

People are unique, yet most products and services operate under the assumption that we are all the same. Let's design for the average of some known physical measure, you might think, a one-size-fits-all solution. For instance, one could determine the average foot size and design a sock based on that measurement. This approach could be effective if the sock is made from stretchable material, allowing it to accommodate a range of foot sizes comfortably. This approach, known as designing for the average user, raises important questions: How effective is it? Can one size truly fit everyone? While this method may seem simple, its success largely depends on adaptive design: solutions that adjust to individual needs through flexible materials or innovative features, like socks, provide a customized fit for a diverse range of users. In such cases, while the fit may improve, performance can be compromised, as design elements that enhance one aspect may unintentionally hinder another. For instance, the foot bones may not have the flexibility to adapt easily, potentially impacting overall comfort and functionality.

People's minds work in all sorts of ways, and that's why some product features end up gathering dust or being used in ways no one expected. It is not that the product was bad; it just was not designed with different users in mind. To truly connect with everyone, designers need to embrace the differences instead of trying to squeeze everyone into the same mold, even though it is easier said than done.

Designing for a wide audience means trying to meet the needs of as many people as possible, which sounds simple but is not always easy. It is tempting to focus on the majority and overlook the outliers, but that is where things get tricky. Trying to embrace the full spectrum of human diversity started with making public buildings accessible to everyone, regardless of any limitations, and the idea has grown from there. "Design for all" is about including the full spectrum of our diversity so that everyone from every walk of life can get things done. It is about pushing the boundaries of design to break down barriers, whether we are using a gadget to complete a task or relying on technology to connect with others.

3.2 EXISTING FRAMEWORKS

There are several established approaches to designing for human diversity and accessibility. In this section, we will focus on three key frameworks: universal design, inclusive design, and design for all, each of which offers a distinct perspective on creating environments, products, and systems that accommodate a wide range of users. These approaches differ in origin, methodology, and emphasis, but they all share the common goal of reducing barriers and enhancing usability for as many people as possible (Persson et al., 2015).

3.2.1 UNIVERSAL DESIGN

In the past, add-ons were designed for individuals with disabilities, but these were often an afterthought rather than an integral part of the design. Designs catered to those who were able, and *assistive devices or technologies* were provided for those with difficulty using them.

Universal design is about one product that most people can use without needing any adaptation (Connell et al., 1997). A ramp to access buildings is not limited to wheelchair users. It can be used by many people who use trolleys and bicycles and even those who have difficulty climbing stairs. Universal design, though, has an issue: it is *one product that fits most* without adaptation. Ronald Mace (www.britannica.com/biography/Ronald-L-Mace) was the visionary behind universal design. He became wheelchair-bound at the age of nine. Despite the challenges he faced in accessing the buildings at North Carolina State University, Mace persevered and earned a degree in architecture. His experiences fueled his passion for creating inclusive environments, and he went on to spearhead the Universal Design movement, advocating spaces that accommodate people of all abilities.

Universal design is based on seven principles (Connell et al., 1997; Mace et al., 2024):

1. *Equitable use*: Ensure inclusivity by offering identical or equivalent methods of use for everyone, regardless of ability, while avoiding any form of segregation or stigma. Prioritize equal access to privacy, security, and safety features, ensuring the product is functional and appealing to diverse users. A website created to be accessible to everyone, including individuals who are blind, is an example of this principle.
2. *Flexibility during use*: Accommodate diverse preferences and abilities, allowing users to choose their preferred methods, whether right- or left-handed. Enhance user experience by facilitating accuracy, precision, and adaptability to individual pacing. An educational app that allows users to read or listen to lesson materials is an example of this principle.
3. *Simple and intuitive use*: Allow users of varying experience levels and language skills to navigate easily. Remove complexity and align with user expectations, ensuring clear information communication and providing helpful prompts and feedback. An example of this is

an electronic product with functions that are intuitive and easy to use without the need to read a user manual.

4. *Perceptible information*: Ensures that all information is communicated clearly and effectively, using multiple formats, like visuals, text, and touch, to accommodate different sensory abilities and environmental conditions. Design elements are easily distinguishable and compatible with various assistive devices, making instructions straightforward and accessible to everyone. An example is a smoke alarm that uses loud sound and flashing lights to alert users, ensuring the warning is effective regardless of noise levels or a person's hearing ability.
5. *Tolerance for error*: Reduce the risk of errors and accidents, with frequently used elements placed for easy access and hazardous elements removed or protected. Include clear warnings, fail-safe mechanisms, and features that discourage automatic actions in tasks needing attention, ensuring user safety at all times. An example is a power tool with a safety guard that automatically shuts off when not in use and a two-step activation process to prevent accidental startup, reducing the risk of injury and ensuring user safety.
6. *Low physical effort*: Minimize physical strain, enabling users to maintain a neutral body position using reasonable forces. Reduce repetitive actions and sustained effort, ensuring the product can be used comfortably and efficiently. An example is a smart chair that adjusts automatically, reducing the need for constant repositioning and/or adjustment to sit comfortably (Gross et al., 1994).
7. *Size and space for approach and use*: Ensure that all users, regardless of their body size, posture, or mobility, have comfortable access to key elements. Provide a clear line of sight and easy reach for both seated and standing users, accommodate different hand and grip sizes, and offer enough space for assistive devices or personal assistance. An example is a height-adjustable table with spacious under-counter clearance, allowing both seated and standing users, including those in wheelchairs, to access and use the workspace comfortably.

Many products have been developed using universal design principles, and one example is the OXO Good Grips peeler. Inspired by Sam Farber's wife, who suffered from arthritis and needed a peeler that wouldn't hurt her hands, this design addressed her need and resulted in a product that is not only safer and more comfortable for people with arthritis but also easier to use for everyone. Another example is the Xbox Adaptive Controller, developed after Spencer Allen, a passionate gamer, was left paralyzed from the chest down due to an accident (Dalgleish, 2024). Determined to play again, he created his own custom rig with added buttons and joysticks, allowing him to regain the precise control at the level he once did. The Alexa Show and Tell (https://vi.ie/alexa-accessibility-show-and-tell-on-the-echo-show/) uses a camera to identify objects. Simply hold a product in front of the camera and ask Alexa to identify it. Alexa will then tell you what you're holding and provide a brief description. While designed primarily for blind and low-vision users, it is a feature that benefits everyone.

3.2.2 INCLUSIVE DESIGN

Even though universal design may address some of the issues, it may not be a solution for every situation because of the diversity of the human race. ***Inclusive design***, on the other hand, provides ***multiple solutions or variations*** to account for each individual's uniqueness and diversity. It caters to users with specific needs, referred to as "extreme users," with a focus on functional interactions (Coleman, 1994; Clarkson et al., 2003). It is defined as a "... methodology that enables and draws on the full range of human diversity" (Microsoft Inclusive Design, 2024). Another definition is "The design of mainstream products and/or services that are accessible to, and usable by, as many people as reasonably possible ... without the need for special adaptation or specialized design." (British Standards Institution, BS 7000-6, 2005).

Everyone deserves the same level of comfort and support when seated. Providing one type of chair for larger individuals and another for smaller ones is not equitable. The Aeron chair addresses this by offering three different sizes, S, M, and L, ensuring that everyone can enjoy the same seating experience, tailored to their specific needs.

Inclusive design also benefits anyone facing temporary or situational challenges. For example, someone with temporary hearing impairment due to loud noise would find inclusive design particularly helpful. Our senses shape our interactions with technology—what we see, hear, say, touch, learn, and remember. As technology advances, particularly with Generative AI like ChatGPT, Gemini, Pi, and Claude, a fascinating question emerges: Are we supposed to adapt to technology, or is technology evolving to fit us? Not everyone can use these large language models (LLMs) effectively, as crafting the right prompts is crucial. This challenge has led to the rise of prompt engineering (Knoth et al., 2024). Another example is Industry 4.0, which was supposed to be an integration of automation and data exchange in manufacturing. Many manufacturers have realized that total automation is challenging, leading them to reintroduce human workers into the process (World Economic Forum, 2016). This has paved the way for Industry 5.0 (Adel, 2022; Ghobakhloo et al., 2023), a new concept that emphasizes collaboration between humans and machines as discussed later in this chapter.

3.2.3 DESIGN FOR ALL

Design for all (D4All) is an inclusive approach that caters to the needs of all users, whatever their abilities or disabilities (Clarkson et al., 2003). The European

Institute for Design and Disability (EIDD) now referred to as Design for All Europe (2004), with their motto "Good design enables, bad design disables" has defined design for all as *"design for human diversity, social inclusion and equality"*. The philosophy of this framework is that design is a human right; it must ensure participation and equality in all aspects of life for everyone. It includes policy-level thinking, not just design execution. It is about building a society where design considers all citizens.

To successfully implement D4All, it is important to conduct user research involving a broad spectrum of users, including various user groups and potential users. With the gathered insights, you can then develop a wide range of personas, after which the design principles in Chapters 2 and 3 can be used.

Products should be designed around people rather than having to train or teach people how to use the products (Harley, 2015). This is the basis of user-centered design (UCD). Do we really need a user manual for a product or instructions for a service? Why can't the product use be intuitive? Gone are the days of technology-based design that did not cater to overcoming human limitations.

The designer must also consider a product or service's value to users. De Couvreur and Goossens (2011) effectively illustrate why people buy and use products through their five-level strategy, which is rooted in Boztepe's (2007) user-value adaptation theory (Figure 3.2). This framework offers a comprehensive overview of the key factors that most matter to users. It covers many aspects related to Design for X (DFX) (Chapter 9). How do we design a product that may have all these different needs for different people?

To design a product right, the user needs must be captured and not the solutions. Imagine an elderly person trying to reach a utensil from the top shelf of a cupboard. The obvious solution might be to provide a stool or ladder, but that only addresses the surface problem. The real need is accessibility—how can we ensure that this person can comfortably and safely retrieve the item without additional tools? Designers must dig deeper, going beyond the quick fix to uncover the core issue. The techniques and tools for achieving this are explored in Chapter 4. Now, consider the individual: Can they see clearly? Do they have full mobility? Can they stand steadily without assistance? (Figure 3.3). These are the critical questions that shape design.

People are unique, with different shapes, sizes, capabilities, limitations, and needs. To create a product or service that addresses these differences, we must design for a wide range of needs that accommodate a larger population unless we aim for a personalized product. Determining personas is one way to group people based on common characteristics (Section 3.2.3.1). While Chapter 2 discussed the capabilities and limitations, remember that even a capable user may not be motivated to use a product. The design should also address behavioral barriers (Section 3.2.3.2). Finally, to ensure the product or service is accessible to all, we must consider the physical characteristics of users, ensuring it is easy and comfortable for everyone to use (Section 3.2.3.3).

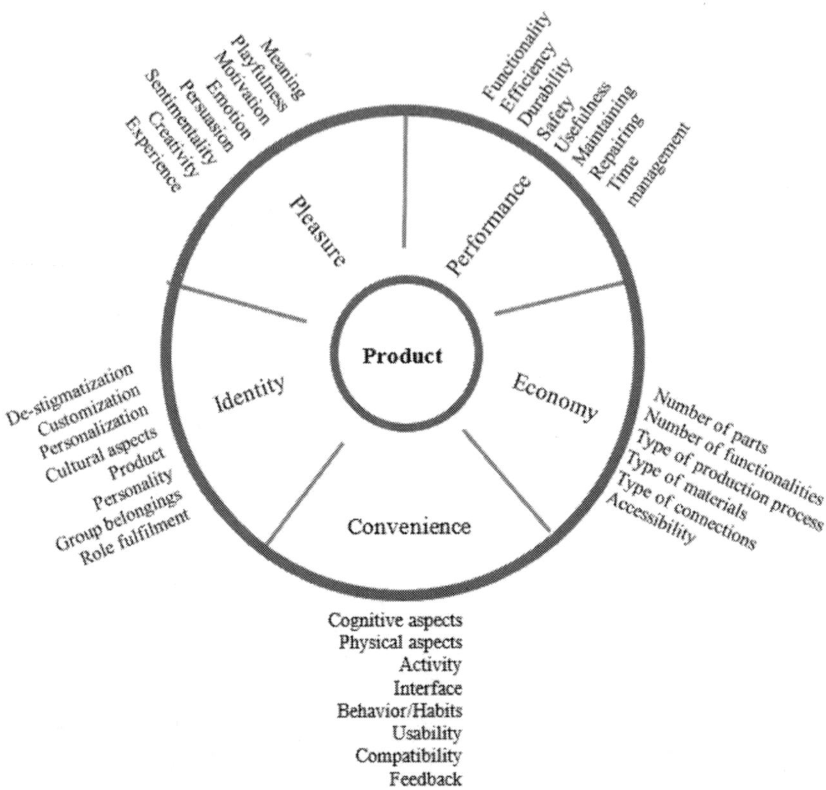

FIGURE 3.2 Product purchase and use considerations.

3.2.3.1 PERSONAS

A persona, as opposed to a person, is a fictional character. Some argue against using personas because they do not relate to a real person (LeRouge et al., 2013; Nielsen, 2019). Even though it may be fictional, it is a realistic description of a product's target user or user group. When we empathize with people, different people may have different needs, and for one, it may not be possible to cater to all the needs with one product. Some of the needs may fit in better with others. For this reason, personas are created to cater to this mix and match.

Defining user groups or market segments is like sketching the outline of a crowd, useful but a bit too vague when it comes to truly connecting with the users. The broad categories rely on statistics to capture the essence of a group, but the result can feel impersonal and abstract, making it hard to design for the ultimate user. Personas, however, turn the "faceless" numbers into a vivid character, representing a specific user who embodies the most important traits of the need group. This transformation from data to narrative creates a story that sticks in the designer's mind, guiding the design process with a human touch.

FIGURE 3.3 A wheelchair user struggles to access items from an overhead rack beyond reach.

Such an approach allows for creating a relatable and detailed narrative that is easier to remember and apply throughout the design phase, making it more likely to influence decisions and outcomes.

Personas should be created based on user research (Chapter 4) rather than dreamed-up information to be accurate and representative of user needs (Pruit & Grudin, 2003). It should be done as early as possible after all the user research is conducted using techniques given in Chapter 4. It is best to create personas as a team, as the contributions from each team member will help with the use of personas, even though they may be fictitious.

Creating personas starts with identifying key user traits and grouping similar ones into clusters. Those that overlap can be merged, and the less important groups can be dropped. Once the distinct traits are identified, the details that bring each persona to life are added to make them believable and memorable. The idea is to turn raw data into vivid characters that resonate, ensuring they guide the design authentically and positively. Creating a persona aims to develop a real and relatable character without getting lost in unnecessary details. A name and photo are important in making the persona stick in everyone's mind. At the same time, it is important to note that too much irrelevant information can dilute the key insights, making it harder to focus on what really matters. The trick is to craft a persona that is both memorable and directly tied to the design objectives. Some components or elements of a persona are shown below (Jansen, 2021), though not all of them may be applicable in every context or scenario. When analyzing the various user personas, ensure they share at least one common metric. Without it, you will end up comparing abstract elements like bios, keywords, or other qualitative descriptions, making meaningful analysis difficult.

Eleanor

Age: 65 years

Location: Boise, Idaho

Salary: $35,000 (pension and social security)

Status: Widowed

Job: Retired School Teacher

Education: Bachelor's Degree in Education

Years of Experience: 30 years in public education

Time Management: Skilled at managing her daily schedule, balancing social activities, and community involvement.

"I am motivated by a desire for autonomy and the ability to manage my own affairs. I value my independence and seek solutions that enhance my mobility without compromising my safety."

Personality

Extrovert (80%)

Logical

Independent

Skills

Smart devices for communication and information gathering.

Familiar with social media platforms

Open to learning about new technologies

Core Needs

Maintain independence and mobility without relying on family or friends for transportation.

Stay connected with her grandchildren and participate in their lives.

Instant availability, seamless and efficient commute that reduces travel time with high safety standards and no accidents

Secure data usage: User-controlled privacy policy

Influencers

Family who are concerned for her well-being and mobility.

Friends who share experiences and recommendations about transportation options.

Local community organizations that provide resources and information about mobility solutions

Frustration and Pain Points

Inconsistent ride-hailing experience – surge pricing, driver variability, unpredictable wait times

Privacy and data Security concerns – skeptical about how her data is used by companies

Vehicle availability – worried that vehicles may not available and affordable

Defining Traits

Caring: She often looks out for others in her community, volunteering at local charities.

Tech-Savvy: comfortable using technology to stay informed and connected.

Inquisitive: enjoys learning about new trends and innovations, particularly those that can enhance her lifestyle.

Preferred Apps/Channels

YouTube Uber

Preferred Brands/Trusted Sources

VISA AMERICAN EXPRESS

Determining the ideal number of personas is not an easy task. Although no empirical study has conclusively established the optimal number, existing research generally suggests a range of five to ten personas (Marsden et al., 2015). In industrial design, the typical practice is to use six or fewer, with projects rarely involving more than 12 personas (Nielsen et al., 2015). A key reason for this limitation is that managing too many personas can overwhelm users, making it harder to remember and distinguish between them (Nielsen, 2019). Even when focusing on just one to three personas, it is still crucial that these personas collectively represent the broader end-user population. For data-driven personas that capture diverse user groups, such as social media or online audiences, Salminen et al., (2022) suggests using more than ten personas to account for the wide range of behaviors and demographics in the data.

Once personas are established, the designer can explore the relevant physical variables to ensure the product or service delivers an optimal user experience. These personas may differ significantly in their physical characteristics. If the design principles outlined in Chapter 2, such as maintaining a neutral posture are to be followed, it is important to understand the structure of the people. This allows the designer to account for the full range of users, whether within a single persona or across multiple personas, ensuring that the design accommodates diverse needs and body types.

Universal adoption is not guaranteed even with well-defined personas and a product or service tailored to meet their needs. At this point, the role of behavioral factors becomes essential in understanding user engagement.

3.2.3.2 OVERCOMING BEHAVIOR

If you do not exercise regularly, think of what comes to mind. You know very well that exercise is good for health but will find enough reason not to be involved in exercise. If a product or service to make people exercise is to be designed, then it is important to consider behavior. The Fogg behavior model (2009) provides a framework for analyzing human behavior. It suggests that people act only when three key elements, motivation, ability, and a prompt are present simultaneously. Motivation drives the desire to change, while ability ensures the individual can perform the act. A prompt, whether from a person or device, triggers the action. The required behavior will not occur if any of these elements are missing.

$$\textbf{Behavior} = \textbf{Motivation} \cup \textbf{Ability} \cup \textbf{Prompt}$$

When motivation and ability surpass the activation threshold, a prompt will lead to the desired behavior (Figure 3.4). Conversely, if a person falls below this threshold, a prompt will fail to initiate the behavior.

Today, a teenager with a smartphone is highly motivated and fully capable of using it. A simple notification, a quick "ting" is all they need to reach out and engage with the device. On the other hand, an elderly person with lower

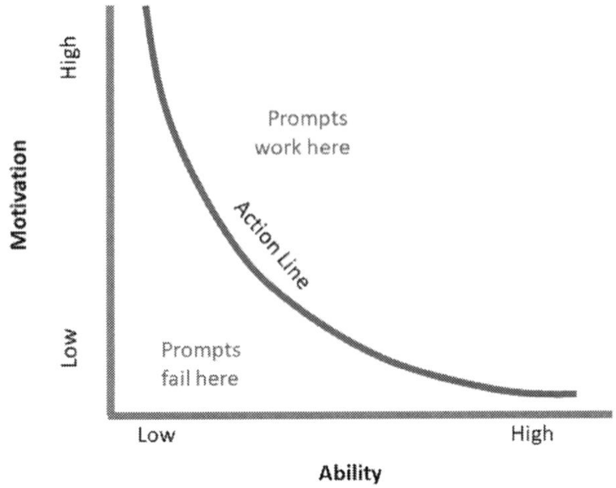

FIGURE 3.4 The Fogg behavioral model.

confidence or ability to use a smartphone might not feel the same urge. That "ting" likely won't prompt them to check the notification. This is why delivering the right triggers for the right users is essential when designing products and services. And this is where personas become crucial in design for all (D4All). Understanding user diversity ensures that our designs speak to everyone, regardless of their abilities or motivations. A sense of belonging drives ***motivation***, as people naturally seek acceptance and avoid rejection. It also stems from anticipation, whether it is the promise of pleasure or the desire to avoid pain. These are core motivators that influence our choices and actions. Fogg breaks down ***ability*** into six key factors: the time it takes, the financial cost, the physical and mental effort required, social acceptance, and how habitual the action is. These elements determine whether a person feels capable of performing the act. If any of these barriers are high, the likelihood of the action will drop. Lastly, Fogg identifies three types of ***prompts***. The first is a "spark," designed for those who have the ability but need a motivational push like an alarm clock jolting you awake. For those who lack the ability, a "helper" or facilitator such as a guide dog for a blind person to guide them. The third type is a "signal," or alert for individuals with both the ability and motivation but need a cue to act. Each prompt plays a vital role in turning intention into action.

Thus, when designing a product or service, it is important to ensure that motivation, ability, and effective prompts are aligned, as only when these three elements are in place can the desired behavior be successfully triggered.

3.2.3.3 VARIABILITY: A STATISTICAL APPROACH TO QUANTIFICATION

One fundamental statistical concept is the central limit theorem, which explains how the average (or sum) of a large number of random variables tends to follow

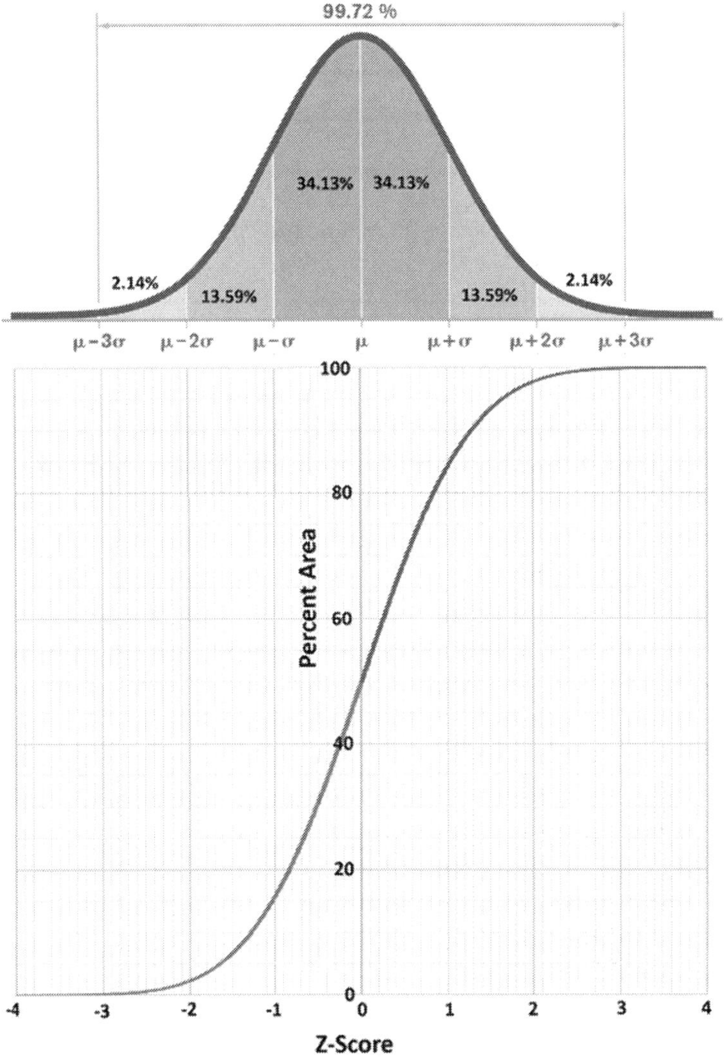

FIGURE 3.5 Area under the normal curve corresponding to Z-score value.

a normal (bell-shaped) distribution, even if the original data doesn't. The upper part of Figure 3.5 shows such a curve. Because a variable can have any value, it is transformed into what is referred to as a Z-score, which tells you how far a particular data point is from the average (mean) of a dataset, measured in terms of standard deviations.

$$Z = \frac{\text{Variable} - \text{Mean}}{\text{Standard Deviation}} = \frac{X - \mu}{\sigma}$$

If a *Z*-score is 0, the data point is exactly at the average. If a *Z*-score is +1, the data point is 1 standard deviation above the average. If a *Z*-score is –1, it is 1 standard deviation below the average. In simple terms, the *Z*-score shows how unusual or typical a value is compared to its average.

In the context of the bell curve (Figure 3.5), the area under the curve represents the probability or percentage of data points that fall within a given range. For instance, the area between a Z-score of –1 and +1 covers 68.26 percent of the data, meaning 68.26 percent of data points are within one standard deviation of the mean. Similarly, between *Z*-scores of –2 and +2, 95.44 percent of the data is included. For precise percentages in other ranges, consult any statistics textbook. The area under the curve illustrates the likelihood of a data point falling within a specific range based on its *Z*-score. Conversely, if a known percentage of data points needs to be covered, the corresponding *Z*-score range can be found using tables or Figure 3.5. Let us say we want to design for the middle 20 percent of people. That is between 40 percent and 60 percent. This we have to consider those with a *Z*-score between –0.255 and +0.255.

3.2.3.4 INDIVIDUAL DIFFERENCES: ACCOMMODATING VARIABILITY IN PEOPLE

People are of different sizes and shapes and have varying physical characteristics, so designing products that fit everyone is no small feat. Balancing fit, functionality, and aesthetics is crucial. A product that is too loose, like a shoe bigger than the foot, simply won't perform as intended. On the other hand, a shoe that's too tight and poorly designed to align with the user's anatomy and needs not only fails to function properly but also causes discomfort. Both extremes compromise usability and comfort. Getting the balance right is key to creating a product that not only looks good but works as it should for every user. Good design is essential for both performance and comfort.

There are different strategies to accommodate different sized people. These include the following:

(a) Choosing the people who fit the job or the product. This is a poor option as it goes against the concept of D4All.
(b) Building-in adjustability in the product. This is a good option and Section 3.10 on Design for a range will illustrate how to determine the adjustability. An example is a height-adjustable chair.
(c) Allowing a choice of sizes that suit individuals by employing sizing systems or through Mass Customization (Section 3.11).
(d) Optimally-sized with respect to the fixed dimensions of the individual for whom the product is made, referred to as personalized.

$$\text{Arithmetic Grading: } Y = X + a$$

$$\text{Geometric Grading: } Y = aX$$

In shoe design, it all starts with a mold called a "last." The base size last, typically US size 9 for men, is first created and then scaled to produce other sizes, a process known as grading. For men's shoes, each half size up in the US system adds 4.25 mm (1/6 inch) in length, meaning a size 9.5 would be 280.25 mm long compared to size 9's 276 mm (Rossi 2000). But length isn't the only thing that changes. Critical dimensions like the ball girth, which increases by 6 mm (1/4 in) per size. While the shoe length increases by 8.5 mm per size, the collar and back heights increase by just 1 mm per size. The grading is an additive system from one size to the next and is referred to as arithmetic grading as opposed to geometric grading, where the increments/decrements of dimensions from one size to another are a constant percentage of the original dimensions (Miller 1976).

Shape-related dimensions in a person don't always scale proportionally. For instance, a longer foot does not mean it is wider (Cabrera et al. 2004; Kouchi & Yamazaki, 1992)). This is one reason that consumers prefer certain brands of footwear because the width and lengths match their feet among other characteristics such as the stiffness and resilience of the midsole. The disproportionate relationship between measures has been described using allometry. Mathematically, $Y = aX^b$ where X and Y morphological measurements and a and b are scaling constants (Huxley 1932). When b equals 1, the relationship between X and Y is considered isometric, meaning the object maintains geometric similarity (Reiss, 1989). If b differs from 1, Y is said to scale allometrically with X, indicating a lack of geometric similarity. Specifically, if b is less than 1, X has a negative allometric effect on Y (i.e., Y increases at a slower rate than X). Conversely, when b exceeds 1, X exerts a positive allometric effect on Y. Even though allometric models have been shown to be true for the various measurements of the human body (Xiong et al., 2008), clothing and shoes rely on a simpler sizing system.

Allometric Scaling: $Y = aX^b$

$b > 1$ *positive allometry;*

$b < 1$ *negative allometry*

$b = 1$ *is Isometric or Geometric Similarity*

Sizing and grading (Luximon et al., 2012) are vital processes in developing wearable products, including footwear, headwear, helmets, glasses, wearable electronics, and clothing, that cater to a wide range of body dimensions. By accurately measuring and adjusting designs, these products can achieve proper fit and comfort, promote inclusivity, and enhance consumer satisfaction across various sizes and shapes. Recent research moves beyond traditional anthropometric-based methods to consider the complete 3D shape of body parts that interact with these products (Luximon & Goonetilleke, 2019).

3.2.3.5 DESIGN FOR A RANGE

Anthropometric data are available from several key sources, such as:

- ISO 7250: A standard that provides essential body measurement definitions. Part 1 covers definitions and landmarks, Part 2 provides statistical body measurement data from multiple countries, and Part 3 offers design ranges for use in product standards (International Organization for Standardization, 2015, 2017, 2024).
- CAESAR 3D Database: This database (www.caesar-database.com/) features 3D anthropometric data from North America (2,400 subjects), the Netherlands (1,250 subjects), and Italy (800 subjects) (Robinette et al., 2002; Harrison & Robinette, 2002).
- DINED: The Database for Industrial Design from TU Delft (https://dined.io.tudelft.nl/en/about), which can generate mannequins based on personas or specific body measurements.
- PeopleSize: A source covering nine nationalities, nine adult age groups, child data by year, and infant data (www.openerg.com/psz/).

When designing for a range of users, one powerful tool is the concept of percentiles. A percentile tells you the value below which a certain percentage of people fall. For instance, if a measurement is at the 80th percentile, it means 80 percent of people have a smaller dimension. Designers often aim to accommodate 95 percent of the population to keep costs manageable—but this approach may leave out a small portion of users. Striking the right balance is key, and Table 3.1 highlights some of the most useful percentiles to consider when designing for a wide range of people.

How do you use these values? Let's take an example. One critical dimension for seat design is popliteal height, which is the distance from the floor to the back of the knee. For the average male in the United States, this height is 445 mm with a standard deviation of 29 mm (Pheasant & Haslegrave, 2018). Assuming that this is the only value that is known, to design for the 5th percentile, we would use a Z-score of –1.64 to calculate the appropriate height, ensuring the seat is comfortable for the lower end of the population range. We use equation 3.1 which gives $X = \mu + Z\sigma$.

$$\begin{aligned}\text{5th percentile popliteal height of US males} &= 445 - 1.64\,(29)\ \text{mm}\\ &= 397.4\ \text{mm}\end{aligned}$$

TABLE 3.1 The Z-score corresponding to the xth percentile (refer to Figure 3.5)

Percentile	2.5th	5th	50th	95th	97.5th
Z-Score	–1.96	–1.64	0	1.64	1.96

Similarly, the 95th percentile US male popliteal height = 445+1.64 (29) mm = 492.6

Assuming a shoe height of 25 mm, if the seat height is to accommodate 95 percent of the US male population, the seat should be adjustable from 422.4 mm (397.4 + 25) to 517.6 mm (492.6 + 25). To cover the range from the 5th to the 95th percentile, representing 90 percent of the population, the required adjustability would be 95.2 mm (517.6 – 422.4). If a non-adjustable seat is what is required, then one solution could be to design it with a fixed height of 517.6 mm and provide an adjustable footrest, allowing for a height range from 0 to 100 mm. This way, users of different heights can still find comfort and support. Readers are urged to integrate ergonomic principles into their design processes for optimal results (Kroemer, 2017; Pheasant & Haslegrave, 2018).

Think of hopping onto a bicycle designed just for you where every component is perfectly set (Figure 3.6). Your ride will be enjoyable. That is what happens when you dial-in the ideal cycling posture. For a smooth, efficient ride, your body needs to form precise angles: your elbows should be at 150–160°, your hips at 55–65°, knees flexed at 25–35°, and your ankles flexing between 5° and 15° (Holliday & Swart, 2021; Burt, 2022). These positions ensure maximum power, comfort, and reduced risk of injury. To achieve such a configuration, manufacturers can manipulate the distance from the saddle tip to the handlebars, saddle height, and handlebar height, and center tube height. Custom-made bikes make this easy, but designing one that fits a wide range of people needs to account for the natural variability in the population to create a bike that fits most riders comfortably!

When designing for a range there are some basic principles to follow. Generally, the chosen values are the 5th percentile or the 95th percentile. Figure 3.7 illustrates the difference between these two. In this figure, the 5th

FIGURE 3.6 Posture adapted when riding a bicycle.

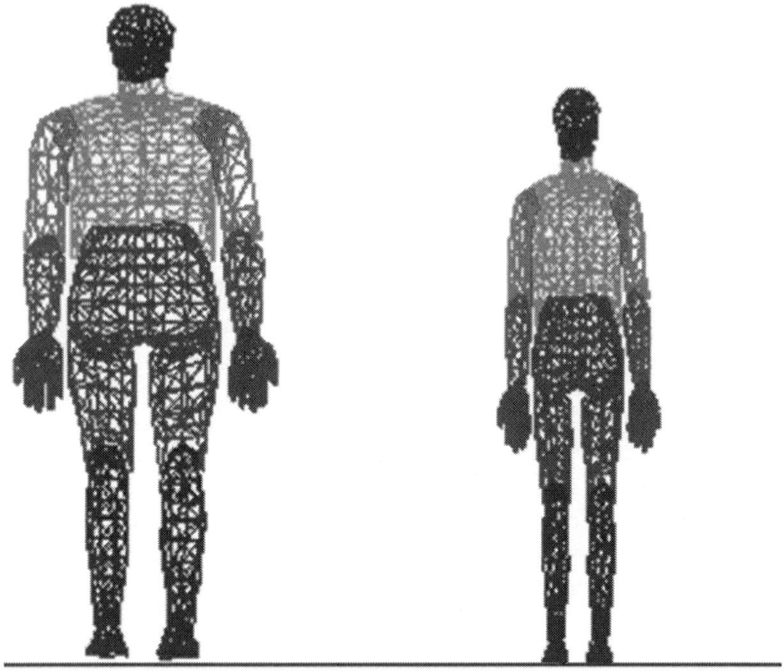

FIGURE 3.7 A 95th percentile person and a 5th percentile person as depicted in the Mannequin software.

percentile person has all limb lengths corresponding to the 5th percentile. This may not always be true. Hence, it is necessary to make adjustments depending on the context in which the relevant dimension is used.

Principles related to anthropometry in design are as follows:

1. Define the target population for which the product or service is intended.
2. When designing for clearance, use the 95th percentile to ensure that 95 percent of the target population will have adequate space.
3. When designing for reach, consider smaller individuals by using the 5th percentile.
4. Statistics can be used to prove that an unknown dimension that is equal to the sum of two known dimensions (x, y), then the mean (m) and standard deviation (s) of that unknown dimension can be found using:

$$m_{x+y} = m_x + m_y$$

$$s^2_{x+y} = s^2_x + s^2_y + 2r\, s_x\, s_y$$

where r is correlation coefficient between variable x and variable y.

5. Statistics can be used to prove that an unknown dimension that is equal to the sum of two known dimensions (x, y), then the mean (m) and standard deviation (s) of that unknown dimension can be found using:

$$m_{x+y} = m_x - m_y$$

$$s^2_{x-y} = s^2_x + s^2_y - 2r\, s_x\, s_y$$

6. When exact population dimensions are unavailable, they can be estimated through ratio scaling by using the dimensions of a similar population matched for factors like age, gender, and ethnicity (Pheasant & Haslegrave, 2018):

$$\left(\frac{m_{xA}}{m_{yA}}\right) = \left(\frac{m_{xB}}{m_{yB}}\right) \text{ and } \left(\frac{s_{xA}}{s_{yA}}\right) = \left(\frac{s_{xB}}{s_{yB}}\right)$$

   where m_{xA} and s_{xA} is the mean and standard deviation of variable x in population A and m_{yA} and s_{xA} is the mean and standard deviation of variable y in population A. Similarly, m_{xB}, m_{yB} s_{xB}, and s_{yB} correspond to population B. Variable y can represent a measurement, such as stature. As noted by Pheasant and Haslegrave (2018), while there is no mathematical proof for these equations, they are useful for expediency when collecting the necessary data is impractical.
7. After determining the design dimensions, use 3D drawings, mock-ups, or prototypes to assess functionality, with particular attention to posture as outlined in Chapter 2.

Relying solely on 2D techniques will eventually fall short of achieving the desired accommodation percentages, as individuals do not possess exact percentiles for every measurement of the body. Consequently, the level of accommodation decreases. To address this limitation, researchers have employed advanced statistical methods, such as principal component analysis (PCA) (Hudson et al., 2006; Zehner et al., 1993). There are numerous reports showcasing the use of databases, 3D models, and CAD design software (Hudson et al., 2006; Luximon et al., 2016; Pastura et al., 2012; Robinette et al., 2024). It is not easy to fit a body part such as a head using 2D anthropometry. In many cases, 3D accommodation is required.

3.3 MASS CUSTOMIZATION AND PERSONALIZATION

Today's customers value a personal connection with the products they use—whether it's a shirt embroidered with their initials or a scarf featuring their name.

Customization empowers users by giving them control over their experience from a variety of options, allowing them to shape products to fit their unique needs. Personalization, on the other hand, tailors an experience based on the user's preferences.

In an effort to become more customer-centric, many companies introduced new programs and procedures to address individual demands. However, balancing efficiency with the diverse needs of customers has led companies worldwide to adopt mass customization, delivering unique value while avoiding the inefficiencies of one-off solutions. Mass customization is a game-changer for companies looking to maximize profits while offering personalized experiences to customers. By tailoring products and services to individual preferences, companies can keep production costs low yet deliver high value (Tseng, Jiao, & Merchant, 1996). Surprisingly, the idea of mass customization isn't new—Ford was an early pioneer of the concept. It is defined as "a strategy for fulfilling individual customer desires while maintaining mass production efficiency" (Tseng & Jiao, 2001, p 685). Mass customization is employed across various industries, including financial and investment services, retail, home goods such as furniture and modular homes, and software, among others.

Mass customization can be categorized into four primary types: collaborative customization, adaptive customization, transparent customization, and cosmetic customization (Gilmore & Pine, 1997). Each type represents a distinct approach to tailoring products and services to meet individual customer needs. Collaborative customizers engage in a dialogue with individual customers to help them express their needs, identify the exact solution that meets them, and create personalized products accordingly. LEGO gives fans the opportunity to propose their own designs for new sets. Through an online platform, users can share their concepts, gather support, and participate in discussions with the community. Designs that receive sufficient votes may be turned into official LEGO sets, allowing fans to play an active role in the development process (ideas.lego.com). In adaptive customization, users can modify a standard product to suit their needs. This approach is ideal for businesses where technology allows for easy self-customization. The DEWALT® Cordless Configurator (www.dewalt.com/products/power-tools/cordless-kit-configurator), allows customers to customize features based on their preferences and project needs. Users can choose from various voltage options to suit tasks ranging from light household repairs to heavy-duty construction. Additionally, they can personalize their drill setup by selecting attachments such as drill bits, impact drivers, and battery types, making it adaptable for different applications like drilling into wood, metal, or masonry. This customizable approach ensures a tailored solution for every user's specific requirements.

Cosmetic customizers tailor the presentation of a standard product to suit different customers. This approach is ideal when customers use the product similarly but prefer distinct packaging or presentation. Instead of altering the product itself, cosmetic customizers adjust how the product is displayed, marketed, or personalized, such as through customized packaging, varied promotional messaging, or personalized names on items.

During the "Share a Coke" campaign, Coca-Cola replaced its iconic logo on bottles with popular names and phrases, creating a personalized and engaging experience for customers (www.youtube.com/watch?v=4JmKVXgAFNA). While the beverage inside remained unchanged, the packaging was tailored to resonate with individual consumers (www.youtube.com/watch?v=0wnJpTpbJPE). Customers could find bottles featuring their names or the names of friends and family, adding a personal touch that encouraged sharing. This approach demonstrated how cosmetic customization could deepen customer engagement by creating a sense of connection and individuality without altering the core product.

In transparent customization, unique goods or services are tailored to individual customers without explicitly revealing the customization. Such an approach is suitable when customers' needs are predictable or easily inferred and when they prefer not to communicate their preferences repeatedly. By observing customer behavior discreetly, transparent customizers adjust their standard offerings to meet individual needs.

Netflix (www.netflix.com) leverages advanced algorithms to analyze user viewing habits, preferences, and ratings, delivering highly personalized content recommendations. Although the platform offers a standard library of movies and shows, it subtly tailors the user experience by suggesting titles that align with the user's viewing history. For example, a user who frequently watches romantic comedies will see similar films prioritized in their recommendations, all without needing to explicitly share these preferences. This seamless and intuitive approach enhances user satisfaction by creating a tailored experience that feels effortless. Similarly, smart technologies like the Nest Learning Thermostat elevated personalization by observing users' heating and cooling preferences over time. By learning from behaviors and routines, the thermostat could automatically adjust settings to optimize comfort and conserve energy, though it required time to adapt to individual habits. Despite its innovative capabilities, Google discontinued all support for the product in April 2024 due to reliability and other issues.

Today, iconic brands like Nike and Apple have taken this concept to the next level. Customers can design products to their exact tastes, choosing everything from color and material to personalized features like engravings and monograms (Figure 3.8). This blend of individuality and efficiency continues to revolutionize industries, giving companies a competitive edge in the modern market. Another example of mass customization is in the automotive industry, where many car manufacturers allow customers to personalize their vehicles to exact specifications. From choosing exterior colors and interior finishes to selecting advanced features and technology packages, drivers can create a car tailored to their preferences. This customization extends to choosing the make and model, exterior and interior colors, and engine type, enabling customers to own a uniquely tailored vehicle while maintaining cost efficiency.

FIGURE 3.8 Personalized Soccer cleats from Nike (www.nike.com/gb/nike-by-you).

Mass customization also thrives in the food industry, where many restaurants let customers personalize their meals. Patrons can select ingredients and adjust portion sizes to suit their preferences, creating an individual and cost-effective dining experience.

Some use the term "individualization" (Riemer & Totz, 2003), for offering the right product and service to the right customer at the right time and the right place (Sunikka & Bragge, 2012). Personalization uses insight based on each customer's personal and behavioral data to deliver a superior experience. More often than not, personalization requires customer engagement to cocreate a personalized experience, which can occur through customer reviews, purchase data, and social media interactions, among others (Berger & Piller, 2003; Lim et al., 2022). The Adidas customization platform, MiAdidas, launched in 2000, was phased out to bring in a new service with co-creation or customer co-design. This process merges the creativity of designers with the insights of customers—who, despite lacking formal design training, contribute valuable perspectives to the design process (Sanders & Stappers, 2008; Reichwald & Piller, 2009).

Customization and personalization are crucial for enhancing user satisfaction and creating a deeper connection between customers and products. By offering both the ability to personalize experiences and the option to customize products, companies can effectively address individual preferences and needs, ultimately leading to a more engaging and tailored consumer experience.

3.4 DESIGN TRENDS

In recent years, with improvements in computational power and technology, many "Assistants" and robots have been designed to enhance people's capabilities and overcome limitations in diverse populations.

3.4.1 AI ASSISTANTS

AI and virtual assistants, such as Google Assistant, Siri, Alexa Show and Tell, enhance accessibility by enabling users to interact with technology through natural language (Sciarretta & Alimenti, 2021; van Heijster et al., 2022). This capability is especially valuable for individuals with varying levels of digital literacy or physical disabilities that make traditional input methods difficult. By supporting voice commands, these assistants offer an intuitive and inclusive way to access information and perform tasks, effectively reducing barriers to technology use and fostering greater inclusivity.

AI assistants that learn from interactions can simplify tasks, and build rapport and trust. For users who may feel marginalized or overlooked by traditional technology, having an AI feature that understands their needs can greatly improve their experience. This personalized engagement fosters a sense of belonging and empowerment, making technology more accessible and inviting. As AI assistants collect data on user preferences, they can continuously refine their services, aligning with iterative design. Whether explicit or implicit, feedback enables these systems to evolve and better meet the needs of all users. This adaptability is essential for creating relevant and effective solutions as user needs evolve over time.

Some AI assistants (Danaher, 2018) excel in personalization by adapting to the unique preferences and habits of users. This adaptability is key in design for all, as it acknowledges the diverse needs of individuals. For example, users with cognitive disabilities may benefit from simplified interactions and tailored reminders to help manage daily tasks more effectively. The ability of personal AI assistants to learn and evolve over time enables them to cater to a wide range of preferences, ultimately improving the user experience for everyone. However, it is important to address the ethical considerations in the personalization process.

3.4.2 SOCIAL ROBOTS AND EXOSKELETONS

Social robots (Ramírez-Duque et al., 2021) and exoskeletons (Søraa & Fosch-Villaronga, 2020) are transformative technologies that offer essential support to individuals with diverse needs. Social robots are designed to engage with humans in a friendly and interactive way, providing companionship and emotional support, especially for vulnerable groups like the elderly or those with disabilities. For example, robots such as Paro (www.parorobots.com), a therapeutic robotic seal, help reduce feelings of loneliness and enhance emotional well-being in healthcare settings. These robots also serve therapeutic purposes, aiding children with autism in developing social skills through interactive play. Meanwhile, exoskeletons are wearable robotic devices that improve physical capabilities, enabling individuals with mobility impairments to regain their independence. They, like the EksoGT (eksobionics.com) are used in rehabilitation to help patients stand and walk, promoting recovery and strengthening muscles.

Exoskeletons also assist workers in physically demanding jobs by reducing strain and the risk of injury, ultimately boosting productivity. Both technologies not only enhance the quality of life for users but also empower them to participate more fully in daily activities. As these technologies continue to evolve, social robots and exoskeletons have the potential to revolutionize support systems, fostering greater independence and improving overall well-being in various aspects of life.

3.4.3 COLLABORATIVE ROBOTS

Companies are redefining the way humans and machines collaborate. Collaborative robots, or cobots are game-changing robotic partners designed not to replace workers but to amplify people's strengths and compensate for their limitations. The result? A workspace that's not just more efficient, but also more inclusive—where technology empowers every worker to perform at their best.

Unlike their traditional counterparts, which are caged off for safety, cobots work right alongside people, seamlessly adapting to human needs (Zacharaki et al., 2020; International Organization for Standardization, 2011). Cobots allow humans and robots to seamlessly share tasks and exchange information, boosting efficiency and productivity (Kolbeinsson et al., 2019). A simple example is their use in assembly lines, where they work safely alongside humans, matching their pace and handling ergonomically challenging tasks. Too heavy to lift? A cobot will do it. Need precision beyond human capability? That is where cobots shine. Whether it is handling intricate assembly tasks, managing repetitive motions, or providing the extra muscle for heavy lifting, these robots are designed to complement the diversity of human abilities—regardless of size, strength, or endurance.

As mass customization drives the need for agile production lines, manufacturers are embracing collaborative workspaces where humans and cobots team up, each bringing their unique skills to the table (Sadrfaridpour, & Wang, 2018). Unlike traditional robots, cobots complement rather than replace human workers—this cooperation is at the heart of Industry 5.0 (Grabowska et al., 2022). By taking over repetitive, tedious, and hazardous tasks, cobots will also allow people to focus on more meaningful work, as discussed in Chapter 2. Trust is the foundation of human–robot collaboration. Studies reveal that for humans to trust robots, they need confidence in the robot's abilities (cognitive trust), such as consistent behavior, reliability, and predictability. Equally important are the robot's traits (affective trust), like how close it operates and the "personality" it portrays. When these elements align, true human–robot teamwork becomes possible. (Hancock et al., 2011).

To ensure safe human interaction, collaborative robots must incorporate built-in safety mechanisms, such as speed and motion limits. Their design focuses on compactness and lightweight construction, with safety features like soft coverings, rounded edges, and flexible joints (Bauer et al., 2016).

Optimizing task allocation between humans and robots is essential for enhancing human–robot collaboration (HRC) (IJtsma, et al., 2019; Parasuraman et al., 2000). Begin by decomposing the work and evaluating it quantitatively, taking into account both time efficiency and ergonomic factors (Annett, 2003). Effective task allocation should capitalize on the strengths of both humans and robots, assigning tasks based on their skill sets, offloading tasks to robots when feasible, and allowing for dynamic reassignment if delays occur. The workload should be distributed according to each agent's physical abilities and skill levels to maximize efficiency and minimize human strain (Lamon et al., 2019; Pearce et al., 2018). A capability-based approach is key, tailoring job assignments to the strengths of either the human or the robot while considering human worker preferences and work allocation.

Improving teamwork and collaborative interaction requires addressing human factors affecting task performance, such as task type, fatigue, skill level, and environment. For optimal productivity, robots should avoid disrupting the human's physical or perceptual focus (McGhan & Atkins, 2014). To effectively reduce human workload while maintaining performance, autonomously controlled physical and social human–robot interactions (pHRI and sHRI) that have built-in trust and emotion aspects tend to outperform manually controlled robots (Sadrfaridpour & Wang, 2018).

By carefully optimizing task allocation and integrating human factors, organizations can significantly enhance the effectiveness of human–robot collaboration. Embracing advanced interaction strategies and ensuring robots complement rather than disrupt human focus will lead to more productive and harmonious teamwork.

3.5 CONCLUSION

Designing for diversity requires more than just adhering to universal design principles; it necessitates a comprehensive approach that includes a design-for-all philosophy and an understanding of user behavior. While universal design provides a baseline, inclusive design offers multiple solutions to cater to a wide range of needs, acknowledging the diversity of human experiences. Effective design involves not only creating adaptable products but also engaging in thorough user research and employing personas to address specific needs. The integration of mass customization, AI, and robots further illustrates the shift toward personalized, efficient solutions. Embracing these approaches enables the creation of products and services that are not only functional but also inclusive, promoting accessibility and enhancing user experience and satisfaction across diverse populations.

> Never underestimate the ability of a small group of committed individuals to change the world.
>
> —Margaret Mead

REFERENCES

Accessible Canada Act (2019). Ministry of Justice, Canada. https://laws-lois.justice.gc.ca/PDF/A-0.6.pdf. Last accessed 21 December 2024.

Adel, A. (2022). Future of industry 5.0 in society: human-centric solutions, challenges and prospective research areas. *Journal of Cloud Computing*, 11, 40. https://doi.org/10.1186/s13677-022-00314-5

Annett, J. (2003). Hierarchical task analysis. In *Handbook of Cognitive Task Design* (pp. 17–36). CRC Press.

Bauer W., Bender, M., Braun, M., Rally, P., & Scholtz, O. (2016). *Lightweight Robots in Manual Assembly—Best to Start Simply!*. Fraunhofer IAO.

Berger, C. & Piller, F. (2003). Customers as co-designers. *Manufacturing Engineer*, 82(4), 42–45. doi:10.1049/me:20030407

Boztepe, S. (2007). Toward a framework of product development for global markets: A user-value-based approach. *Design Studies*, 28(5), 513–533. https://doi.org/10.1016/j.destud.2007.02.010

British Standards Institution. (2005). Design Management Systems: Managing Inclusive Design (BS 7000-6:2005), BSI Standards. https://knowledge.bsigroup.com/products/design-management-systems-managing-inclusive-design-guide.

Brumberg, J. S., Pitt, K. M., Mantie-Kozlowski, A., & Burnison, J. D. (2018). Brain–computer interfaces for augmentative and alternative communication: A tutorial. *American Journal of Speech-Language Pathology*, 27(1), 1-12 https://doi.org/10.1044/2017_AJSLP-16-0244.

Burt, P. (2022). *Bike Fit: Optimise Your Bike Position for High Performance and Injury Avoidance*. 2nd edition. Bloomsbury Sport.

Cabrera, J., Tsui, K.L., & Goonetilleke, R.S. (2004). A scale model for fitting object shapes from fixed location data. *IIE Transactions*, 36, 1099–1105.

Clarkson, P. J., Keates, S., Coleman, R., & Lebbon, C. (2003). Inclusive design. *Design for the Whole Population*. Springer. https://doi.org/10.1007/978-1-4471-0001-0

Coleman, R. (1994). The case for inclusive design: an overview. 12th Triennial Congress. International Ergonomics Association and the Human Factors Association of Canada, Toronto, Canada.

Connell, B.R., Jones, M., Mace, R., Mueller, J., Mullick, A., Ostroff, E., Sanford, J., Steinfeld, E., Story, M., & Vanderheiden, G. (1997). *The Principles of Universal Design*. The Center for Universal Design, N.C. State University. https://design.ncsu.edu/wp-content/uploads/2022/11/principles-of-universal-design.pdf. Retrieved 10 Aug 2024.

Courtney, A., & Chow, H. M. (2000). A study of tile design for tactile guide pathways. *International Journal of Industrial Ergonomics*, 25(6), 693–698. https://doi.org/10.1016/S0169-8141(99)00051-7.

Dalgleish, M. (2024). Who can play? Rethinking video game controllers and accessibility. In: Spöhrer, M., Ochsner, B. (eds), *Disability and Video Games. Palgrave Games in Context*. Palgrave Macmillan. https://doi.org/10.1007/978-3-031-34374-2_3.

Danaher, J. (2018). Toward an ethics of AI assistants: An initial framework. *Philosophy & Technology*, 31(4), 629–653.

De Couvreur, L., & Goossens, R. (2011). Design for (every)one: co-creation as a bridge between universal design and rehabilitation engineering. *CoDesign*, 7(2), 107–121. https://doi.org/10.1080/15710882.2011.609890

Design for All Europe (2004). The EIDD Stockholm Declaration 2004. Adopted on 9 May 2004, at the Annual General Meeting of the European Institute for Design and Disability in Stockholm. https://dfaeurope.eu/what-is-dfa/dfa-documents/the-eidd-stockholm-declaration-2004/. Last accessed 6 August 2024.

Disability Discrimination Act (1992). Federal Register of Legislation, Australian Government. www.legislation.gov.au/C2004A04426/2018-04-12/text. Last accessed 22 December 2024.

Dudek, M., Eisenbeis, L., Alem, N., Ronaghan, J., MacDonald, D. & Reeves, K. N. (2022). The importance of being equitable in product design. www.bcg.com/publications/2022/the-importance-of-equitable-products. Last accessed 1 September 2024.

Elmannai, W., & Elleithy, K. (2017). Sensor-based assistive devices for visually-impaired people: current status, challenges, and future directions. *Sensors* (*Basel*), 17(3), 565. https://doi.org/10.3390/s17030565.

European Accessibility Act (2019). *Official Journal of the European Union*. https://eur-lex.europa.eu/eli/dir/2019/882/oj.

Fitzpatrick, R. B. (1991). Americans with Disabilities Act of 1990. *Journal of the National Association of Administrative Law Judiciary*, , 11, 13.

Flores Ramones, A., & del-Rio-Guerra, M. S. (2023). Recent developments in haptic devices designed for hearing-impaired people: A literature review. *Sensors*, 23(6), 2968. https://doi.org/10.3390/s23062968.

Fogg, B. J. (2009). A behavior model for persuasive design. Persuasive '09: Proceedings of the 4th International Conference on Persuasive Technology. Article No. 40, pp. 1–7. https://doi.org/10.1145/1541948.1541999.

Ghobakhloo, M., Iranmanesh, M., Tseng, M. L., Grybauskas, A., Stefanini, A., & Amran, A. (2023). Behind the definition of Industry 5.0: a systematic review of technologies, principles, components, and values. *Journal of Industrial and Production Engineering*, 40(6), 432–447. https://doi.org/10.1080/21681015.2023.2216701

Gilmore, J. H., & Pine II, B. J. (1997). The four faces of mass customization. *Harvard Business Review* 75 (1): 91–101.

Grabowska, S., Saniuk, S., & Gajdzik, B. (2022). Industry 5.0: improving humanization and sustainability of Industry 4.0. *Scientometrics* 127, 3117–3144. https://doi.org/10.1007/s11192-022-04370-1

Gross, C., Goonetilleke, R.S., Banaag, J., & Nair, C. (1994). Feedback system for load bearing surfaces. U.S. Patent Number 5,283,735; 1 February 1994.

Hancock, P. A., Billings, D. R., Schaefer, K. E., Chen, J. Y. C., de Visser, E. J., & Parasuraman, R. (2011). A meta-analysis of factors affecting trust in human–robot interaction. *Human Factors*, 53(5), 517–527. https://doi.org/10.1177/0018720811417254

Harley, A. (2015). Personas make users memorable for product team members. Nielsen Norman Group. www.nngroup.com/articles/persona/. Last accessed 9 Aug 2024.

Harrison, C. R., & Robinette, K. (2002). CAESAR: Summary Statistics for the Adult Population (Ages 18-65) of the United States of America (Technical Report AFRL-HE-WP-TR-2002-0170). United States Air Force Research Laboratory, Human Effectiveness Directorate, Crew System Interface Division. https://apps.dtic.mil/sti/tr/pdf/ADA406674.pdf

Holliday, W., & Swart, J. (2021). Anthropometrics, flexibility and training history as determinants for bicycle configuration. *Sports Medicine and Health Science*. 4;3(2):93–100. https://doi.org/10.1016/j.smhs.2021.02.007.

Hudson, J. A., Zehner, G. F., Parakkat, J., & Choi, H. J. (2006). A Methodology for Evaluating Advanced Operator Workstation Accommodation (Technical

Report AFRL-HE-WP-TR-2007-0016). Air Force Research Laboratory, Human Effectiveness Directorate.

Huxley, J. S. (1932). *Problems of Relative Growth*. Methuen.

IJtsma, M., Ma, L. M., Pritchett, A. R., & Feigh, K. M. (2019). Computational methodology for the allocation of work and interaction in human-robot teams. *Journal of Cognitive Engineering and Decision Making*, 13(4), 221–241.

International Organization for Standardization. (2011). Robots and robotic devices—Safety requirements for industrial robots Part 2: Robot systems and integration (ISO 10218-2:2011). https://committee.iso.org/sites/tc299/home/projects/ongoing/iso-10218-2.html

International Organization for Standardization. (2015). Basic human body measurements for technological design. Part 3: Worldwide and regional design ranges for use in product standards. (ISO 7250-3:2015). www.iso.org/standard/64237.html.

International Organization for Standardization. (2017). Basic human body measurements for technological design. Part 1: Body measurement definitions and landmarks. (ISO 7250-1:2017). www.iso.org/standard/65246.html.

International Organization for Standardization. (2024). Basic human body measurements for technological design. Part 2: Statistical summaries of body measurements from national populations. (ISO/TR 7250-2:2024). www.iso.org/standard/84822.

Jansen, J. (2021). Elements of a persona profile. https://persona.qcri.org/blog/elements-of-a-persona-profile. Last accessed 1 Sept 2024.

Knoth, N., Tolzin, A., Janson, A., & Leimeister, J. M. (2024). AI literacy and its implications for prompt engineering strategies. *Computers and Education: Artificial Intelligence*, 6, 100225. https://doi.org/10.1016/j.caeai.2024.100225

Kolbeinsson, A., Lagerstedt, E., & Lindblom, J. (2019). Foundation for a classification of collaboration levels for human-robot cooperation in manufacturing. *Production & Manufacturing Research*, 7(1), 448–471.

Kouchi, M. & Yamazaki, N., (1992). Allometry of the foot and the shoe last. *Journal of the Anthropological Society of Nippon*, 100 (1), 101–118 (in Japanese).

Kroemer, K. H. E. (2017). *Fitting the Human*. CRC Press.

Kuriakose, B., Shrestha, R., & Sandnes, F. E. (2020). Tools and technologies for blind and visually impaired navigation support: A review. *IETE Technical Review*, 39(1), 3–18. https://doi.org/10.1080/02564602.2020.1819893.

Lamon, E., De Franco, A., Peternel, L., & Ajoudani, A. (2019). A capability-aware role allocation approach to industrial assembly tasks. *IEEE Robotics and Automation Letters*, 4(4), 3378–3385.

LeRouge, C., Jiao, M., Sneha, S., & Tolle, K. (2013). User profiles and personas in the design and development of consumer health technologies. *International Journal of Medical Informatics*, 82(11), e251–e268. doi: https://doi.org/10.1016/j.ijmedinf.2011.03.006.

Lim, J. S., & Zhang, J. (2022). Adoption of AI-driven personalization in digital news platforms: An integrative model of technology acceptance and perceived contingency. *Technology in Society, 69*, 101965. https://doi.org/10.1016/j.techsoc.2022.101965

Luximon, Y., Ball, R. M., & Chow, E. H. C. (2016). A design and evaluation tool using 3D head templates. *Computer-Aided Design and Applications*, 13(2), 153–161. https://doi.org/10.1080/16864360.2015.1084188

Luximon, A., & Goonetilleke, R. S. (2019). Foot size and foot shape of children, adults and elderly. In *DHM and Posturography* (pp. 295–319). Amsterdam: Elsevier. https://doi.org/10.1016/B978-0-12-816713-7.00023-4

Luximon, A., Zhang, Y., Luximon, Y., & Xiao, M. (2012). Sizing and grading for wearable products. *Computer-Aided Design*, 44(1), 77–84.

Mace, R. L., Hardie, G. J., & Place, J. P. (2024). Accessible environments: toward universal design. North Carolina State University: The Center for Universal Design. https://mn.gov/mnddc/parallels2/pdf/90s/90/90-AEN-CAH.pdf. Last Accessed 21 Dec 2024.

Marsden, N., Link, J., & Büllesfeld, E. (2015). Geschlechterstereotype in Persona-Beschreibungen. In: S. Diefenbach, N. Henze & M. Pielot (Hrsg.): Mensch und Computer 2015 Tagungsband, Stuttgart: Oldenbourg Wissenschaftsverlag, 2015, S. 113-122. https://doi.org/10.1515/9783110443929-013

Mashiata, M., Ali, T., Das, P., Tasneem, Z., Badal, M. F. R., Sarker, S. K., ... Das, S. K. (2022). Towards assisting visually impaired individuals: A review on current status and future prospects. *Biosensors and Bioelectronics: X*, 12, 100265. https://doi.org/10.1016/j.biosx.2022.100265.

McGhan, C. L., & Atkins, E. M. (2014). Human productivity in a workspace shared with a safe robotic manipulator. *Journal of Aerospace Information Systems*, 11(1), 1–18.

Microsoft Inclusive Design (2024). www.microsoft.com/design/inclusive/. Accessed 15 August 2024

Miller, R. G., (1976). *Manual of Shoemaking*. Clarks Training Department.

Nielsen, L. (2019). *Personas—User Focused Design*. Springer.

Nielsen, L., Hansen, K. S., Stage, J., & Billestrup, J. (2015). A template for design personas: analysis of 47 persona descriptions from danish industries and organizations. *International Journal of Sociotechnology and Knowledge Development (IJSKD)*, 7(1), 45–61.

Okolo, G.I., Althobaiti, T., & Ramzan, N. (2024). Assistive systems for visually impaired persons: challenges and opportunities for navigation assistance. *Sensors (Basel)*, 24(11), 3572. https://doi.org/10.3390/s24113572.

Parasuraman, R., Sheridan, T. B., & Wickens, C. D. (2000). A model for types and levels of human interaction with automation. *IEEE Transactions on Systems, Man, and Cybernetics—Part A: Systems and Humans*, 30(3), 286–297. https://doi.org/10.1109/3468.844354.

Pastura, F. C. H., Guimarães, C. P., Zamberlan, M. C. P., Cid, G. L., Santos, V. S., Streit, P., ... & Batista, D. S. (2012). 1D and 3D anthropometric data application on public transport vehicle layout and on oil and gas laboratories work environment design. *Work*, 41, 4618–4625. https://doi.org/10.3233/WOR-2012-0078-4618

Pearce, M., Mutlu, B., Shah, J., & Radwin, R. (2018). Optimizing makespan and ergonomics in integrating collaborative robots into manufacturing processes. *IEEE Transactions on Automation Science and Engineering*, 15(4), 1772–1784.

Perikangas, S., Määttä, A., & Tuurnas, S. (2023). Ensuring social equity through service integration design. *Public Management Review*, 1–21. https://doi.org/10.1080/14719037.2023.2246962

Persson, H., Åhman, H., Yngling, A.A., & Gulliksen, J. (2015). Universal design, inclusive design, accessible design, design for all: different concepts—one goal? On the concept of accessibility—historical, methodological and philosophical aspects. *Universal Access in the Information Society*, 14(4), 505–526. https://doi.org/10.1007/s10209-014-0358-z

Pheasant, S., & Haslegrave, C.M. (2018). *Bodyspace*. CRC Press.

Pruitt, J., & Grudin, J. (2003). Personas: practice and theory. In Proceedings of the 2003 Conference on Designing for user experiences, June, 1–15.

Reichwald, R., & Piller, F. (2009). Interaktive Wertschöpfung in der Innovation: Open Innovation. In: Interaktive Wertschöpfung. Gabler. https://doi.org/10.1007/978-3-8349-9440-0_4

Reiss, M. J. (1989). *The Allometry of Growth and Reproduction*. Cambridge University Press.

Ramírez-Duque, A. A., Aycardi, L. F., Villa, A., Munera, M., Bastos, T., Belpaeme, T., ... & Cifuentes, C. A. (2021). Collaborative and inclusive process with the autism community: a case study in Colombia about social robot design. *International Journal of Social Robotics*, 13, 153–167.

Riemer, K., & Totz, C. (2003). The many faces of personalization. In *The Customer-Centric Enterprise* (pp. 35–50). Springer. https://doi.org/10.1007/978-3-642-55460-5_3

Robinette, K. M., Blackwell, S., Daanen, H. A. M., Boehmer, M., Fleming, S., Brill, T., Hoeferlin, D., & Burnsides, D. (2002). Civilian American and European Surface Anthropometry Resource (CAESAR) Final Reports, Volume I: Summary [Technical Report]. Air Force Research Laboratory, Human Effectiveness Directorate. https://apps.dtic.mil/sti/tr/pdf/ADA406704.pdf

Robinette, K.M., Veitch, D., Bredenkamp, K., & Alemany, S. (2024). *Product Fit and Sizing: Sustainable Product Evaluation, Engineering, and Design*. CRC Press.

Rossi, W.A. (2000). *Professional Shoe Fitting*. Pedorthic Footwear Association and National Shoe Retailers Association.

Rynarzewska, A. I., Tanner, J. F. (Jeff), & Edmondson, D. R. (2024). An introduction to the "The role of inclusion, diversity, equity, & access (IDEA) in today's global marketing environment" special issue. *Journal of Global Scholars of Marketing Science*, 34(1), 1–4. https://doi.org/10.1080/21639159.2023.2264312

Sadrfaridpour, B., & Wang, Y. (2018). Collaborative assembly in hybrid manufacturing cells: An integrated framework for human–robot interaction. *IEEE Transactions on Automation Science and Engineering*, 15(3), 1178–1192. doi: 10.1109/TASE.2017.2748386.

Salminen, J., Jung, S., Nielsen, L., Şengün, S., & Jansen, B. J. (2022). How does varying the number of personas affect user perceptions and behavior? challenging the 'small personas' hypothesis! *International Journal of Human–Computer Studies*, 168, 102915. https://doi.org/10.1016/j.ijhcs.2022.102915.

Sanders, E. B. N., & Stappers, P. J. (2008). Co-creation and the new landscapes of design. *CoDesign, 4*(1), 5–18. https://doi.org/10.1080/15710880701875068

Sciarretta, E., & Alimenti, L. (2021). Smart speakers for inclusion: how can intelligent virtual assistants really assist everybody?. In *Human–Computer Interaction. Theory, Methods and Tools: Thematic Area*, HCI 2021, Held as Part of the 23rd HCI International Conference, HCII 2021, Virtual Event, July 24–29, 2021, Proceedings, Part I 23 (pp. 77–93). Springer International Publishing.

Søraa, R. A., & Fosch-Villaronga, E. (2020). Exoskeletons for all: The interplay between exoskeletons, inclusion, gender, and intersectionality. *Paladyn, Journal of Behavioral Robotics*, 11(1), 217–227.

Sunikka, A., & Bragge, J. (2012). Applying text-mining to personalization and customization research literature–Who, what and where? *Expert Systems with Applications*, 39(11), 10049–10058.

Tseng, M. M., Jiao, J., & Merchant, M. E. (1996). Design for mass customization. *CIRP Annals*, 45(1), 153–156. https://doi.org/10.1016/S0007-8506(07)63036-4

Tseng, M., & Jiao, J. (2001). Mass customization. In Gavriel Salvendy (ed.), *Handbook of Industrial Engineering: Technology and Operations Management*. Wiley. pp. 684–709

van Heijster, H., Liebrecht, C., Bol, N., van Calis, J., van der Cruijsen, A., Dijkgraaf, M., ... & Antheunis, M. (2022). Inclusive design of a sensitive virtual assistant to support people in vulnerable positions in their access to care. In CONVERSATIONS 2022: the 6th International Workshop on Chatbot Research, Applicatinos and Design, 1–6 November .

Waddell, C. D. (2006). Worldwide accessibility laws and policies. In: *Web Accessibility*. Apress. https://doi.org/10.1007/978-1-4302-0188-5_17

World Economic Forum (2016). https://www.weforum.org/stories/2016/02/this-factory-is-replacing-robots-with-humans/. Last accessed 15 August 2024.

World Health Organization (2013). *A Practical Manual for Using the International Classification of Functioning, Disability and Health* (ICF). https://cdn.who.int/media/docs/default-source/classification/icf/drafticfpracticalmanual2.pdf?sfvrsn=8a214b01_4&download=true.

World Health Organization (2001). International Classification of Functioning, Disability and Health (ICF). www.who.int/standards/classifications/international-classification-of-functioning-disability-and-health. Last accessed 6 August 2024.

World Health Organization (2023). Disability. www.who.int/news-room/fact-sheets/detail/disability-and-health. Last accessed 6 August 2024.

Xiong, S., Goonetilleke, R. S., Witana, C. P. & Au, E. Y. L. (2008). Modelling foot height and foot shape-related dimensions. *Ergonomics*, 51(8), 1272–1289.

Zacharaki, A., Kostavelis, I., Gasteratos, A., & Dokas, I. (2020). Safety bounds in human robot interaction: A survey. *Safety Science*, 127, 104667.

Zallio, M., & Clarkson, P. J. (2021). Inclusion, diversity, equity and accessibility in the built environment: A study of architectural design practice. *Building and Environment*, 206, 108352. https://doi.org/10.1016/j.buildenv.2021.108352

Zehner, G. F., Meindl, R. S., & Hudson, J. A. (1993). *A Multivariate Anthropometric Method for Crew Station Design*: Abridged (Technical Report AL-TR-1993-0054). Armstrong Laboratory, Air Force Material Command. https://apps.dtic.mil/sti/tr/pdf/ADA270652.pdf

4 Empathy in Design

> The goal of a designer is to listen, observe, understand, sympathize, empathize, synthesize, and glean insights that enable him or her to "make the invisible visible".
>
> ——Hillman Curtis

4.1 EMPATHY IN DESIGN

Empathy is the ability to understand and share the feelings, emotions, and perspectives of others, and this is the foundation of human-centered design. It empowers designers to craft solutions that emotionally, psychologically, and practically connect with users. Without empathy, even the most sophisticated designs risk failure, as they may overlook critical aspects of the actual user experience.

An example is Heathrow Airport's Terminal 5 launch in 2008. Despite a £4.3 billion investment aimed at revolutionizing travel with advanced technology, the terminal's opening quickly devolved into chaos, cancellations, and lost luggage (Kennedy, 2008). The root issue was a lack of empathic considerations: insufficient staff preparation, inadequate testing of realistic passenger scenarios, and overlooked practical and emotional traveler needs. Had empathic practices been integral to the design process, many operational crises might have been anticipated, emphasized, and mitigated.

In today's product development, user voices exert significant influence, showing the value of empathic approaches. Online reviews and user feedback shape design strategies, emphasizing the necessity for user-centric methods in an evolving technological landscape (Miao et al., 2023; Stagg et al., 2023). Designers utilize various empathic methods, including contextual inquiry, storytelling, ethnographic research, and Kansei engineering, to gain comprehensive insights into user needs, behaviors, and emotions (Han & Moghaddam, 2021; Twomlow et al., 2022). Integrating

DOI: 10.1201/9781003476177-4

empathic metrics from the outset ensures design decisions remain authentic, informed, and impactful.

4.2 NEED FOR EMPATHY

Beyond defining empathy as foundational, it is essential to clarify why empathic design is critically important. Empathy ensures designs extend beyond basic functionality, genuinely addressing the diverse emotional, psychological, and cultural needs of users. Design for All (introduced in Chapter 3) reaches its fullest potential when guided by empathic insights, allowing designers to move beyond mere compliance and continuously refine accessibility features through sustained user engagement (Dong & Clarkson, 2005; Naranjo-Rojas et al., 2022). Empathy thus plays a pivotal role in shifting from compliance-driven to user-centered design. When designers fully understand and embrace users' unique needs and experiences, they can develop solutions that authentically connect with users and drive meaningful innovation.

Moreover, empathy acts as a bridge to cultural understanding as well. By thoughtfully addressing cultural nuances and sensitivities, designers can craft products that effectively engage people from diverse cultural contexts, achieving broader relevance and deeper inclusivity (Hussain et al., 2012). In this way, empathic approaches enable designers to embed cultural relevance directly into their products, enriching the overall user experience.

An empathic approach also allows designers to recognize and accommodate the full spectrum of human diversity, including sensory sensitivities and neurodivergent cognitive styles. Understanding neurodivergent experiences—such as sensory sensitivities or distinctive processing styles—helps designers create more accommodating and inclusive products (Heylighen & Dong, 2019). Techniques such as "empathic persona-based scenarios" further equip designers with meaningful insights into users' emotional and situational realities, fostering designs that are inclusive and universally accessible (Cooper et al., 2014; Stagg et al., 2023).

Ethical considerations further highlight the significance of empathy. By integrating fairness, respect, and social responsibility into processes, designers ensure that solutions uphold human dignity and contribute positively to society (Brown & Katz, 2019). In essence, empathic design is not merely advantageous; it is indispensable for creating products that genuinely honor and support the diversity and complexity of human users.

4.3 EMPATHY AS A DESIGN DRIVER

Empathy is a fundamental principle that supports human-centered design and drives every phase of the design process. By leveraging empathic approaches, designers effectively inform and guide the entire design journey, from understanding users' cognitive landscapes (Kleinbub et al., 2019) to addressing their psychological, emotional (Desmet et al., 2019), and cultural dimensions (Heylighen & Dong, 2019).

Through nuanced awareness of users' cognitive and emotional states, designers continually refine their products. For example, mood-driven design methodologies incorporate users' emotional states into design decisions, adapting features dynamically to meet users' current feelings or moods. Applications such as automotive interiors with mood-responsive ambient lighting illustrate how mood-driven design directly influences user experience, ensuring sustained relevance and satisfaction (Desmet et al., 2019; Tholen et al., 2020).

Continued engagement with users not only refines the design but also strengthens the alignment between user needs and design solutions. In this process, empathic communication, characterized by actively listening, demonstrating genuine understanding, and responding sensitively to users' expressed thoughts and feelings becomes essential. Empathic communication helps designers build trust, enabling users to share honest and insightful experiences and preferences.

Complementing empathic communication, effective communication involves clarity, precision, and openness, ensuring that information exchange between designers and users is both productive and purposeful. Effective communication includes clearly articulated questions, transparent discussions about design constraints, and structured interactions that systematically elicit meaningful user feedback. Together, empathic and effective communication foster a robust feedback loop, guiding the iterative development of the design and ensuring that user insights consistently inform every stage of the design process (Surma-Aho & Hölttä-Otto, 2022).

Moreover, empathy humanizes technological solutions by shifting the design focus from purely technical functionality toward addressing users' emotional, psychological, and social needs. When designers embed empathy into technology, they ensure that the solutions align with emotional needs and become intuitively meaningful to users, rather than appearing abstract or disconnected from real-world experiences. For example, Twomlow et al. (2022) demonstrated how empathic data visualization grounds technological interventions firmly within users' lived realities and emotional responses. Instead of merely presenting abstract statistical information, such empathically designed visualizations highlight personal and community-level implications, making complex data accessible, understandable, and personally relevant. This human-centered approach ensures that technological advancements remain relatable, build stronger emotional connections, and ultimately foster greater user engagement and trust (Miao et al., 2023).

Empathy serves as a medium for design innovation, enhancing user experiences by creating interfaces, interactions, and environments that support the emotional needs (Concannon and Tomalin, 2023; Ferreira et al., 2023). It allows designers to uncover intangible user needs that might otherwise remain overlooked, fostering innovation and iterative refinement that lead to impactful solutions across a wide range of domains. For example, Ferreira et al. (2023) illustrated how user-centered design improved satisfaction in real-estate customer relationship management (CRM) systems by aligning functionality with user emotions and goals. Similarly, Tholen et al. (2020) reinforced the role of empathic analysis, using neural and physiological data to tailor design solutions to users' cognitive and emotional states. Incorporating empathic insights into the design process ensures that even underrepresented user groups feel seen and understood.

4.4 EMPATHY-FOSTERING METHODS AND DATA

Now we introduce key methods to generate empathic data, paired with practical principles for their effective use, enabling designers to develop actionable insights and to define impactful design challenges. The empathic data collected will serve as input for organizing and interpreting user insights (Section 4.5).

4.4.1 METHODS FOR EMPATHIC UNDERSTANDING

Empathy-fostering methods capture user behaviors, preferences, emotions, and contexts, enabling designers to deeply understand user experiences. Table 4.1 summarizes key methods, how to effectively apply them, and their associated benefits. These methods enable designers to generate comprehensive empathic data.

4.4.2 EMPATHIC DATA

Empathic data captures users' subjective experiences, which require careful interpretation. Table 4.2 categorizes three main types of empathic data with their sources and principles for effective use of the data. These principles guide designers to ensure that the subjective empathic data collected remains accurate, meaningful, and insightful.

TABLE 4.1 Methods for Empathic Understanding with their Potential Benefits.

Methods	Description	Benefits	References
Observational studies	Directly observes users in their natural environments to note behaviors and interactions without interference.	Provides authentic, context-rich insights into user behaviors and preferences.	Baxter et al. (2015); Hanington and Martin (2019)
Contextual inquiry	Observes and interviews users as they perform specific tasks within their everyday settings.	Captures detailed insights into users' goals, challenges, and decision-making processes.	Baxter et al. (2015); Holtzblatt and Beyer (2017)
Ethnographic interviews	Conducts open-ended interviews within users' natural environments to deeply understand their experiences and perspectives.	Generates detailed, contextual narratives revealing underlying motivations and cultural influences.	Hanington and Martin (2019); Ladner (2014)
Storytelling	Creates narratives that vividly illustrate users' experiences, emotions, and interactions.	Engages stakeholders emotionally, enhancing empathy and deeper understanding of user experiences.	Quesenbery and Brooks (2010); Gruen et al. (2002)
User journey mapping	Visually documents and analyzes users' interactions with products or services across multiple stages.	Highlights pain points and emotional highs and lows, identifying opportunities to enhance user experience.	Kalbach (2016); Stickdorn et al. (2018)
Design immersion	Actively participates in users' environments or experiences, temporarily adopting their roles.	Strengthens designers' intuitive understanding by experiencing users' perspectives firsthand.	Fulton Suri (2003); Brown and Wyatt (2010)
Personal inquiry	Encourages users to reflect through journals, diaries, or self-assessment exercises.	Facilitates deeper self-expression, revealing users' internal thoughts, motivations, and emotional states.	Hanington and Martin (2019); Sanders and Stappers (2012)
Task analysis	Breaks down complex tasks into smaller, manageable steps to examine users' cognitive processes and decision-making.	Identifies inefficiencies and pain points, pinpointing specific areas for design improvements.	Kirwan and Ainsworth (1992); Stanton (2006)
User personas	Creates detailed, research-based profiles representing key user segments.	Helps designers maintain a clear focus on diverse user needs, guiding empathic and targeted design decisions.	Cooper et al. (2014); Pruitt and Adlin (2006)
Kansei engineering	Captures and analyzes users' emotional responses to specific design attributes through structured evaluations.	Reveals emotional drivers behind user preferences, informing emotionally resonant design choices.	Nagamachi (2002); Schütte et al. (2004)

TABLE 4.2 Empathic Data Type and the Principles for Effective Use.

Type	Description of the Data Type	Generated from (Methods)	Principles for Effective Use
Self-reported data	Users explicitly express emotions, perceptions, and experiences.	Ethnographic interviews Storytelling Personal inquiry User personas	(a) Ensure clear, unbiased questions are used in the empathic methods (b) Validate collected data through multiple sources (c) Address potential response biases by ensuring user comfort and confidentiality
Behavioral data	Observable user actions and interactions that require subjective interpretation for empathic understanding.	Observational studies Contextual inquiry Task analysis	(a) Clearly define observable behaviors (b) Complement with qualitative context (c) Maintain ecological validity by observing natural user interactions.
Ethnographic data	Rich, detailed qualitative insights from naturalistic user observations and interactions.	Observational studies Ethnographic interviews Design immersion	(a) Immerse deeply in user contexts (b) Practice reflexivity to minimize researcher bias (c) Use triangulation for enhanced validity

4.4.2.1 PRINCIPLES FOR SELF-REPORTED DATA

Self-reported assessments may be unreliable due to social desirability bias and limited self-awareness, leading individuals to overstate their empathic tendencies, emphasizing a need for multi-method evaluation strategies. To prevent such biases,

(a) Ensure clear, unbiased questions are used in the empathic methods
- Develop interview guides or survey questions with neutral wording to avoid influencing users' responses. Avoid leading questions or assumptions.
- Conduct pilot tests or mock interviews to verify the clarity and neutrality of questions before actual data collection.
- Balance open-ended and structured questions to gather rich qualitative insights without restricting users' genuine expression of thoughts and emotions.

(b) Validate collected data using multiple sources
- Use multiple empathic methods (e.g., interviews, personal journals, user personas) to corroborate insights. This approach helps identify consistent themes and reduces biases from any single data source.
- Cross-reference self-reported data with behavioral observations or user-generated artifacts (photos, diaries, or social media posts) to validate authenticity.
- Seek patterns or discrepancies across different data sources and address them through follow-up interviews or supplementary methods.

(c) Address potential response biases by ensuring user comfort and confidentiality
- Clearly communicate the confidentiality and anonymity of users' responses, ensuring users feel safe and comfortable sharing honest insights.
- Select neutral and comfortable environments for interviews or interactions, minimizing distractions or environmental pressures that might bias responses.
- Establish rapport and trust with participants through respectful communication, empathy, and patience, allowing them sufficient time to articulate their experiences authentically.

4.4.2.2 PRINCIPLES FOR BEHAVIORAL DATA

Behavioral data offers a more objective and ecologically valid measure of empathy by capturing how individuals actually respond in real-world or simulated contexts. Unlike self-reports, it reflects situational empathy and reveals discrepancies between claimed attitudes and observable actions (Decety & Cowell, 2015). These are more resource-intensive, context-dependent, and may still fail to capture the internal emotional or cognitive processes underlying empathic behavior. To prevent such issues,

(a) Clearly define observable behaviors
- Specify precisely what behaviors will be observed and how they will be documented before starting the study. For example, clearly define "user frustration" through observable signs (e.g., repeated clicks, facial expressions, sighs).
- Develop standardized behavior coding or logging protocols to ensure consistent and reliable data collection across different observers or sessions.
- Regularly calibrate and train observers to ensure consistency in identifying and recording defined behaviors.

(b) Complement with qualitative context
- Record additional qualitative data (such as contextual notes, verbalized thoughts, or environmental conditions) alongside behavioral observations to enrich interpretation.
- Conduct post-observation interviews or brief conversations with users to clarify behaviors and confirm the underlying reasons or motivations behind observed actions.
- Combine observational data with narrative descriptions or storyboards to visualize user journeys and interactions comprehensively.

(c) Maintain ecological validity by observing natural user interactions
- Conduct observations in environments and situations that closely reflect real-world contexts, ensuring that user behaviors are genuine and not influenced by artificial conditions.
- Minimize observer intrusiveness (e.g., use unobtrusive video recording or passive observation techniques) to reduce the Hawthorne effect, where participants alter behavior due to awareness of observation.
- Extend observation periods or multiple sessions to ensure the captured behaviors are representative and not merely momentary or situational.

4.4.2.3 PRINCIPLES FOR ETHNOGRAPHIC DATA

Ethnography allows researchers to observe empathic interactions as they naturally occur, especially in culturally embedded practices. However, it is time-consuming, relies heavily on the researcher's interpretive lens, and is subject to observer bias and challenges in replicability (Pink, 2020). To mitigate these issues,

(a) Immerse deeply in user contexts
- Spend sufficient time within users' natural settings (e.g., homes, workplaces, community spaces) to experience and observe interactions firsthand, building a nuanced understanding of their lived experiences.
- Adopt a learner's stance, observing and participating without immediate judgment or assumptions to accurately capture authentic user behaviors and interactions.

(b) Use participant-observation methods, engaging directly in user activities and routines, to enhance contextual understanding and empathic connection.

(c) Practice reflexivity to minimize researcher bias
- Regularly reflect on your own assumptions, biases, and emotional reactions throughout the research process, documenting these reflections to ensure they don't distort interpretations.

- Involve multiple researchers in the analysis of ethnographic data, facilitating critical discussion and diverse perspectives that balance out potential biases.
- Employ structured analytic tools such as reflective journals, memos, or peer reviews that promote continuous critical self-examination and transparency.

(d) Use triangulation for enhanced validity

- Combine multiple ethnographic methods (such as observational studies, contextual inquiries, and interviews) to cross-verify insights and reinforce the reliability of your findings.
- Engage in data triangulation, methodological triangulation (multiple research methods), and investigator triangulation (multiple researchers) to ensure robust validation of data patterns.
- Integrate findings across qualitative and quantitative data (where available) to strengthen the overall interpretive rigor and deepen understanding of user experiences.

4.5 ORGANIZATION OF EMPATHIC DATA

Empathic data can be represented through frameworks and visualizations, each serving distinct purposes, but they work cohesively. Empathy frameworks guide both the collection and organization of empathic data by providing structured, systematic ways to capture, categorize, and interpret empathic data, ensuring clarity, consistency, and comprehensiveness. These frameworks range from practical templates directly used to document data, such as empathy maps and journey maps, to conceptual models like Activities, Environments, Interactions, Objects, and Users (AEIOU) (Martin & Hanington, 2012), or People, Objects, Environments, Messages, and Services (POEMS) (Kumar, 2012), guiding deeper analysis of user experiences. Visualizations translate organized data into graphic formats leading to insights by highlighting patterns and trends. These visualizations range from visual storytelling and graphical representation to facilitate the communication of abstract concepts to diverse stakeholders.

4.5.1 FRAMEWORKS FOR EMPATHIC DATA

The frameworks listed in Table 4.3 include both practical templates and conceptual models, each tailored for distinct stages and objectives within the user-centered design process. The summary of these frameworks (Table 4.3) facilitates informed selection based on specific design goals and contexts. By choosing appropriate frameworks, designers can effectively navigate the complexities of user insights, ensuring clarity and depth in understanding users' experiences, emotions, and motivations.

TABLE 4.3 Description, Benefits and Design Focus of Empathic Methods.

Framework	Description	Benefits	Design Focus	References
Empathy map	Organizes user insights into four quadrants: what users see, hear, say, and do.	Helps designers capture users' thoughts, feelings, and behaviors effectively.	Enhances empathic understanding of user experiences.	Gray (2017); Osterwalder and Pigneur (2010)
User story	Provides short, structured narratives of user needs and goals from their viewpoint.	Enables clear, communication of user requirements and functionalities.	Guides precise and user-focused design development.	Cohn (2004); Patton (2014)
User personas	Creates detailed profiles representing user segments, including demographics, goals, motivations, and pain points.	Supports empathic visualization of diverse user types for tailored design solutions.	Facilitates targeted understanding of user groups.	Cooper et al. (2014); Pruitt and Adlin (2006)
Customer profile	Details customer attributes by identifying their jobs, pains, and gains systematically.	Provides deep insights into customer motivations and challenges for improved outcomes.	Enables targeted design of empathic and user-centered solutions.	Osterwalder et al. (2014)
Five human factors	Examines five key dimensions of user behavior: cognition, emotion, perception, physiology, and social interactions.	Ensures comprehensive consideration of cognitive, emotional, and physical user requirements.	Supports holistic integration of human-centered aspects into the design.	Wickens et al. (2015); Salvendy (2012)
POEMS	Categorizes user experience into five dimensions: People, Objects, Environments, Messages, and Services.	Captures multidimensional user insights systematically to identify improvement opportunities.	Enables comprehensive evaluation of user interactions across contexts.	Kumar (2012)

(continued)

TABLE 4.3 (Continued)

Framework	Description	Benefits	Design Focus	References
AEIOU	Analyzes user experiences by organizing data into Activities, Environments, Interactions, Objects, and Users.	Provides structured and detailed insights into user contexts and behaviors.	Facilitates systematic exploration and analysis of user interactions.	Wasson (2000); Martin and Hanington (2012)
HEART	Assesses user experience quantitatively through Happiness, Engagement, Adoption, Retention, and Task Success metrics.	Allows designers to measure and enhance user satisfaction and effectiveness using clear indicators.	Supports evaluation and optimization of usability and user engagement.	Rodden et al. (2010)
Kano model	Organizes product features into Basic (must-have), performance (expected), and delight (unexpected and pleasing).	Helps prioritize design features effectively by assessing their impact on user satisfaction.	Facilitates balance between essential requirements and innovative features.	Kano et al. (1984); Sauerwein et al. (1996)
Emotional design	Considers three experiential layers: visceral (initial impression), behavioral (functionality), and reflective (meaning).	Enhances emotional engagement between users and products, strengthening overall user satisfaction.	Creates intuitive and emotionally resonant user experiences.	Norman (2004)

4.5.2 EMPATHIC DATA VISUALIZATION

Empathy frameworks and visualization tools work together. While empathy frameworks provide structured methodologies for systematically collecting and categorizing user data, visualization tools (Table 4.4) dynamically interpret and translate these insights into intuitive visual formats. This integrated approach bridges data collection and practical application, empowering design teams to make informed, empathic, and user-centered decisions. Effective visualization tools transform empathic data into comprehensible insights.

4.5.3 CONSTRUCTING USER INSIGHTS

Empathy-driven design processes culminate in clearly defined insights about target populations, essential for effective problem framing, the focus of Chapter 5. Identifying the target population's distinct characteristics, preferences, behaviors, and emotional drivers through empathic methods and frameworks ensures accurate representation of user needs. By using structured frameworks and visualization methods outlined in this chapter, designers can comprehensively articulate the nuanced requirements and expectations of specific user segments.

The effectiveness of frameworks and visualizations for empathic data depends on the specific research context and the unique requirements of each design endeavor. Selecting the most appropriate tools requires a clear understanding of the design focus and the intended outcomes. When carefully chosen and integrated, these elements collectively enable designers to embed empathy into the design process effectively. This structured approach not only facilitates the identification of problems but also lays a strong foundation for innovative and impactful solutions. Ultimately, the integration of these tools enhances the quality of design outcomes.

Table 4.5 shows a case study on selecting the appropriate method and visualization in seat design. Seat comfort involves understanding users' physical, emotional, and situational needs. The right combination of frameworks and visualization methods (Figure 4.1) ensures a user-centered approach to obtaining insights to meet practical and emotional needs.

Accurately capturing and clearly articulating target population insights through empathy-driven methods establishes a solid foundation for framing meaningful and precise design problems. The transition to Chapter 5 will leverage these comprehensive insights, guiding designers toward focused problem identification, enabling innovative solutions genuinely aligned with user needs. The following are examples of empathy-driven Insights for the Seating case study:

(a) *Physical comfort needs*: From Heatmaps, designers identified pressure points and discomfort zones when users sat for extended periods, highlighting the need for variable seat cushioning and adjustable lumbar support.

TABLE 4.4 Description, Benefits, and Design Focus of Visualization Methods.

Visualization Method	Description	Benefits	Design Focus	References
Storyboards	Presents sequential visual narratives illustrating users' experiences and interactions with a product or service.	Provides comprehensive views of user journeys, highlights pain points, and reveals improvement opportunities.	Explores user interactions, experiences, and behaviors in design contexts.	Quesenbery and Brooks (2010); Truong et al. (2006)
Heatmaps	Presents visual representations showing intensity and frequency of user interactions with design elements.	Identifies areas of high and low engagement rapidly, clarifies user behavior patterns, and supports targeted improvements.	Examines user engagement levels and interaction patterns effectively.	Albert and Tullis (2013); Pernice and Nielsen (2009)
Word clouds	Presents textual data visually by sizing words according to frequency or significance within a dataset.	Highlights dominant themes or keywords effectively, offers snapshots of user sentiment, and identifies preference trends.	Analyzes textual feedback and identifies key user-expressed themes clearly.	Viégas and Wattenberg (2008); McNaught and Lam (2010)
Network diagrams	Illustrates visual relationships and connections among various elements or entities in a graphical form.	Visualizes complex relationships clearly, identifies dependencies quickly, and reveals areas for potential improvements.	Explores connections, relationships, and dependencies between design elements clearly.	Freeman (2004); Hansen et al. (2011)
Emotional charts	Displays visual representations of user emotional responses using color-coded scales or graphical elements.	Highlights emotional patterns clearly, identifies emotional triggers effectively, and supports emotionally engaged design.	Analyzes user emotions, reactions, and affective responses systematically.	Plutchik (2001); Desmet (2003)

Infographics	Represents complex information, data, or knowledge in visually engaging graphical formats.	Simplifies complex content clearly, improves comprehension significantly, and increases audience engagement effectively.	Communicates complex information, data, and insights visually and accessibly.	Lankow et al. (2012); Smiciklas (2012)
User flow diagrams	Maps detailed visual pathways of user interactions within digital or physical design environments.	Identifies workflow bottlenecks efficiently, optimizes user journeys clearly, and supports improved interaction design.	Maps user workflows, identifies usability gaps, and reveals opportunities for enhancements clearly.	Garrett (2010); Kalbach (2016)
Comparative analysis charts	Displays visual comparisons of different design options, emphasizing strengths, weaknesses, and variations.	Facilitates informed decisions clearly, contrasts design alternatives effectively, and supports iterative refinement processes.	Compares design options systematically to inform decision-making clearly.	Albert and Tullis (2013); Few (2004)
Sentiment analysis charts	Presents visual analyses of the sentiment or emotional tone from user feedback using graphical charts.	Captures user sentiment trends clearly, identifies prevalent attitudes effectively, and supports design refinement processes.	Analyzes user sentiment, attitudes, and feedback systematically to enhance user satisfaction.	Pang and Lee (2008); Liu (2015)

TABLE 4.5 Framework and Visualization Methods Selection for Seats.

Phase	Objective	Tool	Why Appropriate?
Empathic framework (Data organization)	Organize and categorize the data collected for systematic analysis.	Empathy map	Helps document users' feelings, thoughts, and challenges related to their seating experiences, emphasizing emotional and practical needs.
		User personas	Personalizes the design process by detailing demographic, ergonomic, and comfort preferences for target users.
		Kano model	Prioritizes features into basic (must-have), performance (enhance satisfaction), and delight (unexpected exciting features).
		POEMS framework	Analyzes user interactions with People, Objects, Environments, Messages, and Services, identifying opportunities for enhancement.
Visualization (Data communication)	Communicate insights and design concepts effectively to stakeholders and the design team.	Heatmaps	Highlights areas of pressure or discomfort, providing insights into physical usability and guiding ergonomic improvements.
		Sentiment Analysis charts	Captures emotional responses, highlighting areas needing emotional satisfaction enhancements.
		Storyboards	Depicts user scenarios in different environments (e.g., office, home), helping visualize real-world applications.
		User flow diagrams	Maps user interactions (e.g., adjusting the chair or switching positions), illustrating usability concerns and opportunities.

(b) *Emotional and psychological needs*: Sentiment analysis charts revealed users associated positive feelings with seating that made them feel "secure" and "relaxed," suggesting that aesthetic elements like color, material softness, and ergonomic curves strongly influence emotional satisfaction.

(c) *Behavioral patterns*: User flow diagrams showed frequent micro-adjustments (e.g., shifting posture, adjusting seat height), indicating a need for intuitive, easily reachable adjustment controls without interrupting the user's workflow.

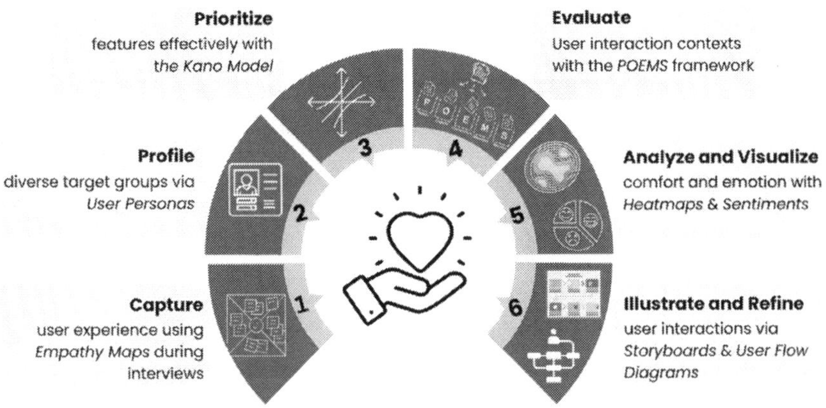

FIGURE 4.1 The integrated framework and visualization for sitting.

(d) *Diverse demographic needs*: User personas based on interviews and surveys captured distinctions between different user groups (e.g., office workers needing dynamic support during long work hours vs. elderly users prioritizing ease of getting up from the seat), leading to differentiated design strategies.

(e) *Latent needs and unspoken expectations*: Empathy maps uncovered that users valued "invisible comfort." They preferred chairs that required no conscious thought to remain comfortable, emphasizing the need for automatic ergonomic adaptation rather than manual adjustments.

(f) *Contextual usage insights*: Using the POEMS framework, designers observed that seating comfort was influenced not just by the chair itself but also by environmental factors (e.g., desk height, lighting, ambient noise), suggesting that seat design must consider environmental adaptability.

4.6 CONCLUSION

In conclusion, this chapter emphasizes the critical role of empathy in design, highlighting the importance of deeply understanding users' emotions, experiences, and perspectives. By examining various methods such as observational studies, contextual inquiries, ethnographic interviews, and Kansei engineering, designers can systematically capture rich empathic data. The structured use of frameworks and visualizations further enhances the clarity and actionability of these insights, promoting effective communication and collaboration across multidisciplinary teams.

The insights derived from empathic practices lay a strong foundation for effective problem-framing, discussed further in the next chapter. By clearly articulating and defining user challenges through empathy, designers enhance their ability to identify and address real-world problems with precision and clarity. This empathic foundation ensures that subsequent problem-solving strategies remain firmly rooted in a comprehensive understanding of human experiences.

> You never really understand a person until you consider things from his point of view... until you climb into his skin and walk around in it.
>
> —Harper Lee

REFERENCES

Albert, W., & Tullis, T. (2013). *Measuring the User Experience: Collecting, Analyzing, and Presenting Usability Metrics.* Morgan Kaufmann.

Baxter, K., Courage, C., & Caine, K. (2015). *Understanding Your Users: A Practical Guide to User Research Methods.* Morgan Kaufmann.

Brown, T., & Wyatt, J. (2010). Design thinking for social innovation. *Stanford Social Innovation Review,* 8(1), 31–35.

Brown, T., & Katz, B. (2019). *Change by Design: How Design Thinking Creates New Alternatives for Business and Society.* HarperBusiness.

Cohn, M. (2004). *User Stories Applied: For Agile Software Development.* Addison-Wesley Professional.

Concannon, S., & Tomalin, M. (2023). Measuring perceived empathy in dialogue systems. *AI and Society,* 39, 2233–2247.

Cooper, A., Reimann, R., Cronin, D., Noessel, C., & Csizmadi, D. (2014). *About Face: The Essentials of Interaction Design.* Wiley.

Decety, J., Cowell, J. M. (2015). Empathy, justice, and moral behavior. *AJOB Neuroscience.* 2015; 6(3):3–14. https://doi.org/10.1080/21507740.2015.1047055.

Desmet, P. M. A. (2003). Measuring emotion: development and application of an instrument to measure emotional responses to products. In M. A. Blythe et al. (Eds.), *Funology: From Usability to Enjoyment,* 111–123. Springer.

Desmet, P. M. A., Xue, H., & Fokkinga, S. F. (2019). The same person is never the same: introducing mood-stimulated thought/action tendencies for user-centered design. *She Ji,* 5(3), 167–187.

Dong, H., & Clarkson, J. (2005). "Barriers to inclusive design in the built environment: Empathy, expertise, and cultural factors." *Design Studies, 26(3).*

Ferreira, M. S., Antão, J., Pereira, R., Bianchi, I. S., & Tovma, N. (2023). Improving real estate CRM user experience and satisfaction: A user-centered design approach. *Journal of Open Innovation: Technology, Market, and Complexity,* 9(2), 100076.

Few, S. (2004). *Show Me the Numbers: Designing Tables and Graphs to Enlighten.* Analytics Press.

Freeman, L. C. (2004). *The Development of Social Network Analysis: A Study in the Sociology of Science.* Empirical Press.

Fulton Suri, J. (2003). Empathic design: informed and inspired by other people's experience. *Empathic Design,* 51(4), 51–58.

Garrett, J. J. (2010). *The Elements of User Experience: User-Centered Design for the Web and Beyond* (2nd ed.). New Riders.

Gray, D. (2017). *Gamestorming: A Playbook for Innovators, Rulebreakers, and Changemakers*. O'Reilly Media.

Gruen, D., Rauch, T., Redpath, S., & Ruettinger, S. (2002). The use of stories in user experience design. *International Journal of Human–Computer Interaction*, 14(3–4), 503–534.

Han, Y., & Moghaddam, M. (2021). Analysis of sentiment expressions for user-centered design. *Expert Systems with Applications*, 171, 114604.

Hanington, B., & Martin, B. (2019). *Universal Methods of Design: 125 Ways to Research Complex Problems, Develop Innovative Ideas, and Design Effective Solutions.* Rockport Publishers.

Hansen, D. L., Shneiderman, B., & Smith, M. A. (2011). *Analyzing Social Media Networks with NodeXL*. Morgan Kaufmann.

Heylighen, A., & Dong, A. (2019). To empathise or not to empathise? Empathy and its limits in design. *Design Studies*, 65, 107–124.

Holtzblatt, K., & Beyer, H. (2017). *Contextual Design: Design for Life*. Morgan Kaufmann.

Hussain, S., Sanders, E. B. N., & Steinert, M. (2012). Participatory design with marginalized people in developing countries: Challenges and opportunities experienced in a field study in Cambodia. *International Journal of Design*, 6(2), 91–109.

Kalbach, J. (2016). *Mapping Experiences: A Complete Guide to Creating Value through Journeys, Blueprints, and Diagrams*. O'Reilly Media.

Kano, N., Seraku, N., Takahashi, F., & Tsuji, S. (1984). Attractive quality and must-be quality. *Journal of the Japanese Society for Quality Control*, 14(2), 39–48.

Kennedy, D. (2008, March 28). Disastrous opening day for Terminal 5. *The Independent*. www.the-independent.com/news/uk/home-news/disastrous-opening-day-for-terminal-5-801376.html

Kirwan, B., & Ainsworth, L.K. (1992). *A Guide to Task Analysis*. Taylor & Francis.

Kleinbub, J. R., Palmieri, A., & Benelli, E. (2019). Measuring empathy: A statistical physics grounded approach. *Physica A: Statistical Mechanics and Its Applications*, 526, 120979.

Kumar, V. (2012). *101 Design Methods: A Structured Approach for Driving Innovation in Your Organization*. Wiley.

Ladner, S. (2014). *Practical Ethnography: A Guide to Doing Ethnography in the Private Sector*. Routledge.

Lankow, J., Ritchie, J., & Crooks, R. (2012). *Infographics: The Power of Visual Storytelling*. Wiley.

Liu, B. (2015). *Sentiment Analysis: Mining Opinions, Sentiments, and Emotions*. Cambridge University Press.

Martin, B., & Hanington, B. (2012). *Universal Methods of Design: 100 Ways to Research Complex Problems, Develop Innovative Ideas, and Design Effective Solutions*. Rockport Publishers.

McNaught, C., & Lam, P. (2010). Using Wordle as a supplementary research tool. *The Qualitative Report*, 15(3), 630–643.

Miao, W., Lin, K. C., Wu, C. F., Sun, J., Sun, W., Wei, W., & Gu, C. (2023). How could consumers' online review help improve product design strategy? *Information (Switzerland)*, 14(8), 434.

Nagamachi, M. (2002). Kansei Engineering as a powerful consumer-oriented technology for product development. *Applied Ergonomics*, 33(3), 289–294.

Naranjo-Rojas, A., Perula-de-Torres, L. Á., & Molina-Recio, G. (2022). Patients, caregivers, and healthcare professionals' needs when designing the content of a mobile application for the clinical monitoring of patients with chronic obstructive pulmonary disease and home oxygen therapy: A user-centered design. *Internet Interventions*, 29, 100552.

Norman, D. A. (2004). *Emotional Design: Why We Love (or Hate) Everyday Things.* Basic Books.

Osterwalder, A., & Pigneur, Y. (2010). *Business Model Generation: A Handbook for Visionaries, Game Changers, and Challengers.* Wiley.

Osterwalder, A., Pigneur, Y., Bernarda, G., & Smith, A. (2014). *Value Proposition Design: How to Create Products and Services Customers Want.* Wiley.

Pang, B., & Lee, L. (2008). Opinion mining and sentiment analysis. *Foundations and Trends in Information Retrieval*, 2(1-2), 1–135.

Patton, J. (2014). *User Story Mapping: Discover the Whole Story, Build the Right Product.* O'Reilly Media.

Pernice, K., & Nielsen, J. (2009). *Eyetracking Web Usability*. New Riders Press.

Pink, S. (2020). *Doing Sensory Ethnography* (2nd ed.). SAGE Publications.

Plutchik, R. (2001). The nature of emotions. *American Scientist*, 89(4), 344–350.

Pruitt, J., & Adlin, T. (2006). *The Persona Lifecycle: Keeping People in Mind Throughout Product Design*. Morgan Kaufmann.

Quesenbery, W., & Brooks, K. (2010). *Storytelling for User Experience: Crafting Stories for Better Design*. Rosenfeld Media.

Rodden, K., Hutchinson, H., & Fu, X. (2010). Measuring the user experience on a large scale: User-centered metrics for web applications. *Proceedings of the SIGCHI Conference on Human Factors in Computing Systems (CHI '10)*, 2395–2398. Atlanta, GA, USA, April 10–15. New York: ACM. https://dl.acm.org/doi/10.1145/1753326.1753687

Salvendy, G. (Ed.). (2012). *Handbook of Human Factors and Ergonomics* (4th ed.). Wiley.

Sanders, E. B. N., & Stappers, P. J. (2012). *Convivial Toolbox: Generative Research for the Front End of Design*. BIS Publishers.

Sauerwein, E., Bailom, F., Matzler, K., & Hinterhuber, H. H. (1996). The Kano model: how to delight your customers. Preprint from the 9th International Working Seminar on Production Economics, pp. 313–327. https://www.researchgate.net/publication/240462191_The_Kano_Model_How_to_Delight_Your_Customers

Schütte, S., Eklund, J., Axelsson, J. R., & Nagamachi, M. (2004). Concepts, methods and tools in Kansei Engineering. *Theoretical Issues in Ergonomics Science*, 5(3), 214–231.

Smiciklas, M. (2012). *The Power of Infographics: Using Pictures to Communicate and Connect with Your Audiences*. Que Publishing.

Stagg, B. C., Tullis, B., Asare, A., Stein, J. D., Medeiros, F. A., Weir, C., Borbolla, D., & Hess, R. (2023). Systematic user-centered design of a prototype clinical decision support system for glaucoma. *Ophthalmology Science*, 3(3), 100279.

Stanton, N. A. (2006). Hierarchical task analysis: Developments, applications, and extensions. *Applied Ergonomics*, 37(1), 55–79.

Stickdorn, M., Hormess, M., Lawrence, A., & Schneider, J. (2018). *This is Service Design Doing: Applying Service Design Thinking in the Real World*. O'Reilly Media.

Surma-Aho, A., & Hölttä-Otto, K. (2022). Conceptualization and operationalization of empathy in design research. *Design Studies*, 78, Article 101075. https://doi.org/10.1016/j.destud.2021.101075

Tholen, M. G., Trautwein, F. M., Böckler, A., Singer, T., & Kanske, P. (2020). "Functional magnetic resonance imaging (fMRI) item analysis of empathy and theory of mind." *Human Brain Mapping,* 41(10), 2611–2628. https://doi.org/10.1002/hbm.24966

Truong, K. N., Hayes, G. R., & Abowd, G. D. (2006). Storyboarding: an empirical determination of best practices and effective guidelines. *Proceedings of the 6th Conference on Designing Interactive Systems,* 12–21.

Twomlow, A., Grainger, S., Cieslik, K., Paul, J. D., & Buytaert, W. (2022). A user-centred design framework for disaster risk visualisation. *International Journal of Disaster Risk Reduction,* 77, 103067. www.sciencedirect.com/science/article/pii/S2212420922002862

Viégas, F. B., & Wattenberg, M. (2008). TIMELINES: Tag Clouds and the Case for Vernacular Visualization." *Interactions,* 15(4), 49–52.

Wasson, C. (2000). Ethnography in the field of design. *Human Organization,* 59(4), 377–388.

Wickens, C. D., Hollands, J. G., Banbury, S., & Parasuraman, R. (2015). *Engineering Psychology and Human Performance* (4th ed.). Psychology Press.

5 Framing the Problem

> If I had an hour to solve a problem, I'd spend 55 minutes thinking about the problem and five minutes thinking about solutions.
>
> —Albert Einstein

5.1 THE PROBLEM

In Chapter 4, we learned the art of uncovering valuable insights and focusing on what truly matters. Now, let's embark on a journey to define the right problem.

A problem can be defined as a significant and unclear situation requiring fresh solutions and a well-thought-out plan for successful resolution (Treffinger, 1995). It is important to recognize the ambiguity of the situation, which needs clarification as a first step. Design problems are often ill-structured or inadequately defined (Simon, 1995), as designers initially grapple with incomplete and imprecise mental representations. To bridge this gap, the problem and solution spaces must co-evolve, leading to the refinement and completion of these representations (Dorst & Cross, 2001). Just as there are many pairs of numbers that add up to 10, design problems often have numerous possible solutions depending on how well we understand the problem. That is why they are "open-ended" and "wicked" (Churchman, 1967; Rittel & Webber, 1973, 1984). Wicked problems are complex, multifaceted, and difficult to define and solve due to their interconnected nature, uncertain outcomes, and involvement of multiple stakeholders with conflicting interests and differing views and preferred solutions. What one individual may find attractive, another may perceive as unappealing. Such problems often lack

DOI: 10.1201/9781003476177-5

clear solutions and can evolve over time. Wicked problems are like those tricky puzzles that refuse to be solved with ordinary methods. You cannot follow a straight path; you need to think outside the box, try different angles, and even break a few rules. It is like trying to untangle a knot—sometimes, you have to twist and turn it until you find the right approach. That is why we may need to go beyond just defining or refining a problem and frame and reframe it. Framing is essentially "selection and salience" (Entman, 1993), guiding us toward pinpointing the most impactful problem.

5.2 DEFINE AND REFINE THE PROBLEM

Defining the problem is establishing an understanding of the nature, scope, and context of the problem. It may involve getting to some of the potential root causes, boundaries, and key stakeholders involved, with the purpose of gaining clarity and focusing on what the problem entails.

Refining, on the other hand, is enhancing or sharpening the definition of the problem through iterative analysis and exploration. It involves clarifying uncertainties, narrowing the scope, identifying specific aspects to address, and removing bias and false assumptions. The purpose is to deepen the understanding of the problem and identify key challenges or opportunities.

5.3 GENERATING A PROBLEM STATEMENT

The 5W method (who, what, why, where, and when) is attributed to Kipling's poem (1902) (pp. 157–158), a timeless tool for unraveling the understanding of anything. His words echo through the ages:

I KEEP six honest serving-men
They taught me all I knew;
Their names are What and Why and When
And How and Where and Who.

Understanding and generating a problem statement lies in addressing five fundamental questions related to who, what, why, where, and when:

Who has the problem?
What is the specific *issue* they are currently facing?
Where/when does the problem happen (Context)?
Why does it happen?
Why is this problem important (Relevance)?

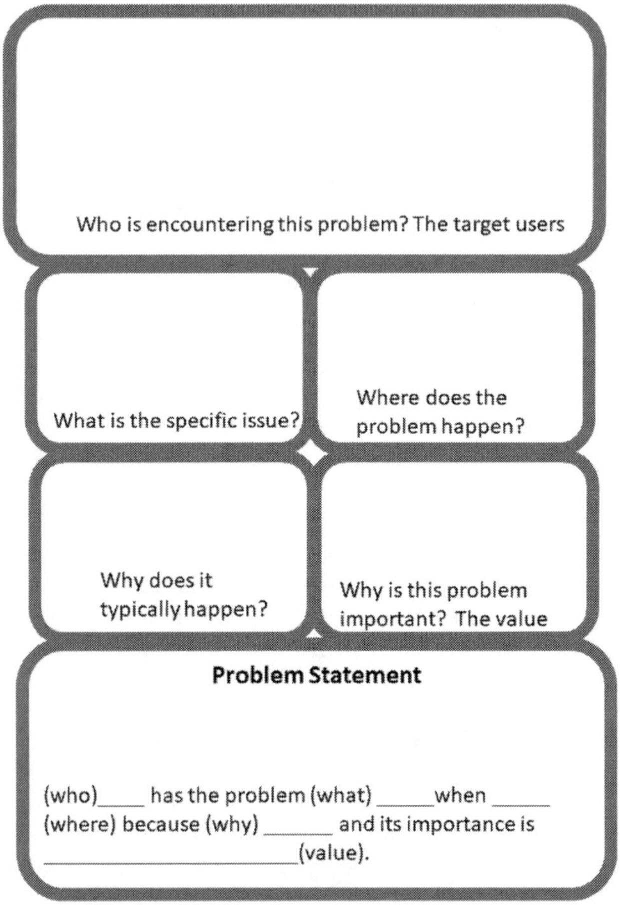

FIGURE 5.1 Worksheet for generating the initial problem statement.

The template in Figure 5.1 is useful for generating the initial problem statement. While crafting a problem statement might seem simple at first glance, it can become complex, especially when multiple opinions are involved. The real challenge lies in avoiding ambiguity and pinpointing the root problem. The 40–20–10–5 rule is an approach to zoom in on the core issue (Cantero-Gomez, 2019). Begin by describing the problem in 40 words, then distill it to 20, and finally, 5–10 words, ensuring clarity and focus so *the design challenge is short and easy to remember.*

5.4 PROBLEM DECOMPOSITION

When we confront a challenge or hear the pain points of people, our reflex is to leap straight into finding a fix, the one right answer (von Oech, 2008) (Figure 5.2).

FIGURE 5.2 Traditional problem solving.

As Pablo Picasso said, "Every act of creation begins with an act of destruction." Similarly, in problem-solving, there is a technique known as problem decomposition. Ergonomists refer to it as *task analysis*. The process entails dissecting intricate problems into smaller, more digestible parts, like dismantling a machine into its constituent elements. Such a method offers several benefits:

1. Break down the complex problem into manageable parts.
2. Facilitates the identification of critical issues.
3. Enables prioritization of discovered issues.
4. Fosters a solution-oriented approach for addressing them.

Amidst our quest for the ultimate solution, we must not overlook the value of addressing pain points incrementally. Picture it as optimizing each component in the machine before reassembling it for peak performance like a Formula-1 car. So, while the allure of a single, comprehensive solution may be strong, do not underestimate the power of incremental advancements through problem decomposition. Take, for instance, the birth of ride-hailing companies like Uber, Lyft, Didi, Careem, and Grab. They revolutionized the transportation landscape by tackling the pain points associated with traditional taxis. As illustrated in Table 5.1, dissecting challenges and addressing the pain points in some components (specifically Sub-tasks 1, 3, and 4) laid the groundwork for the revolutionary taxi services we now enjoy.

In quality circles, an approach similar to problem decomposition is known as the cause-and-effect diagram, or more vividly, the fishbone or Ishikawa diagram (Ishikawa, 1986). Much like a doctor diagnosing an illness by examining symptoms across different systems, this diagram sorts potential causes into key categories like materials, machines, methods, environment, and people, helping to pinpoint where the real issue lies. It is a method commonly used to find the root cause of a problem and jump into improvements in a product, service, or process. In all these cases, the investigator pursues the most accessible solution to the potential pain point or cause, which may not necessarily be the best one.

One way to get over the hurdle of stopping after the one right solution is to go beyond the first and find the second, third, fourth, and beyond answers. It is just like a chess master looking for a better move even though he/she sees a good move. Doubts will creep in when we find ourselves questioning whether we are making the right move in tackling the right problem. We start pondering the very essence of our endeavor, asking ourselves, "Why are we even doing this?"

TABLE 5.1 Dissecting a Journey from Point X to Point Y Through the Traditional Taxi Model. Addressing the Pain Points Paved the Way for a Ride-hailing Model.

Sub-Task	Potential Pain Points	Potential Solution
1. Go to taxi stand	Walking to taxi stand	Getting vehicle to person's location
2. Get into taxi	Difficult for person with impairment	Ramp for person in wheelchair
3. Give destination to driver	Language and communicating	Translator or position location on e-map
4. Pay for ride	No cash or change	E-payment
5. Get off at destination	Destination may not be exact.	Audio guidance or warnings.
	Difficult for person with impairment	Ramp for person in wheelchair

5.5 LEVERAGING SYSTEM 2 THINKING FOR CREATIVE BREAKTHROUGHS

We are hardwired to be solution-seekers from our many years spent studying and acing exams fixated on finding the one right answer. We define the problem with the tools and solutions that are available. It is in our DNA to crave closure, to tie up loose ends, and to move on to the next thing, which stems from our reliance on System 1 thinking: a very rapid, intuitive process based on past experiences (Kahneman, 2011), like the solutions in Table 5.1. These are essentially System 1 thinking. A problem is not just a hurdle to overcome; it is an opportunity to put your thinking to the test and dive into the messy, beautiful world of problem-solving. What is required is engaging in System 2 thinking: a slower, more logical one that requires more effort. The way to do that is by questioning and scrutinizing whether the aspired value we aim for is truly the most impactful in terms of viability, feasibility, and desirability.

Addressing the pain points in our insights will not propel us toward groundbreaking innovation. To create products that truly delight us, we must aim to pinpoint the exact problem. Altering the wording in a question can unlock many answers. Consider the stark contrast between "what chair to design" and "what sitting experience to design." The change in viewpoint upon hearing each of these statements is remarkably distinct. The former adopts a rather narrow, close-ended perspective, whereas the latter embraces a broader, open-minded stance. Embracing an open-ended challenge is like unlocking the floodgates of creativity, allowing many ideas to come in. The way we frame the problem shapes the solutions we find. Restricting ourselves to a single statement puts us at risk of overlooking numerous possibilities. This should not be the norm. Instead, we should try to dig deeper into the initial insight and explore many problems. This is precisely why the first half of the first diamond in the design process diverges (Figure 1.4). It is similar to embarking on a treasure hunt for innovation. We explore a wide range of problems, seeking the one shining gem that not only captivates our imagination but also holds the promise of transforming lives. Ask yourself: Does the challenge to make a *big impact* have many solutions, unlike those in Table 5.1, considering the right context? Focus on those that will make it a challenge that one is eager to take on.

Exploring "what-if" scenarios is like opening a door to a world of possibilities, steering us away from a solitary answer toward many potential outcomes. Designers constantly ask themselves, "What if I did this?" and then assess the idea independently. Diverse perspectives are the fuel for innovative thinking. While one person may originate a question, involving a group of six to eight people with complementary skills can enrich and elevate that question to new heights (Brett & Wang, 2020). Large teams focus on solving problems, while small teams excel at discovering new problems because small teams are good at being disruptive.

In the rest of this chapter, we will leverage the most promising insight found in the previous chapter to craft a problem statement that unlocks a worthwhile challenge to address.

5.6 PROBLEM FRAMING AND REFRAMING

You may capture countless photographs, but only the finest are chosen to be framed and displayed. The same principle applies here in problem identification. Framing helps refine and highlight what truly matters, bringing focus and clarity to the work. The purpose of framing is to engage in reframing (Beckman & Barry, 2007). Unfortunately, framing and reframing are elements of design that tend to be omitted in many situations as we tend to jump into a solution for a problem that was defined by a stakeholder in a "fuzzy" manner (Quinlan, 2014). The real problem may not be what was said. This is where problem framing becomes useful. Frames are "structures of belief, perception and appreciation" (Gamson & Modigliani, 1989; Schön & Rein, 1994). Framing involves highlighting certain aspects of perceived reality to emphasize specific problem definitions, causal interpretations, moral evaluations, or treatment recommendations for what is under discussion (Entman, 1993). It looks at how people understand tricky situations and figure out what to do to solve a problem (Rein & Laws, 2000; Grainger et al., 2019).

Problem reframing is all about finding a problem worth solving or a better one to solve rather than the most direct one (IDEO, 2015). It is often hidden from everyone when the design process begins. The concept of reframing is well done by the military, where different perspectives, contexts, and interpretations are used to broaden the understanding of the problem and uncover new insights. A large group of stakeholders with intense engagement is required for this purpose, so the process is similar to co-design (Beckman, 2020; Bradwell &

Marr, 2008; Dorst, 2015). Reframing is all about creating new perspectives for the problem (Senge 2006).

5.6.1 CHALLENGES TO FRAMING

There can be many challenges to framing a problem. Functional fixedness is one where a mental obstacle of being unable to envision an object's alternative uses to solve a problem (Duncker, 1945). It is like the old saying, "If all you have is a hammer, everything looks like a nail." Overcoming functional fixedness allows one to unlock the potential in a situation, enabling one to frame a problem without being constrained by preconceived notions about the solution.

Another hurdle is the *Einstellung* (German word for attitude or setting) effect (Luchins, 1942), which is where we stop considering alternatives after finding what we believe is the correct "answer." While functional fixedness hampers our ability to see new problem-solving approaches based on past experiences, the *Einstellung* effect restricts us from exploring other potential frames once a frame is identified.

5.6.2 THE ABC OF FRAMING

Dorst (2011) illustrates a way to show the conventional problem-solving approach compared to the ideal one using the analogy of 10 = 5 + 5 or the **A** = **B** + **C** (ABC) analogy. Designers are driven by the pursuit of value through innovative exploration, a facet often overlooked by many. Today, the emphasis lies not solely on functionality but on the value designers can bring into people's lives, the society, and the world. As Steve Jobs famously emphasized, "Think different," highlighting the important role of those who challenge norms in propelling progress. Hence, it becomes imperative to chart a course toward desired outcomes that unlock numerous potential creative solutions. Ultimately, after the ideation phase, our aim is to land on the most suitable thing, **C** (Figure 5.3).

The HOW together with the ASPIRED VALUE is the FRAME (i.e. the combination of A and B). In most engineering design problems, we know the result we are seeking (A) and the working principle (B) of the device we want to build,

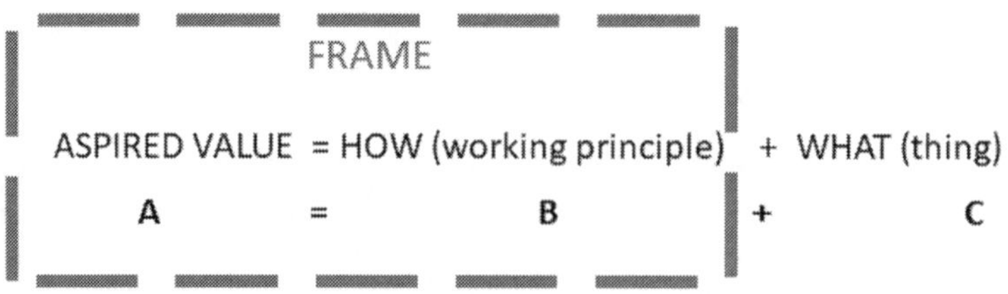

FIGURE 5.3 The framing concept.

and then we make the device, service, or system (C) that is necessary. This results in incremental innovation and has limited creativity, even though it may generate new business models, as with the taxi service (Table 5.1). When our goal is simply a chair, the "what" (C) is quite constrained. However, when we aim for a complete sitting environment (A), the possibilities for "C" become boundless. That is thinking differently! So, it is important to frame the challenge well. That is capturing the most promising AB combination. One useful way is to use a starter phrase such as *How Might We* (HMW) together with an insight (Kelley & Kelley, 2015). Each of the insights can be further explored, as illustrated in Section 5.7 of this chapter. The "how" gives some confidence that there are solutions. "Might" means we can try ideas, irrespective of whether they work or not, and "we" emphasizes that we work together to improve each other's ideas. But here's where things get tricky, and where design thinking sometimes falls short: who exactly is "we"? Does it only refer to the people in the room, the "designers"?

If our goal is to design for humanity, then "we" must be more inclusive. It should encompass not just the designers, but also the people we are designing for. This calls for participatory design or co-design approaches, where users and communities are not just considered but actively involved in shaping outcomes. Their lived experiences, values, and insights are critical to the design process. In this sense, "we" includes people outside the room, those who must be consulted, engaged, and included from the start. As emphasized earlier, achieving meaningful, human-centered innovation requires a large and diverse group of stakeholders engaged deeply throughout the process.

HMW statements turn the challenge into an actionable question. If we are looking for multiple solutions, an appropriate starter can be in the form of "In what ways might we ..." and other similar statements.

5.6.3 CASE STUDY 1: CHECK ASSUMPTIONS

In the early days of NASA's space program, one of the biggest challenges was figuring out how to shield the space capsule from burning due to the intense heat generated by friction during re-entry (Launius & Jenkins, 2011, p. 22). The challenge was stated as "find a metal capable of withstanding the heat of reentry". In the first framing, using the ABC analogy:

prevent capsule from burning (A) = metal that will withstand heat (B) + shell (C)

We are designing C, assuming A and B are known. Does B exist, or can we find a metal that will withstand the heat of reentry? So, the objective can be to find a new metal. Instead, NASA's primary objective was to prevent the capsule from burning rather than find a new metal. *Checking assumptions* is always necessary to be able to reframe a problem. A reframed question by Generative

AI is: HMW "protect spacecraft during reentry into the Earth's atmosphere?" In other words, we seek an alternative B that allows more creative solutions to the problem.

prevent capsule from burning (A)= protect spacecraft (B) + shell (C)

5.6.4 CASE STUDY 2: BROADEN HORIZONS THROUGH THEME EXPLORATION

We can be more creative if A can also be modified, resulting in disruptive innovation. For this, we ought to dig deeper into the aspired value. Let us illustrate using the case presented by Ackoff & Greenberg (2008). In the hustle and bustle of old New York City, there's a classic tale of office life: customer service flooded with complaints about the dreaded slow elevator. One way to state this problem is that it is a technical snag: "Elevators are slower than molasses!" Unfortunately, the consultants brought in to solve the issue looked exactly at those—faster elevators, more elevators, computer-controlled elevators, you name it. With the ABC analogy,

faster movement of people = increased speed + elevator

faster movement of people = higher number + elevator

To truly reframe, we must challenge assumptions. Can we feasibly install more or faster elevators? However, the harsh truth dawns: none of these options are practical in an aging building. This is where many falter, abandoning the problem rather than considering a reframed perspective by reassessing the desired outcome. In this scenario, the researcher overlooked the elevator users—a second theme slipped through the cracks of evaluation. *Why* was there a rush to speed up movement? Because folks were fed up with twiddling their thumbs while waiting. What we truly seek is to eliminate boredom while waiting for the elevator. Unsurprisingly, this can be achieved affordably and effectively. Now, if we had the aspired value as "enhance the experience and satisfaction of tenants within the building" or "faster in/out access of people" we may have captured the second theme and found many Cs and Bs that satisfy such an A. In this case, B would be how might we minimize boredom. Even though solutions (C) are too early to be discussed in this book, the thing (C) that the building management opted for in this case was a low-key solution: mirrors. Yes, just mirrors. Suddenly, the wait became a chance for some impromptu self-reflection—and the complaints vanished into thin air. And there you have it—a tale of turning a mundane annoyance into a moment of reflection, where a simple idea of asking *WHY* people were complaining saves the day without breaking the bank. Adjusting the aspired value (A) required good framing of the original problem to capture a second theme.

5.6.5 TAKE GIANT LEAPS WITH HMWS

Craft the HMW questions with a positive spin to ignite creativity. Instead of asking, “HMW prevent glass from breaking?” consider asking, “HMW enhance glass care?” Also, refrain from proposing specific solutions like “HMW use a protective coating to enhance glass care,” as there are many ways to improve glass care beyond a protective coating. Writing these HMW statements is an iterative process that will improve with practice. Be ambitious and take a giant leap (moonshot) when writing these statements to aim for significant progress. Instead of saying, “HMW reduce maintenance cost,” say, “HMW make maintenance cost to be zero?” so that the ambiguity in the statement helps to broaden the exploration space, allowing for a diverse range of potential solutions. One may say a maintenance cost of zero is not achievable. It is worth quoting Nelson Mandela, “It always seems impossible until it is done.” If you set it to zero, you will start exploring how to make it maintenance-free or self-maintaining. Put the HMW questions to the test with input from the design team or stakeholders, refining them based on feedback and insights. It is important to involve users, designers, and domain-knowledge experts in the HMW generation process, as the questions provide a roadmap for moving forward to the ideation stage. Domain-knowledge experts may have very conservative views, and so it is important to question them in a non-committal way. Engaging participants from diverse backgrounds helps bring out the various issues beyond mere user needs.

5.7 TECHNIQUE FOR PROBLEM FRAMING AND REFRAMING

If the challenge seems complex, use problem decomposition to split the problem into five to ten smaller and relevant sub-challenges that can be made actionable using HMWs. Having good HMWs help generate more impactful solutions when ideating.

Whenever a question is posed, the natural tendency is to seek solutions. These solutions may come very quickly but they die down as quickly. Another issue is that people participating in such an exercise may have social anxiety to give out their solution not knowing how others will judge him/her. Shifting the focus to brainstorming questions instead of answers facilitates overcoming cognitive biases and exploring unexplored terrain more effectively without any of the aforementioned issues (Gregersen, 2018).

Question bursts allow anyone to pose their questions without any fear. Descriptive questions, such as what’s working and what is not working, and why, are much better than speculative ones, such as what if, what might not be, or why not? As Gregersen (2018) suggests, a good rule of thumb is to set 4 minutes for about 15 questions and do this exercise about two to three times possibly with different people. The process can be made effective by having these question bursts center around certain themes (Van Manen, 1990, p. 89).

The question bursts will result in many questions, some of which may be related, and some questions may generate more questions. Asking "why," similar to the Five Why technique developed by Sakichi Toyoda, the founder of Toyota in the 1930s, enables the discovery of a wider spectrum of perspectives and frames for a problem (Ohno, 1978; Serrat, 2017).

How would this process unfold for the New York City office elevator problem mentioned before? Starting with the central issue of "slow in/out access to tenants," we dig deeper by asking why it is slow and strategies to make it fast (Figure 5.4). The themes for this case center around the elevator, the building, and the tenant. The first answer may highlight the elevators' sluggishness because of their age. Alternatively, it may point to a limited number of elevators, potentially caused by space constraints, and so forth. Each of these reasons can be taken further as necessary. These were the aspects first investigated when the management was looking for solutions. What they did not look at was the theme related to the tenants of the building. If the framing was around reducing the waiting time to reduce boredom and thereby reduce complaints, many feasible solutions could have been developed. The reduced boredom resulted in the viable, feasible, and desirable solution (#2 in Figure 5.4).

An effective method for visualizing these questions and their associated HMW frame is illustrated in Figure 5.4. The lower box is used to frame the problem related to the cause stated above it. Additional boxes can be added as needed, especially if the five-why technique is used. The key is to understand that each distinct reason is examined by posing "why" questions and placed in the adjacent box in the same row related to one theme. Boxes within the same row are somewhat interconnected through the theme, whereas those in different rows may exhibit a degree of independence from one another or belong to another theme. The corresponding HMW for each result is then written in the box below it.

5.8 THE END GAME

The process of framing and reframing can seem endless, but eventually, the design team must stop and transition to the next phase of ideation. The critical question then becomes when is it time to conclude the reframing process? Various approaches exist for making this determination. One straightforward method involves assessing each option using an impact-difficulty matrix, offering a quick and efficient evaluation. Alternatively, reflection-in-action (Schön, 1983) and reflection-on-action or established techniques like the analytic hierarchy process (Saaty, 1970) provide a structured framework for decision-making. Moreover, advancements in AI technology offer additional avenues for weighing the pros and cons of each option, contributing to more informed and objective decision-making processes.

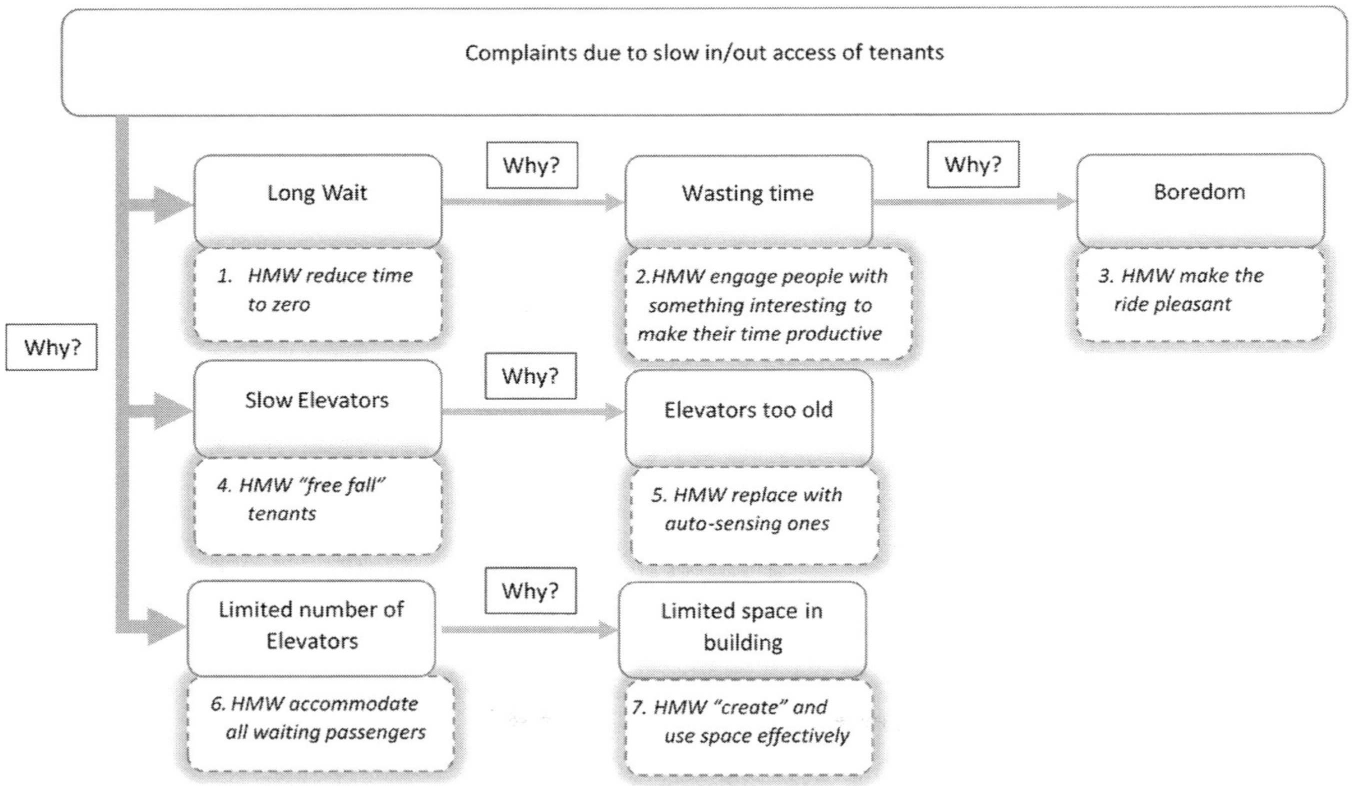

FIGURE 5.4 Structured question bursts to formulate the related moonshot frame. In the Five Why technique, there will be more boxes for the five-whys in each row.

5.8.1 IMPACT-DIFFICULTY MATRIX

In Section 5.7, we have determined various frames, each representing different aspects of the problem. To decide where to focus our efforts, we can assess each frame based on two factors: how hard it is to tackle and how important it is (Luma Institute, 2024). Depending on the problem, you can add or change these criteria to suit the problem. Today, impactful is valued more than important. Let's assume you and your team have identified the seven key frames (Figure 5.4) through the previously mentioned procedures. Now, make informed yet bold estimates about the level of difficulty involved in solving each problem and the potential impact if it is solved. These crude assessments will serve as a basis for evaluation. A simple visualization for the evaluation uses a four-quadrant matrix (Figure 5.5) that plots difficulty against impact.

The visual helps to prioritize the tasks so that conflicts within the design team can be reduced and agreement on which problem to tackle first can be expedited. The potential locations of each frame depicted in Figure 5.4 are illustrated in Figure 5.5. The top two quadrants represent high-impact frames, with the left side being easier to address (for example, frames 2 and 4). The focus should be on these high-impact–low-difficulty frames. On the other hand, the lower quadrants show frames with lower impact (for example, frames 3 and 7), which we can choose to avoid. It is clear from Figure 5.5 that frame 2

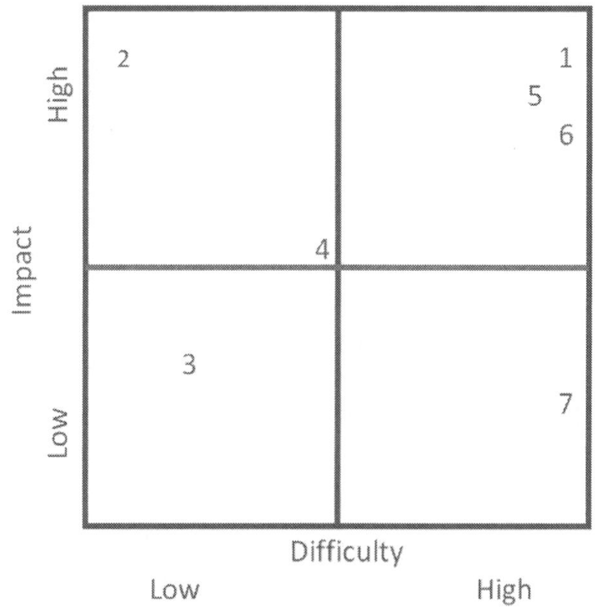

FIGURE 5.5 The impact-difficulty matrix. The labels 1, 2, ...7 indicate the relative impact and difficulty of each frame shown for the elevator issue in Figure 5.4.

(Figure 5.4), *HMW engage people with something interesting so their waiting time is productive* is the one with the greatest impact and lowest difficulty. By applying this technique, the case study presented in Section 5.6.4 could have been approached more systematically.

5.8.2 ANALYTIC HIERARCHY PROCESS

Analytic Hierarchy Process (AHP) (Saaty, 1970) is another approach for handling the decision-making process, especially when subjective as well as objective variables exist for decision-making. It helps identify and prioritize selection criteria, analyze the collected data, and expedite the decision-making process. It is a multi-criteria decision-making tool.

5.8.3 "REFLECTION-IN-ACTION" AND "REFLECTION-ON-ACTION"

Reflection is a good way to know whether the right frame has been chosen. Some people can think about their actions while engaged in them, a concept termed "reflection-in-action" (Schön, 1983). Equally important is "reflection-on-action," a process of thinking back on what has been done to understand and learn from the experience. The criteria for determining whether the right frame has been chosen are:

1. Am I capable of solving the issue?
2. Will I be satisfied with the outcome upon solving this problem?
3. Have I ensured coherence?
4. Does the solution align with my fundamental values and theories?
5. Is the inquiry progressing?

Schön (1983) suggests testing the frame through research or prototyping (Chapters 7 & 8). Reflecting on the test results during and after actions will determine if the correct frame has been selected. If not, revisiting and reframing are necessary to complete the reflective loop and ensure effective problem-solving.

5.9 ETHICAL CONSIDERATIONS

When designing, ethical considerations are important when creating and selecting the framed problem. Designers bear the responsibility of ensuring that the problems they identify and address align with ethical standards and social responsibility. This involves critically examining the potential consequences of their aspired values on various stakeholders, including communities, environments, and future generations. Ethical implications arise when defining design problems, as designers must navigate complex issues such as environmental sustainability, social justice, and inclusivity. By scrutinizing problem statements through an ethical lens, designers can attempt to address the right problems that uphold principles of fairness, equity, and respect for human rights. Ultimately, ethical problem framing serves as a guiding principle for designing products, services, and systems that contribute positively to humanity while minimizing harm.

5.10 PROBLEM FRAMING AND REFRAMING IS ITERATIVE

Problem framing is not an activity that is done just once. It is an ongoing, iterative process that should be present throughout the design journey, and hence the concept of reframing (Dorst, 2015). Our frame of reference is dynamic as we gather more information and develop more understanding, resulting in reframing the problem. For instance, consider a handheld vacuum cleaner. Traditionally, vacuum cleaners are designed with the assumption that people will manually remove dust and small particles. But why do we need them? It is simply to eliminate dust—holding the device isn't necessary. This insight has led to the development of robot cleaners that automate tasks without requiring human intervention. Yet, the problem is not fully solved. How can dust be prevented at the source? Is it brought in by footwear and the garments we wear, or does it enter through windows and doors? If so, we must consider relevant "How Might We" (HMW) questions to address these underlying causes. The ability to question and adjust one's frame of reference is important for expanding creativity as it unveils entirely new perspectives (Seelig, 2012). This makes it necessary for the problem-solver to be open and flexible toward new information and the implications and reframe the original problem. Reframing is not a replacement of the existing frame but rather one of integrating the new into the old (Wang, 2022). The creation of new frames is critical to the development of problem definition.

5.11 DESIGN BRIEF

Once the problem has been reframed, outline the problem's scope to ensure the design process remains focused and feasible. Highlight the constraints, such as budget, materials, environmental impact, or regulatory requirements. A design brief lays the foundation for a project by clearly defining the problem. It serves as a roadmap, guiding the design process by setting objectives, constraints, and key considerations while leaving room for creative exploration and strategic problem-solving (Phillips, 2012).

The contents of a design brief largely depend on the specific needs of the project (Best, 2015). But, a well-crafted brief should include a project overview that has a concise problem statement with desired outcomes, and a list of must-have features or functions. Others to be included are the target population, brand guidelines if there are any, timelines, deliverables, budget, communication, and feedback as rules of engagement.

The design brief plays an important role in stimulating creativity and shaping the direction of outcomes by influencing how designers engage with the problem statement. Research suggests that a *well-framed problem statement* with *action-oriented verbs* lead to more ideas that exhibit higher novelty and flexibility, though they may be less immediately feasible (Kapkin & Joines, 2021). The design brief and the way problems are articulated can serve as powerful catalysts for creativity in the early stages of the design process. As a design brief becomes more concrete and detailed, it increasingly restricts the generation of highly novel ideas. Overly rigid constraints and specific requirements can hinder creative exploration, making unconventional or groundbreaking solutions less likely. Finding the right balance between structure and creative freedom is a delicate and challenging process.

The design brief should encourage iteration, leaving room for prototyping and testing as part of the design process. By striking a balance between specificity and flexibility, the brief serves as a guiding document that aligns all stakeholders and keeps the project on track from concept to execution.

5.12 CONCLUSION

In summary, defining a problem establishes its basic parameters, refining it hones in on specific details and nuances, framing it to explore different perspectives and contexts, and reframing it challenges existing assumptions and opens up new possibilities for innovative solutions. Each step contributes to a more comprehensive understanding of the problem and guides the development of effective problem-solving strategies in design. Framing problems is inherently iterative, requiring continuous refinement and adjustment as new insights emerge. Maintaining a user-centric focus ensures that our problem statements resonate with the needs and experiences of those we aim to serve. The most impactful frame is the foundation for effective problem-solving, guiding us toward impactful solutions as we transition from the design brief to ideation.

> A problem well stated is a problem half-solved.
>
> —Charles Kettering (1876–1958), founder of Delco and Head of Research at General Motors

REFERENCES

Ackoff, R., & Greenberg, D. (2008). *Turning Learning Right Side Up, Putting Education Back on Track*. Wharton School Press.

Beckman, S. (2020) To frame or reframe: where might design thinking research go next?, *California Management Review*, 62(2), 144–162. https://escholarship.org/content/qt5zx787h7/qt5zx787h7.pdf.

Beckman, S., & Barry, M. (2007). Innovation as a learning process: Embedding design thinking. *California Management Review*. 50(1), 25–56.

Best, K. (2015). *Design Management: Managing Design Strategy, Process and Implementation*. Bloomsbury Publishing.

Bradwell, P., & Marr, S. (2008). *Making the Most of Collaboration: An International Survey of Public Service Co-design*, Edited by Julie Pickard. PriceWaterhouseCooper.

Brett, J., & Wang, D. (2020). If you want creativity, keep the team small. *Scientific American*, February 2020. https://www.scientificamerican.com/blog/observations/if-you-want-creative-solutions-keep-your-team-small/.

Cantero-Gomez, Paloma (2019). How to frame A problem to find the right solution. *Forbes*. www.forbes.com/sites/palomacanterogomez/2019/04/10/how-to-frame-a-problem-to-find-the-right-solution/?sh=6e4f3eb35993. Last accessed 25 March 2024.

Churchman, C. West. (1967). Wicked problems. *Management Science* 14(4), B141–B142.

Dorst, K. (2011). The core of 'design thinking' and its application. *Design Studies*, 32(6), 521–532.

Dorst, K. (2015). *Frame Innovation*. MIT Press.

Dorst, K., & Cross, N. (2001). Creativity in the design process: Co-evolution of problem-solution. *Design Studies*, 22, 425–437.

Duncker, K. (1945). On problem-solving (L. S. Lees, Trans.). *Psychological Monographs*, 58(5), i–113. https://doi.org/10.1037/h0093599.

Entman, R. M. (1993). Framing: Toward clarification of a fractured paradigm. *Journal of Communication*, 43, 51–58

Gamson, W. A., & Modigliani, A. (1989). Media discourse and public opinion on nuclear power: A constructionist approach. *American Journal of Sociology*, 95, 1–37.

Grainger, S. Hommes, L., Karpouzoglou, T., Perez, K., Buytaert, W., & Dewulf, A. (2019). The development and intersection of highland-coastal scale frames: A case study of water governance in central Peru. *Journal of Environmental Policy & Planning*, 21(4), 373–390.

Gregersen, H. (2018) Better brainstorming: Focus on questions, not answers, for breakthrough insights. *Harvard Business Review*, 96(2), 64-71.

IDEO (2015). The Field Guide to Human-Centered Design https://www.designkit.org/resources/1.html

Ishikawa, K. (1986). *Guide to Quality Control*. Asian Productivity Organization.

Kahneman, D. (2011). *Thinking, Fast and Slow*. Doubleday Canada.

Kapkın, E., & Joines, S. (2021). The design brief as a creativity catalyst in design education: priming through problem statement. *International Journal of Art and Design Education*, 40, 126–145. https://doi.org/10.1111/jade.12339

Kelley, D., & Kelley, T. (2015). *Creative Confidence: Unleashing the Creative Potential within Us All.* Crown Currency.

Kipling, Rudyard (1902). *Just So Stories*. Macmillan, pp. 157–158.

Launius, R. D., & Jenkins, D. R. (2011). Coming home: reentry and recovery from space (NASA/SP, 2011-593). Available at: www.nasa.gov/sites/default/files/695726main_ComingHome-ebook.pdf.

Luchins, A. S. (1942). Mechanization in problem solving: The effect of Einstellung. *Psychological Monographs*, 54(6), i–95. https://doi.org/10.1037/h0093502

Luma Institute (2024). Importance/Difficulty Matrix. https://designsprintkit.withgoogle.com/methodology/phase1-understand/importancedifficulty-matrix Last accessed 6 April 2024.

Ohno, T. (1978). *Toyota Production System: Beyond Large-scale Production*. Productivity Press.

Phillips, P. L. (2012). Creating the Perfect Design Brief: How to Manage Design for Strategic Advantage. Allworth Press; Design Management Institute.

Quinlan, R. (2014) The design of everyday things (a review). Q Media Solutions, 13 May 2014. https://qmediasolutions.com/the-design-of-everyday-things/, accessed 14 February 2024.

Rein, M., & Laws, D. (2000). Controversy, reframing, and reflection. In W. Salet & A. Faludi (Eds.), *The Revival of Strategic Spatial Planning: Proceedings of the Colloquium* (pp. 93–108). Royal Netherlands Academy of Arts & Sciences.

Rittel, H.W.J., & Webber, M.M. (1973). Dilemmas in a general theory of planning. *Policy Sciences* 4, 155–169. https://doi.org/10.1007/BF01405730

Rittel, H., & Webber, M. M. (1984). Planning problems are wicked problems. In N. Cross (Ed.), *Developments in Design Methodology* (pp. 135–144). Wiley.

Saaty, T. L. (1990). How to make a decision: the analytic hierarchy process. *European Journal of Operational Research*, 48, 9–26. http://dx.doi.org/10.1016/0377-2217(90)90057-I

Schön, D. (1983). *The Reflective Practitioner: How Professionals Think in Action*. Basic Books.

Schön, D. A., & Rein, M. (1994). *Frame Reflection: Toward the Resolution of Intractable Policy Controversies*. Basic Books.

Seelig, T (2012) *inGenius: A Crash Course on Creativity*. HarperOne.

Seelig, T. (2013). How reframing a problem unlocks innovation, innovation engine, *Fast Company*, 19 April, www.fastcompany.com/1672354/how-reframing-a-problem-unlocks-innovation, accessed 14 March 2024.

Senge, P. (2006). *The Fifth Discipline: The Art and Practice of the Learning Organization*. Doubleday.

Serrat, O. (2017). The five whys technique. *Knowledge Solutions* (pp. 307–310). Springer.

Simon, H. A. (1995). Problem forming, problem finding and problem solving in design. In A. Collen & W. Gasparski (Eds.), *Design & Systems* (pp. 245–257). Transaction Publishers.

Treffinger, D. J. (1995). Creative problem solving: Overview and educational implications. *Educational Psychology Review*, 7, 301–311.

Van Manen, M. (1990). *Researching Lived Experience*. The Althouse Press.

von Oech, R. (2008). *A Whack on the Side of the Head: How You Can Be More Creative*. Grand Central Publishing.

Wang, G. (2022) Digital reframing: the design thinking of redesigning traditional products into innovative digital products. *Journal of Product Innovation Management* 39, 95–118.

6

Ideation and Invention

> Creativity is seeing what others see and thinking what no one else ever thought.
>
> —Albert Einstein

6.1 IDEATION

Ideation is a cognitive process that engages higher functions such as thinking and imagination. It is the process of generating, developing, and communicating new ideas, often through collaborative techniques that emphasize quantity and variety. Ultimately, the aim of ideation is to generate original, innovative ideas (Sterberg & Lubart, 1999; Dow, 2022).

6.2 INNOVATION AND INVENTION

Innovation and invention are closely related (Kaufman & Sternberg, 2010; Carayannis, 2020). An innovation is "a new or improved product or process (or combination thereof) that differs significantly from the unit's previous products or processes" (OECD, 2018). An invention is the discovery of "any new and useful process, machine, manufacture, or composition of matter, or any new and useful improvement" (www.uspto.gov/web/offices/pac/mpep/s2104.html). All three are closely linked to bringing something new into the world, addressing specific needs, and harnessing the incredible power of human imagination and ingenuity (Amabile, 2018). Each concept has its own distinct characteristics and qualities.

Imagine being an industrial designer working with a team of stakeholders to create many ideas using different ideation methods. While using human-centered design

DOI: 10.1201/9781003476177-6

principles and constantly exploring different materials, forms, and aesthetics, you craft a chair that not only catches the eye but also stands out from the rest (creativity in action). Innovation is where those creative ideas come alive and transform into practical designs that are marketable and useful. As a designer, you refine the chair's form, dimensions, and structure, considering ergonomic and functional requirements. You introduce innovative features like adjustable elements or use advanced and sustainable materials. But it doesn't stop there. You also consider manufacturing processes, materials sourcing, and cost-effectiveness, ensuring that your creative concepts can be transformed into tangible products that meet the demands of the market. It is an incredibly exciting process of turning ideas into reality. The Aeron Chair, designed by Don Chadwick and Bill Stumpf and sold by Herman Miller, is an icon of design invention and has revolutionized the seating industry with its suspended mesh seating, reinforced plastic frames, and a focus on recyclability (Brown, 2014). It provides the necessary ergonomics (McCauley et al. 2016). Bringing such a chair to life involved years of dedicated research, experimentation, and refinement (Kuang, 2013). The impact of this invention can be seen through the millions of chairs sold in the United States alone.

Every person has the potential to be creative, like a muscle that can be strengthened with training. History has shown us that remarkable inventions can come from unexpected places. A good example is James Murray Spangler, an American janitor who invented a portable electric vacuum cleaner in 1908. It is a testament that anyone can be creative if the need to make a difference and create an impact is strong.

6.3 IDEATION PROCESS

Now we move to the second diamond of GRIP (Figure 1.4). There are many different types and configurations for ideation (Maital & Barzani, 2021; Green et al., 2023; Mumford et al., 2023). For simplicity, the four stages (Figure 6.1) are preparation (Table 6.1), incubation (Table 6.2), illumination (Table 6.3), and implementation (Table 6.4) are discussed (Kilgour, 2022). Each of the four stages has five steps, and it should be used iteratively with principles developed in the other chapters in this book. It is important to note that these stages are not always linear or sequential. The process can be iterative, with individuals moving back and forth between stages, refining their ideas, and incorporating new understandings along the way.

TABLE 6.1 Preparation (Sawyer, 2021).

Defining the objective	This stage involves clearly articulating the purpose and objective of the creative project. It is important to have a well-defined goal or problem that provides direction and focus. (For details, refer to Chapter 5).
Research and exploration	Once the objective is defined, the next stage involves conducting research and exploring relevant information and resources. This includes gathering data, studying the existing solutions or works in the field, and understanding the context in which the creative work will be situated. It may involve reading publications, attending exhibitions and meetings, or exploring online resources.
Idea generation	In this stage, individuals engage in sessions to generate a wide range of ideas. The focus is on encouraging free-flowing and non-judgmental idea generation.
Concept development	Once a pool of ideas has been generated, the next stage involves evaluating and refining those ideas. This includes selecting the most promising or relevant concepts and developing them further.
Set constraints and parameters	This stage involves defining some limitations, boundaries, or criteria that the work should adhere to. Constraints can provide a framework within which solutions can be explored and developed, fostering innovation and problem solving.

6.4 IDEATION METHODS

Ideation methods (Table 6.5) can be categorized into various groups based on their characteristics and approaches (Ross, 2006; Pinkow, 2023). By understanding and utilizing these various ideation methods, individuals and groups can generate innovative ideas and approach problem solving from different perspectives. Ideation methods can be grouped into thinking methods, mind-related methods, brainstorming methods, systematic methods, visual and spatial methods, random methods, analogical methods, metaphorical methods, morphological methods, and synectic methods.

Thinking methods encompass a range of approaches that adopt flexible and unconventional thinking to enable the flames of creativity. These methods include established methods such as lateral thinking (De Bono, 2014), divergent thinking (Reiter-Palmon et al., 2019; Xia et al., 2021), and the Six Thinking Hats (De Bono, 2017). Mind-related methods such as mind mapping harness the immense power of the mind and its cognitive processes to fuel creativity and innovation (Koutstaal & Binks, 2015). Visual methods are powerful tools that employ the potential of visual elements to foster creativity and stimulate idea generation (Ambrose, 2015; Rustler, 2020). Methods such as mind mapping, storyboarding, visual metaphors, visual mapping, and visual analogies are all part of this dynamic category. Brainstorming methods aim to generate a multitude of ideas, whether through group sessions or individual brainstorming

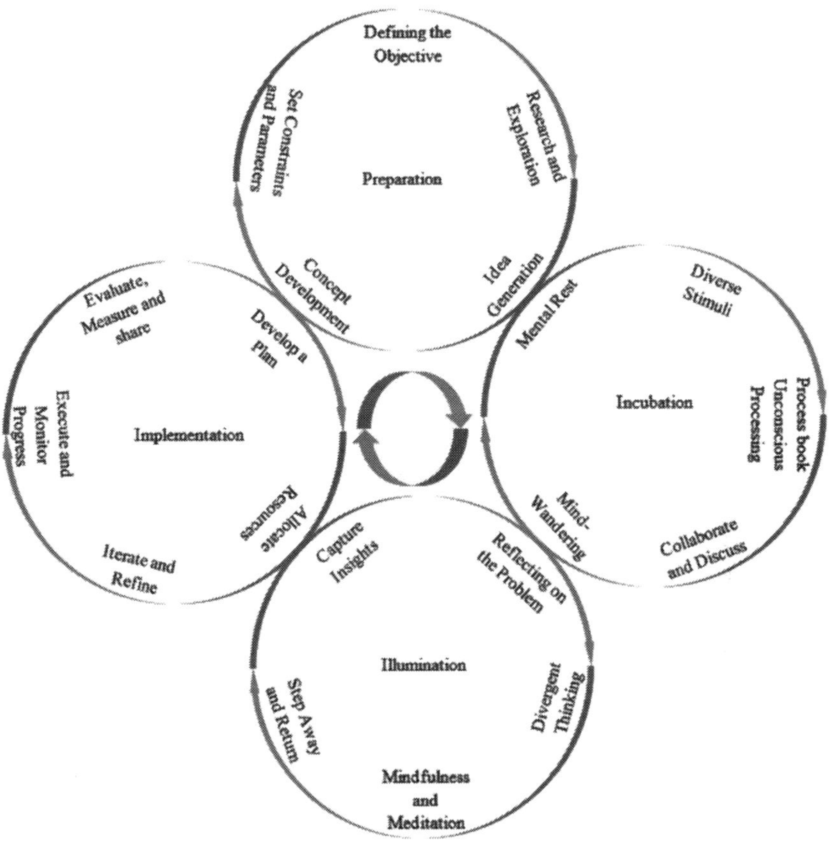

FIGURE 6.1 The ideation process.

exercises (Lewrick et al., 2020; Maaravi et al., 2021; Gong et al., 2022). Commonly employed methods include brainstorming, brainwriting, and brain dump. These exercises can be done using software applications, online platforms, physical aids, and whiteboards. The systematic methods provide valuable frameworks that effectively enhance the creative process. Notable examples of systematic approaches include TRIZ (theory of inventive problem solving) (Gadd, 2011), Osborn-Parnes creative problem solving (CPS) model (Isaksen, 2023), SCAMPER (Substitute, Combine, Adapt, Modify, Put to another use, Eliminate, Reverse) and its variations (Osborn 1953; Suh, 2019). Spatial creativity methods, as described by Suh and Cho (2020), permit individuals to generate and manipulate mental representations of spatial information in innovative and imaginative ways. These methods encompass a variety of methods, including mind mapping, forced connections, reverse thinking, SCAMPER method, analogous thinking, random stimuli, and storyboarding. Random creativity methods (Malycha & Maier, 2017) involve introducing random or unexpected elements to trigger

TABLE 6.2 Incubation (Ritter & Dijksterhuis, 2014).

Mental rest	Mental rest is an essential aspect of the incubation stage. It involves consciously stepping away from the project and giving the mind a break. Mental rest can be achieved through activities like taking a walk, engaging in hobbies, or simply engaging in activities that bring joy and relaxation.
Diverse stimuli	Exposing people to diverse stimuli during incubation can help trigger new intuitions and connections. This involves seeking out various experiences, ideas, and perspectives that are different from the usual routine.
Process book—Unconscious Processing	Keeping a process book or a creative journal can be a valuable tool during the incubation stage. This involves recording the thoughts, observations, sketches, and ideas related to the project or problem people are working on. It serves as a repository for subconscious processing.
Collaborate and discuss	Engaging in collaborative discussions with others can be a fruitful activity during the incubation stage. Sharing ideas, thoughts, and challenges with peers, mentors, or colleagues can provide fresh perspectives and new understandings. It also allows the exchange of ideas and feedback.
Mind-Wandering	Mind-wandering involves letting the thoughts drift without any specific focus or goal. It can happen during idle moments, while engaged in monotonous tasks, or even during activities like taking a shower or going for a run. Mind-wandering allows the subconscious mind to make connections and associations that may not have been apparent during focused thinking.

TABLE 6.3 Illumination (Botella et al., 2018).

Reflecting on the problem	Reflecting on the problem involves actively contemplating and analyzing the problem or challenge at hand. It includes deep thinking, questioning assumptions, and examining the problem from different angles. This stage requires focused attention and critical evaluation of the problem's various aspects.
Divergent thinking	Divergent thinking is a thought process that involves generating a wide range of ideas, possibilities, or solutions. It is the opposite of convergent thinking, which focuses on finding a single correct answer. Divergent thinking encourages expansive and unconstrained thinking, allowing for the exploration of multiple perspectives and unconventional ideas.
Mindfulness and meditation	Mindfulness and meditation practices can enhance the illumination stage by fostering a calm and focused mental state. By cultivating mindfulness and practicing meditation, it can increase self-awareness, quiet the mind, and create a receptive state for insights to emerge.
Step away and return	Stepping away from the problem and taking a break can be an effective strategy to facilitate insights. By temporarily disengaging from the problem, it allows the mind to continue processing information and making connections in the background. Returning to the problem after a break can often lead to fresh insights and new perspectives.
Capture insights	Insights can be fleeting, and it is important to capture them when they occur. This involves having a system in place of recording and documentation. It can be as simple as carrying a notebook or using a digital note-taking app to jot down the thoughts and ideas as they arise.

TABLE 6.4 Implementation (Blocker, 2023).

Develop a plan	Developing a plan is crucial to guide the implementation of creative ideas and solutions. The plan should include clear objectives, milestones, and strategies to achieve the goals within a preset timeline. It should also consider potential challenges and contingencies.
Allocate resources	Allocating resources involves identifying and assigning the necessary people, materials, and finances required for the implementation of the creative project.
Iterate and refine	The implementation stage often involves an iterative process where ideas and solutions are continuously refined and improved. It may involve testing, gathering data, and obtaining feedback from stakeholders. By iterating and refining, people can enhance the effectiveness and quality of the implementation, making necessary adjustments and enhancements along the way.
Execute and monitor progress	Execution is the active phase of implementing creative ideas. During this step, it is crucial to monitor progress regularly. This includes tracking milestones, timelines, and key performance indicators to ensure that the implementation is on track. Monitoring progress allows us to identify any issues or deviations from the plan and take corrective actions as needed.
Evaluate, measure, and share	During this stage, it is important to assess the project's success against defined objectives and desired outcomes. It is essential to document the lessons learned throughout the process and share these insights with the team and stakeholders to encourage knowledge sharing and continuous improvement.

new ideas. Random word, random picture, random stimulus cards or prompts, and random object combination are examples of random creativity methods. Analogical creativity methods (Holyoak & Thagard, 1996; Peters et al., 2021) involve drawing analogies or making comparisons between different domains, situations, or concepts to inspire creative thinking and generate innovative ideas. These methods leverage the power of similarity and transfer of knowledge from one context to another. Some of the analogical creativity methods involve direct analogy, symbolic analogy, personal analogy, historical analogy, cross-domain analogy, functional analogy, and biomimicry. Morphological creativity methods (Geum & Park, 2016) involve breaking down a problem or concept into its constituent parts or dimensions and systematically exploring variations and combinations of those elements to generate creative ideas. These methods allow for a structured and systematic approach to idea generation. Some morphological creativity methods include concept combination, morphological matrix, and cross-pollination. Metaphorical creativity methods (Sanchez-Ruiz et al., 2013) involve using metaphors as a tool for generating creative insights and ideas. Metaphors enable individuals to explore a problem or concept from a different perspective, fostering new connections and sparking innovative thinking. Asking questions like "What is this problem/concept similar to?" or

"What does this problem/concept remind me of?" can serve as a springboard for generating fresh insights and creative ideas during brainstorming and role play. Metaphor exploration methods are used for creative idea generation. Synectic creativity methods (Sahoo et al, 2023) focuses on making connections between seemingly unrelated elements to foster creative insights such as cocklebur seeds for the invention of Velcro by George de Mestral. Methods such as analogy, fantasy analysis, forced connections, reverse thinking, random stimuli, provocation, and attribute listing are employed in this approach.

TABLE 6.5 Selected Ideation Methods.

Method	Description	Sample Scenarios
Analogous thinking	Analogous thinking draws parallels between different fields or situations to generate intuitions. By identifying similarities between unrelated contexts, this method fosters creativity and innovation, allowing individuals to apply solutions from one area to another.	Using principles from nature to inspire a new design for an eco-friendly product.
Brain dump	A brain dump involves quickly writing down all thoughts, ideas, and information related to a specific topic without filtering or organizing them. This method helps clear mental clutter, allowing for a more focused approach to problem solving and creativity.	Organizing a brain dump session to explore all potential features for a new app, capturing every idea without judgment.
Brainstorming	Brainstorming is a group creativity method aimed at generating many ideas in a short period. Participants share their thoughts freely without criticism, encouraging wild ideas and building on each other's contributions to foster innovation.	Hosting a brainstorming session to generate ideas for a community event that engages residents in a community.
Brainwriting	Brainwriting methods include structured methods like the 6–3–5 method (Gong et al., 2023), where six participants write down three ideas each in five minutes. The ideas are then passed around for others to expand upon. This approach promotes collaboration and ensures that everyone's input is valued.	Implementing the 6–3–5 method to generate innovative marketing ideas for a product launch.
Divergent thinking	Encourages generating multiple ideas or solutions to an open-ended question, promoting creativity and exploration without immediate judgment or constraints.	Works well in brainstorming sessions, exploring new possibilities, and innovation processes requiring a variety of ideas.

TABLE 6.5 (Continued)

Method	Description	Sample Scenarios
Forced connections	Forced connections involve combining unrelated ideas or concepts to generate new insights or solutions. By intentionally linking disparate elements, this method encourages creative thinking and can lead to innovative approaches to problem solving	Combining a bicycle and a smartphone to brainstorm ideas for a new fitness tracking device.
Lateral thinking	Approaches problems from new and unexpected angles, breaking away from traditional reasoning to generate innovative solutions. Techniques include brainstorming and using prompts for unconventional ideas.	Useful for brainstorming sessions, overcoming creative blocks, generating unique product ideas, and exploring unconventional solutions.
Mind mapping	Mind mapping is a visual brainstorming technique that involves creating a diagram to represent ideas, concepts, and their relationships. Starting with a central idea, branches are drawn to related topics or subtopics, facilitating organization and enhancing creativity by illustrating connections.	Generating new product ideas by visually connecting customer feedback to potential features.
Osborn-Parnes creative problem solving (CPS) model	The Osborn–Parnes CPS model, updated by Isaksen (2023), is a structured process for fostering creativity and problem solving. It consists of several stages: objective finding, fact finding, problem finding, idea finding, solution finding, and acceptance finding. This model emphasizes divergent and convergent thinking to generate and refine ideas effectively.	Utilizing the CPS model to brainstorm solutions for improving team collaboration and communication.
Random stimuli	Random stimuli involve introducing unrelated words, images, or objects to spark new ideas. This method encourages out-of-the-box thinking by disrupting conventional thought patterns and inspiring fresh perspectives on a problem.	Using a random image to inspire a new advertising concept for a product.

(continued)

TABLE 6.5 (Continued)

Method	Description	Sample Scenarios
Reverse thinking	Reverse thinking involves approaching a problem by considering the opposite of the desired outcome. This method helps identify potential obstacles and reveals new perspectives, encouraging innovative solutions by challenging conventional thinking.	Asking what could make a product fail to uncover hidden weaknesses in the design.
SCAMPER	SCAMPER is a creative thinking method that prompts innovation through seven key actions: substitute, combine, adapt, modify, put to another use, eliminate, and reverse. Developed by Alex Osborn in 1953, SCAMPER encourages individuals to think critically about existing products or ideas and explore new possibilities through systematic questioning.	Using SCAMPER to reimagine a common household item, like transforming a traditional lamp into a multifunctional charging station.
Six Thinking Hats	Developed by Edward de Bono (2017), this method guides group thinking through six perspectives represented by hats: white (facts), red (feelings), black (cautions), yellow (benefits), green (creativity), and blue (process) for structured, balanced, and multi-faceted problem analysis.	Best for group discussions, meetings requiring balanced perspectives, conflict resolution, and exploring problems from multiple viewpoints.
Storyboarding	Storyboarding involves creating a sequence of illustrations or images that outline a narrative or process. Commonly used in film and design, it helps visualize the flow of ideas, allowing for better planning, organization, and communication of concepts or stories.	Creating a storyboard for a new user onboarding process to illustrate each step and interaction.
TRIZ (Theory of Inventive Problem Solving)	TRIZ is a systematic approach to problem-solving and innovation developed by Genrich Altshuller and further explored by Gadd (2011). It identifies and analyzes patterns in technological advancements and inventive solutions. TRIZ provides a set of tools and principles to help inventors and engineers solve problems creatively by overcoming contradictions and optimizing designs.	Using TRIZ principles to identify contradictions in a design process, such as improving usability while reducing costs.

TABLE 6.5 (Continued)

Method	Description	Sample Scenarios
Visual analogies	Visual analogies draw comparisons between two different concepts using visual imagery. This method helps clarify complex ideas by relating them to something more familiar, enhancing comprehension and creativity by highlighting similarities and encouraging innovative thinking.	Using visual analogies to explain a complex technology product by relating it to everyday items familiar to users.
Visual mapping	Visual mapping is a method that organizes information visually, similar to mind mapping but often with more structured layouts. It can include diagrams, flowcharts, or infographics to represent relationships and processes, facilitating clearer comprehension and analysis.	Developing a service blueprint that visually maps out customer interactions and backend processes.
Visual metaphors	Visual metaphors use imagery to convey complex ideas by comparing them to more familiar concepts. This method enhances understanding and creativity by providing a visual representation that simplifies abstract notions, making them more relatable and engaging.	Designing a new product by using visual metaphors to represent key features and user benefits.

6.4.1 SELECTION OF IDEATION METHODS

Selecting ideation methods involves a systematic approach to align methods with project goals and the specific stage of development. Start by clarifying your objectives and assessing the current stage, whether initial, early, middle, late, or final. Evaluate time constraints to determine suitable methods that fit within your timeframe and consider the complexity of each method to ensure they match the participants' skills. Creating a table can help visualize the alignment of various ideation methods with these factors. After piloting your selected methods, gather feedback from participants to understand their effectiveness and engagement, allowing you to refine your approach for future creativity sessions. Table 6.6 shows the different ideation methods and when to consider.

6.5 TECHNOLOGY-DRIVEN METHODS

New emerging methods are being developed due to increasing use of technology to enhance creativity and support the ideation phase of the design process. Some

TABLE 6.6 Relative Complexity and Time Required for the Selected Ideation Methods.

Ideation Method	When to Use	Relative Time	Complexity
Analogous thinking	Early stage for generating insights	**	**
Brain dump	Initial to capture all thoughts quickly	*	*
Brainstorming	To generate many ideas	**	*
Brainwriting	For structured, silent idea generation	**	**
Divergent thinking	For exploring multiple possibilities	**	**
Forced connections	To generate new ideas by linking concepts	**	**
Lateral thinking	To approach problems from new angles	**	****
Mind mapping	To visualize relationships between ideas	*	**
Osborn-Parnes creative problem solving (CPS) model	For structured problem solving	***	***
Random stimuli	To inspire creativity through randomness	*	*
Reverse thinking	To challenge assumptions	**	**
SCAMPER	To innovate by modifying existing ideas	**	**
Six Thinking Hats	To explore ideas from different perspectives	***	***
Storyboarding	To visualize narratives or processes	***	**
TRIZ (Theory of Inventive Problem Solving)	For solving technical problems creatively	****	*****
Visual analogies	To relate new ideas to familiar concepts	**	**
Visual mapping	To represent ideas spatially	**	**
Visual metaphors	To simplify complex ideas visually	**	**

Note: Relative time: * indicates less time, while ***** indicates more time. Complexity: * indicates less complex, while ***** indicates very complex.

of them are human interactive creativity (Tzachrista et al., 2023), design tools and inspiration space (Halskov & Lundqvist, 2021), computer-mediated random leaps (Luximon & Luximon, 2010), AR/VR and knowledge-based design (Rane et al., 2023), social media and crowd creativity (Manghan, 2020), artificial intelligence and smart systems (Shi et al., 2023), and genetic/evolving/self-generated (GES) design (Wang et al., 2022).

6.5.1 HUMAN INTERACTIVE CREATIVITY

Human interaction through brain activity, eye tracking, and gestures plays a significant role in enhancing creativity and ideation processes. These technologies provide valuable insights into human cognitive and physical responses (Kleinmintz et al, 2019; Tzachrista et al., 2023), enabling designers to create more tailored and impactful designs. Brain activity measurement, such as electroencephalography (EEG) or functional magnetic resonance imaging (fMRI), allows designers to analyze cognitive responses and emotional engagement (Dietrich & Kanso 2010; Fink & Benedek 2014; Lebuda & Benedek, 2023). Understanding user preferences and reactions to design elements helps in developing more personalized and effective solutions. Eye tracking technology provides insights into visual preferences and interaction patterns by tracking where people focus their attention. This information helps optimize layouts, improve information hierarchy, and enhance user experiences. Gestures and body movement recognition enable designers to capture and interpret natural human interactions, particularly in interactive design scenarios. By tracking gestures, intuitive and responsive interfaces can be created, fostering user engagement and creativity. By integrating these human interaction technologies, designers gain real-time data and insights that inform decision-making and drive innovation. Leveraging brain activity, eye tracking, and gestures in the creative process leads to personalized experiences, optimized design elements, and more engaging environments. These technologies have the potential to unlock new possibilities in human-centered design, improving the overall creative ideation process and resulting in innovative and impactful designs.

6.5.2 COMPUTER-MEDIATED RANDOM LEAPS

This approach stimulates creativity by injecting randomness and unpredictability into the design process (Luximon & Luximon, 2010). By utilizing parametric design and loosening constraints, designers can generate a wide range of new designs and ideas (Figure 6.2). The icons were created using spline curves and a program written in MATLAB. Closed curves forming mouth, pupil, and eye had different colors and shading. Random leaps have the power to liberate designers from rigid patterns of thinking and ignite exploration of unconventional concepts that may have been overlooked otherwise. In the context of computer-mediated random leaps, humans retain control over the process by defining and implementing various evaluation steps, ensuring that the generated ideas align with desired criteria and objectives.

6.5.3 AR/VR AND KNOWLEDGE-BASED DESIGN

Knowledge-based design is a powerful approach that draws on human knowledge and past design examples to inspire and inform the ideation process (Coburn

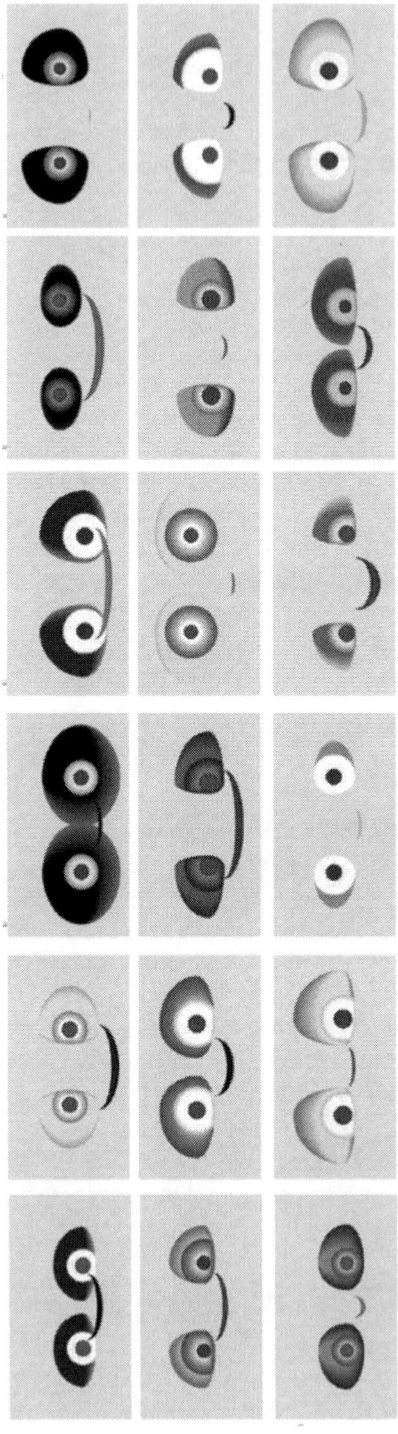

FIGURE 6.2 Icons created using parametric design.

et al., 2017; Wajcman, 2017; Rane et al., 2023). By examining existing designs, designers can gain valuable insights and inspiration that can fuel their creativity. Analyzing successful concepts, approaches, and solutions from the past serves as a foundation for generating new and innovative ideas. This process allows designers to build upon the collective wisdom of their field, learning from what has worked before and identifying opportunities for improvement and advancement. By leveraging human knowledge and studying past design examples, designers can enhance their understanding, broaden their perspectives, and create designs that are informed, effective, and impactful.

6.5.4 SOCIAL MEDIA

Social media has become intertwined in the modern era, leading to a powerful and dynamic approach to innovation (Manghan, 2020). By leveraging the vast and diverse perspectives found on social media platforms, designers can tap into a wealth of experiences and viewpoints, sparking creativity and driving the generation of a wide range of innovative solutions. The collaborative nature of crowd creativity is amplified through social media, fostering dynamic interactions, idea exchange, and constructive feedback among participants. This collaborative process pushes the boundaries of conventional thinking, leading to fresh insights and novel perspectives. For instance, in movies, feedback and discussions on social media platforms can influence the direction of a film's story, with filmmakers considering and incorporating elements or changes based on the conversations happening in the digital sphere. Similarly, in product development, platforms like Kickstarter (www.kickstarter.com) allow a community of supporters to contribute ideas, feedback, and financial support, harnessing crowd creativity to bring innovative ideas to life. By embracing social media and crowd creativity, designers can tap into the collective wisdom of the crowd, resulting in more impactful, inclusive designs, and successful creative endeavors.

6.5.5 ARTIFICIAL INTELLIGENCE AND SMART SYSTEMS

AI systems have made significant advancements in supporting creativity, particularly in the realm of image generation and assisting creative processes (Shi et al., 2023; Choudhury, 2022; Stige et al., 2023; Grech et al., 2023). Some examples include generative adversarial networks (GANs), style transfer algorithms, AI-powered recommendation systems, natural language processing (NLP) based AI systems and music and video composition. GANs are AI systems that can generate realistic images (Figure 6.3). GANs have been used to create stunning visual art, including paintings, photographs, and even original artwork. Style transfer algorithms can analyze the style of one image and apply it to another, creating a new image that combines the content of one image with the aesthetic style of another. AI-powered recommendation systems can analyze vast amounts of

FIGURE 6.3 Images created using AI (using StableDiffusion3.5L).

data, such as user preferences, historical trends, and market insights, to suggest ideas, concepts, or designs to creative professionals. NLP-based AI systems can assist in creative writing, providing suggestions for word choices, sentence structures, or even generating entire paragraphs based on given prompts. Music and video composition AI algorithms can analyze patterns in existing music and generate original compositions based on those patterns. Overall, AI systems are increasingly being used to augment and enhance the creative process across various artistic domains. By leveraging AI-generated insights and capabilities, designers and creative professionals can explore new avenues, overcome challenges, and unlock fresh perspectives that fuel their creativity.

6.5.6 GES (GENETIC/ EVOLVING/SELF-GENERATED) DESIGN

GES design is an iterative approach that leverages the principles of evolution to explore a diverse range of design variations (Kronfeldner, 2014; Capanna, 2015; Wang et al., 2022). This method involves generating, evaluating, and refining design options over time, allowing for the discovery of unexpected and innovative solutions. In GES design, designers create an initial set of design variations, often referred to as a population. Each design variation represents a unique solution to the problem at hand. These design variations can be influenced by various factors, such as environmental conditions, user behavior, or other stimuli that the designer considers relevant. Designs that perform well or exhibit desirable characteristics are selected manually or automatically for further development. Through an iterative process of generating, evaluating, and refining designs, the population evolves over time. This evolution can be guided by mechanisms inspired by natural selection, such as mutation and crossover, which introduce random variations and combine successful design elements from different solutions. As the population evolves, designers can explore a wide range of design possibilities and uncover unexpected and innovative solutions that may not have been initially considered. This approach encourages designers to think beyond conventional boundaries and encourages the exploration of unconventional ideas. GES design is particularly useful when dealing with complex design challenges or when seeking novel and groundbreaking solutions. By tapping into the power of evolutionary processes, designers can harness the potential of natural variation and selection to drive the creative ideation process and arrive at innovative and effective designs.

6.6 INDIVIDUALITY, INDIVIDUAL AND TEAM CREATIVITY

6.6.1 INDIVIDUALITY AND INDIVIDUAL CREATIVITY

Individuality refers to the unique qualities, perspectives, and characteristics that make each person special (Furnham et al., 2008; de Raad, 2020). It is about embracing our personality, beliefs, values, interests, and experiences, recognizing that we have our own way of thinking and behaving. When it comes to creativity, it can manifest at both the individual and team levels, each with its own remarkable aspects and advantages (Walton, 2016; Yuan et al. 2022).

Creativity is a concept that transcends boundaries and finds expression in a variety of contexts. Creativity can be explored and cultivated in art, science, business, education, or everyday life. Amabile's framework (1983) provides valuable insights into the factors contributing to creativity. Three core elements in individual creativity are intrinsic motivation (Deci & Ryan, 1985), creativity-relevant skills or domain knowledge and skills (Weisberg, 1999), and creative thinking (Guilford, 1967; Paulus & Nijstad, 2003; Cropley, 2006). Intrinsic motivation represents the internal drive and passion that fuels creativity. When individuals are genuinely motivated and engaged in their work, their creative potential flourishes, and their ideas flow more freely. Creativity-relevant skills encompass the expertise and abilities that are directly applicable to the creative task at hand. This includes domain knowledge, problem solving skills, and critical thinking. Possessing a strong foundation in these areas significantly enhances one's creative output. Creative thinking or ideation methods individuals employ to generate innovative ideas. It compels individuals to think beyond conventional boundaries, make unexpected connections, and explore novel approaches to problem-solving. In the context of organizations, a supportive work environment plays a pivotal role in fostering creativity. Research by Rainone and colleagues (2021) highlights the significance of creating an atmosphere that nurtures and values creative exploration. This involves providing individuals with autonomy, necessary resources, opportunities for collaboration, and constructive feedback. Such an environment permits individuals and teams to unlock their creative potential, fostering a culture of innovation and growth.

6.6.2 TEAM CREATIVITY

The number of people needed for creative thinking can vary depending on the specific context and goals. Creative thinking can be done individually (Abi Saad & Agogué, 2023) or in groups, and both approaches have advantages (Nguyen & Mougenot, 2022). The optimal number of people for creative thinking depends on various factors, including the complexity of the problem, the diversity of perspectives required, and the dynamics of the group (Ulloa et al., 2023). Sometimes, a small team of two or three individuals may be sufficient, while larger groups may be more effective for tackling complex challenges or fostering collective creativity. Ultimately, the key is to strike a balance between individual and group creative thinking approaches, leveraging the strengths of each to foster a dynamic and productive creative process.

Small group creative thinking (Puccio et al., 2020; Buisine et al., 2017) can be highly effective for generating innovative ideas and solutions. When the group size is limited, it allows for focused and intimate collaboration. While the ideal number of members may vary depending on the specific context, a small group size for creative thinking is typically three to five individuals. Methods used for individual creative thinking can also be used for small group creative thinking. Baruah and Paulus (2008) suggests that going between individual and interactive group sessions may be an effective procedure for group brainstorming. Some benefits of a very small group for creative thinking include intense collaboration, efficient communication, diverse perspectives, rapid iteration, enhanced trust and psychological safety, and flexibility and adaptability.

As highlighted by Paulus and Nijstad (2003), large-group creative thinking has several advantages, including diverse perspectives, collective intelligence, collaboration, and networking. Group brainstorming enhances idea sharing within groups (Osborn, 1953). More recent studies on electronic brainstorming have shown that larger groups of eight or more people can generate more ideas than individuals brainstorming alone (DeRosa et al., 2007).

One effective approach is the fishbowl method, where an inner circle discusses a topic while an outer circle observes (Keck-McNulty, 2004). Participants can rotate to ensure varied input. A study by Shirazi et al. (2020) involving 70 participants, with half using the fishbowl method, found that this method significantly enhances creative thinking. A useful approach in which participants take turns sharing ideas in sequence is the round-robin method. This technique safeguards that everyone can contribute, helping to generate a broad range of perspectives and fostering a sense of inclusion. Another method is role-storming, where team members assume different personas to tackle problems from new angles to generate innovative ideas (McFadzean et al., 1999). The World Café is a structured conversational process where groups discuss various topics in several rounds, moving between tables to build on each other's ideas. This method fosters constructive dialogue, strengthens relationships, promotes collaborative learning, and uncovers new opportunities for action (Ropes et al.,

2020). Tan and Brown (2005) noted that the World Café has been pivotal in Singapore's efforts to cultivate a learning culture and transition from a top-down approach to a more inclusive society. While large group creativity offers many benefits, it also presents challenges such as communication barriers, managing complexity, decision-making difficulties, and ensuring individual contributions are maintained (Paulus et al., 2012).

6.7 CHALLENGES DURING THE CREATIVE PROCESS

Both individual and team creativity face challenges that can hinder their effectiveness and impede the generation of innovative ideas. Individual creativity may encounter obstacles such as self-doubt, lack of inspiration, limited resources, isolation, difficulties managing time, resistance and criticism from others, and the pursuit of perfectionism. Overcoming these challenges involves building self-confidence, seeking feedback, cultivating a supportive community or network, setting realistic goals, and adopting a growth mindset that embraces learning and experimentation.

Team creativity, on the other hand, faces challenges like groupthink, unequal participation, lack of psychological safety, communication and collaboration issues, clashes of creative egos, time and resource constraints, and difficulties aligning objectives and visions. Addressing these challenges requires fostering open communication channels and safe environment (Asencio et al., 2023) by using some of the tools and methods such as multiple communication tools, diverse platforms, share information policy, open door policy, surveys and suggestion boxes, define clear communication norms., regular performance reviews to align goals and objectives, promoting equal participation and inclusion, encouraging constructive feedback and collaboration, and allocating sufficient time and resources for creative activities. Another way to overcome the group-related issues is to use the six thinking hats (de Bono, 2017).

Creating a culture of creative thinking within an organization involves fostering innovation and unlocking individual potential by encouraging idea generation, risk-taking, and experimentation (Table 6.7). It requires challenging the status quo and valuing creative thinking, while promoting open communication, collaboration, and freedom of expression. A new hybrid organizational environment requires a new set of challenges, such as trust and knowledge sharing (Bredberg, 2023). Cultivating a creative environment enables continuous improvement, adaptability to changes, and a competitive edge. Encouraging "impractical" and "bizarre" ideas breaks mental barriers, sparks creativity, and promotes divergent thinking. Overcoming personal inhibitions unleashes creativity and encourages individuals to take risks. Strategies such as cultivating a growth mindset, providing a safe environment, and seeking feedback help overcome these difficulties.

TABLE 6.7 Organizational Environment and Culture.

Creating a supportive environment	▪ Promote open communication, collaboration, and freedom of expression. ▪ Enable continuous improvement and adaptability to changes. ▪ Cultivate a creative environment. ▪ Overcome personal inhibitions. ▪ Unleash creativity. ▪ Cultivate a growth mindset and provide a safe environment. ▪ Foster psychological safety. ▪ Promote learning.
Encouraging creativity and idea generation	▪ Foster innovation. ▪ Encourage idea generation, risk-taking, and experimentation. ▪ Challenge the status quo. ▪ Value creative thinking. ▪ Encourage "impractical" and "bizarre" ideas. ▪ Break mental barriers. ▪ Spark creativity. ▪ Promote divergent thinking.
Nurturing individual potential	▪ Unlock individual potential. ▪ Cultivate a growth mindset and provide a safe environment. ▪ Foster psychological safety. ▪ Promote learning. ▪ Provide mentorship. ▪ Recognize creative contributions. ▪ Lead by example. ▪ Seek feedback.
Supporting organizational success	▪ Enable continuous improvement and adaptability to changes. ▪ Maintain a competitive edge. ▪ Provide necessary resources and tools to support a creative thinking culture. ▪ Encourage collaboration. ▪ Promote balanced risk-taking. ▪ Overcome obstacles. ▪ Find creative solutions to complex problems.

6.8 ORGANIZATIONAL REQUIREMENTS FOR CREATIVE THINKING

Creative thinking in an organization necessitates openness, tolerance, trust, freedom of thought, and freedom of expression (Rainone et al., 2021; Isaksen, 2023; Mumford et al., 2023). Cultivating these elements creates an environment that fosters innovative ideas, empowers individuals to take risks, and generates novel solutions to complex problems.

Openness refers to having an open mindset and being receptive to new ideas, perspectives, and possibilities. Measured using personality inventory subjective scales (Harris, 2004; Costa & McCrae, 2008; Peljko & Auer Antončič, 2022), it involves encouraging curiosity, questioning assumptions, and being willing to explore unconventional approaches. Openness allows for the exploration of diverse viewpoints and helps break free from rigid thinking patterns, fostering a culture of innovation. Personality traits are generally categorized into five groups known as the big five: openness to experience (O), conscientiousness (C), extraversion (E), agreeableness (A), and neuroticism (N). Research has shown a strong relationship between creativity and openness. For example, Hoseinifar et al. (2011) proposed a model to predict creativity based on the five traits: Creativity = 0.35O + 0.21E + 0.17A + 0.19C – 0.23N, with a coefficient of determination (R^2) = 0.48.

Trust is a foundational element for creative thinking within an organization. It involves building a sense of psychological safety, where individuals feel confident sharing their ideas without fear of criticism or negative consequences. Trust enables open and honest communication, promotes risk-taking, and encourages experimentation, all essential for fostering innovation. Trust can be measured using the Trust Inventory and revised Trust Inventory (Couch et al., 1996; Couch & Jones, 1997).

Creative thinking thrives on the freedom to think independently and critically. For instance, Schrödinger's cat thought experiment (Lewis, 2000) is purely conceptual, involving no harm to any animals. Freedom of thought is a fundamental aspect of the international human rights framework and a cornerstone of democratic society (Alegre, 2022). Individuals should be encouraged to challenge established norms, question assumptions, and explore new possibilities. This freedom fosters the generation of unique and imaginative ideas, paving the way for breakthrough innovations.

Alongside freedom of thought, freedom of expression is vital for creative thinking. It involves providing individuals with the opportunity and platform to openly share their ideas, opinions, and perspectives. By allowing diverse voices to be heard, organizations can tap into a wealth of creativity and unlock the potential for innovative solutions.

6.9 EVALUATION METHODS

It is important to note that the assessment of creative ideas is subjective to some degree and can vary based on individual perspectives, preferences, and evaluation criteria. Different evaluators or stakeholders may prioritize different factors and place different weights on each factor based on their specific needs and contexts. Qualitative approaches (expert evaluation, peer review, focus groups, and open-ended surveys or interviews) and quantitative approaches (rating scales, surveys with closed-ended questions, idea ranking or pairwise comparison, and statistical analysis).

6.9.1 THEORETICAL FRAMEWORK FOR CREATIVITY EVALUATION

Diffusion of innovation theory, developed by Everett Rogers (Rogers, 1962), offers a clear understanding of the intricate process of diffusion and adoption over time (Sahin, 2006). Table 6.8 serves as a valuable tool within the theory, showcasing the five progressive levels of innovation. It provides a clear roadmap of how ideas evolve and gain traction within different segments of the population. Moreover, the theory goes beyond mere observation, examining the factors that shape the speed and scope of adoption. In Table 6.9, there is a comprehensive list of these influential factors, providing insightful guidance into the diffusion process. By considering these factors, we can go through the complexities of innovation diffusion and gain a deeper understanding of how and why individuals or groups embrace new ideas, products, or innovations.

Rogers' diffusion of innovation theory provides a framework for understanding the factors that influence the adoption and diffusion of new ideas, products, or innovations (Miller, 2015). It has been widely applied in various fields, including technology, marketing, healthcare, and social sciences, to guide the development and implementation of innovations.

6.9.2 TESTS OF CREATIVE THINKING

The Torrance tests of creative thinking (TTCT), developed by psychologist E. Paul Torrance in 1974, is a widely recognized assessment tool for evaluating

TABLE 6.8 The Progress of Innovations.

Knowledge	The individual becomes aware of the innovation's existence and gains knowledge about its features and benefits.
Persuasion	The individual forms a favorable attitude toward innovation and considers adopting it. Persuasion can occur through various means, such as information, communication, or personal influence.
Decision	The individual engages in a process of evaluating the innovation and its potential benefits and drawbacks. At this stage, the individual decides regarding whether to adopt or reject the innovation.
Implementation	The individual puts the innovation into practice and starts using it. Implementation involves learning and adapting to innovation, overcoming any challenges or barriers, and integrating it into existing routines or systems.
Confirmation	The individual evaluates the outcomes and consequences of adopting the innovation. If the individual perceives positive outcomes, such as improved performance or satisfaction, they are more likely to continue using and endorsing the innovation.

TABLE 6.9 Five Characteristics of Innovations that Influence Their Adoption.

Relative advantage	The extent to which an innovation is perceived as being better than the existing alternatives.
Compatibility	The degree to which innovation is perceived as being consistent with existing values, beliefs, and needs of potential adopters.
Complexity	The perceived difficulty of understanding and using innovation. Innovations that are perceived as simple and easy to use are more likely to be adopted.
Observability	The visibility of the results or benefits of the innovation. Innovations that have observable and tangible benefits are more likely to be adopted.
Trialability	The ability to experiment with innovation on a limited basis before making a full commitment. Innovations that can be tried and tested before adoption are more likely to be adopted.

creativity. TTCT offers a comprehensive framework for understanding the different dimensions of creativity (see Table 6.10). The Scholastic Testing Service administers the test and provides both Individual Student Reports and Group Summary Reports (Torrance, 1974). The Individual Student Report includes scores for creative thinking, a creative index, percentile ranks based on age and grade, and summaries of performance. Meanwhile, the Group Summary Report presents frequency distributions, means, and standard deviations for standard scores, along with standard deviations for all TTCT scores across a class list. It is often used to evaluate creative thinking abilities and identify individuals with high creative potential. Designers often use the four components of TTCT (Torrance-4) to enhance their ideation and concept generation process. By focusing on fluency, designers aim to generate a large quantity of ideas in the early stages of a project (Runco & Jaeger, 2012). They strive to think flexibly by exploring various perspectives, challenging assumptions, and considering alternative approaches (De Bono, 2017). Designers and teams can assess the fluency of ideas generated, considering the quantity and variety of concepts explored. They can evaluate the flexibility of the design, assessing how well it addresses different perspectives and approaches the problem from various angles. Originality is valued as ensuring that the ideas generated are novel and distinguishable from existing solutions (Sternberg & Lubart, 1999; Guilford, 1967). Originality can be assessed by comparing the design to existing solutions and evaluating its uniqueness and novelty. Elaboration comes into play when designers further develop and refine their ideas, expanding upon initial concepts to create detailed design proposals. Elaboration may be influenced by imagination (Weisberg, 1999; Boden, 1990) and curiosity (Sawyer, 2003). Elaboration can be evaluated by considering the level of detail, refinement, and depth of thinking reflected in the design concept. Using the fluency, flexibility, originality and elaboration components, the creative thinking abilities of individuals are assessed and Torrance-4 has often been used to identify individuals with high creative potential.

6.9.3 METRICS FOR THE QUALITY OF IDEAS

When assessing creative ideas, there are several factors to consider, and the specific criteria may vary depending on the context and purpose of the evaluation. These criteria include elements such as novelty, usefulness, feasibility, impact, originality, adaptability, practicality, marketability, sustainability, and alignment with objectives. Each criterion plays a crucial role in evaluating the creative potential of an idea.

TABLE 6.10 Four Components of TTCT.

Component	Description	In Torrance Tests	Interpretation
Fluency	Fluency refers to the ability to generate many ideas or responses.	Individuals are often given a specific prompt and are asked to generate as many ideas or solutions as possible within a given time frame.	Higher fluency scores indicate a greater quantity of ideas.
Flexibility	Flexibility measures the ability to think in diverse ways and approach problems from different angles.	Individuals may be asked to provide multiple solutions or approaches to a given problem.	Higher flexibility scores indicate higher capacity to shift perspectives, consider alternative viewpoints, and adapt thinking strategies to generate more solutions.
Originality	Originality assesses the degree of novelty and uniqueness in one's ideas or responses.	Originality is evaluated based on the rarity and distinctiveness of the generated ideas.	Higher originality scores indicate the high capability to produce ideas that are uncommon, unconventional, or unexpected.
Elaboration	Elaboration refers to the extent to which individuals develop and expand upon their ideas.	Elaboration is assessed based on the level of detail and complexity in the responses.	Higher elaboration scores refer to detailed and rich descriptions, explanations, or extensions of initial thoughts.

To obtain a comprehensive evaluation, a simple method is provided in Table 6.11. In this approach, each criterion is assigned a weight, ensuring that the total weight adds up to 100% (Σweights = 100%). A score ranging from 0 to 100 is then assigned to each criterion based on its performance. The weighted score is calculated by multiplying the weight assigned to each criterion by its corresponding score. In Table 6.11, you'll find examples that utilize weights with higher novelty weight (20%), the Quality of Ideas obtained is 63.5, indicating a highly satisfactory value for the idea being assessed. The weights and scores can be customized and adjusted to align with the specific requirements and priorities of the evaluation process. This flexibility allows one to tailor the assessment to suit the unique context and objectives.

6.9.4 LIMITATIONS

There are many limitations when evaluating creative ideas, and it is necessary to adapt evaluation approaches that mitigate biases and account for uncertainties. A multidimensional evaluation process involving diverse perspectives and considering multiple criteria can help overcome some limitations and provide a more comprehensive assessment of creative ideas. Some limitations are subjectivity and uncertainty, company policies and constraints, linking creativity to product/service/ systems, technological and manufacturing uncertainty, market uncertainty, resource limitations, cultural and contextual factors, and lack of long-term perspective.

In a creative activity, there is a need for an exit strategy, including evaluating the outcome, iterative development, cost constraints, time constraints, resource constraints, transitioning from ideation to the next phase, implementation, and abandonment (Table 6.12).

6.10 USEFUL SOURCES

Designers can seek inspiration from sources such as patents, magazines, company products and reports, social media, movies, and other relevant information available today. Exposure to different cultures and visual stimuli, like images, can enhance the creative process for product design. Including sources showcasing cultural elements, artistic styles, and visual representations can inspire and broaden the perspective when designing products.

Academic journals and conference papers often contain research studies, theories, and methodologies related to product design. They can provide insights into design principles, user experiences, and emerging trends. Some of the well-known databases are PubMed (pubmed.ncbi.nlm.nih.gov), Scopus (www.scopus.com), Web of Science (www.webofscience.com), IEEE Explore (ieeexplore.ieee.org), ACM Digital Library (dl.acm.org), ScienceDirect (www.sciencedir

TABLE 6.11 Scoring the Quality of Ideas.

Criteria	Details	Weight	Score	Weight X score
Novelty	The degree of originality and uniqueness of the idea.	20%	70	14
Usefulness	The practical value or utility of the idea.	10%	60	6
Feasibility	The extent to which the idea can be realistically implemented or executed.	10%	50	6
Impact	The potential effect or influence of the idea including benefits or changes.	10%	60	6
Originality	The extent to which the idea is distinct and not derivative of existing concepts or solutions.	10%	70	7
Adaptability	The potential for the idea to be adapted, modified, or applied in different contexts or situations.	10%	60	6
Practicality	The degree to which the idea is feasible and realistic in terms of implementation, cost, and resources required.	10%	80	8
Marketability	The potential market appeal and commercial viability of the idea.	5%	60	3
Sustainability	Evaluating sustainability involves considering the long-term viability and environmental, social, and economic implications of an idea.	10%	50	5
Alignment with objectives	The extent to which the idea aligns with the goals, values, or objectives of the individual, organization, or project.	5%	70	3.5
Total score				63.5

Categorization of ideas.

0-50	51-60	61-70	71-80	81-90	90-100
Not satisfactory	Satisfactory	Very satisfactory	Good	Very good	Excellent

Note: The weights and scores can be different depending on what is evaluated.

TABLE 6.12 Exit Strategy Considerations.

Evaluating the outcome	Assessing the results of the creative activity to determine its success, impact, and alignment with objectives.
Iterative development	Embracing an iterative approach that involves continuously refining and improving ideas through cycles of testing, feedback, and iteration.
Cost constraints	Considering the financial resources required for implementing and sustaining creative activity, ensuring it remains within budgetary limits.
Time constraints	Considering the time available for the creative process and setting milestones or deadlines to guide the progression of the activity.
Resource constraints	Recognizing the availability and limitations of resources, such as personnel, equipment, materials, or external support, and making decisions accordingly.
Transitioning from ideation to the next phase	Determining when and how to move from the ideation phase to the subsequent stages, such as prototyping, development, or implementation.
Implementation	Executing the selected ideas or solutions and putting them into action, whether through product development, project launch, or implementation of a new process.
Abandonment	Recognizing when an idea or project is not viable or feasible and making a conscious decision to discontinue or abandon it, redirecting resources and efforts elsewhere.

ect.com), JSTOR (www.jstor.org), and ProQuest (www.proquest.com). Google Scholar (scholar.google.com) is not a traditional academic database but a freely accessible web search engine focusing on scholarly literature.

Patents can provide information on existing products, design innovations, and technological advancements. Patent databases such as the United States Patent and Trademark Office (USPTO) (www.uspto.gov), the European Patent Office (EPO) (www.epo.org), or the World Intellectual Property Organization (WIPO) (www.wipo.int) can be explored for relevant patents. Google Patents is a free search engine that allows users to access and search through a vast database of patents and patent applications (patents.google.com).

Exploring company websites, product catalogs, and annual reports can give insights into real-world design practices, market trends, and user needs. Companies often share information about their design processes, philosophies, and the factors influencing their product decisions. Other venues to get information include industry associations and organizations, industry conferences and exhibitions, product reviews, trade publications, trade shows, and press releases.

Platforms like Instagram (www.instagram.com), Pinterest (www.pinterest.com), and Behance (www.behance.net) can be valuable sources of visual inspiration and exposure to diverse design styles. Online design communities and forums can provide an opportunity to engage with other designers, share ideas, and gain insights from the design community.

Films, documentaries, and multimedia content can offer unique perspectives on design, creativity, and cultural influences. They can provide inspiration, storytelling elements, and visual references.

6.11 CONCLUSION

In conclusion, this chapter serves as a comprehensive guide to ideation and creativity, building on our skills from previous chapters. After defining the problem in Chapter 5, we emphasized the significance of individuality, collaboration, and the stages of the creative thinking process. We explored various ideation methods, including emerging techniques like computer-mediated ideation and AI-based approaches. Additionally, we discussed evaluation methods for assessing the quality of ideas and underscored the importance of having an exit strategy for creative activities. By providing valuable visions and tools, this chapter equips readers to foster creativity, generate innovative ideas, and navigate the creative process effectively. As we move forward, we will explore different levels of prototypes, with detailed discussions on prototyping in the next chapter.

> Our dreams may seem impossible, but our creativity can turn them into reality.
>
> —Malala Yousafzai

REFERENCES

Abi Saad, E., & Agogué, M. (2023). Creativity in virtual teams: Systematic review, synthesis and research agenda. *Creativity and Innovation Management*, 32(1), 117–140.

Alegre, S. (2022). *Protecting Freedom of Thought in the Digital Age*. Centre for International Governance Innovation.

Amabile, T. M. (1983). *The Social Psychology of Creativity*. Springer. https://doi.org/10.1007/978-1-4612-5533-8.

Amabile, T. M. (2018). *Creativity in Context: Update to the Social Psychology of Creativity*. Routledge.

Ambrose, G. (2015). *Design Thinking for Visual Communication*. Bloomsbury Publishing.

Asencio, R., Murase, T., Chollet, B., DeChurch, L. A., & Zaccaro, S. J. (2023). Bridging the boundary without sinking the team: Communication, identification, and creativity in multiteam systems. *Group Dynamics: Theory, Research, and Practice*, 27(1), 28.

Baruah, J., & Paulus, P. B. (2008). Effects of training on idea generation in groups. *Small Group Research*, 39(5), 523–541.

Blocker, L. (2023). Achieving implementation: Putting creative ideas to work. In *Handbook of Organizational Creativity* (pp. 133–144). Academic Press.

Boden, M. A. (1990). *The Philosophy of Artificial Intelligence.* Oxford University Press, Inc.

Botella, M., Zenasni, F., & Lubart, T. (2018). What are the stages of the creative process? What visual art students are saying. *Frontiers in Psychology*, 9, 2266.

Bredberg, E. (2023). Beyond the office walls: Exploring trust and knowledge sharing in the hybrid work setting—A qualitative study on employees' perception of the challenges and opportunities when adopting the hybrid work setting. Master Project, University of Gothenburg. https://gupea.ub.gu.se/handle/2077/77865

Brown, A. S. (2014). Iconic design. *Mechanical Engineering*, 136(3), 28–35.

Buisine, S., Guegan, J., & Vernier, F. (2017). Technological Innovation in Group Creativity. In *Creativity, Design Thinking and Interdisciplinarity* (pp. 185–201). Singapore: Springer Singapore.

Capanna, A. (2015). *BiOrganic Design: A New Method for Architecture and the City. Architecture and Mathematics from Antiquity to the Future: Volume II: The 1500s to the Future*, 579–591. Cham: Springer International Publishing.

Carayannis, E. G. (Ed.). (2020). *Encyclopedia of Creativity, Invention, Innovation and Entrepreneurship.* Springer International Publishing.

Choudhury, N. (2022). Can artificial intelligence be used to improve productivity by automating elements of the user experience design processes? *Arts & Humanities*, 57, v1. https://doi.org/10.20944/preprints202202

Coburn, J. Q., Freeman, I., & Salmon, J. L. (2017). A review of the capabilities of current low-cost virtual reality technology and its potential to enhance the design process. *Journal of Computing and Information Science in Engineering*, 17(3), 031013.

Costa, P. T., & McCrae, R. R. (2008). The revised neo personality inventory (neo-pi-r). *The SAGE Handbook of Personality Theory and Assessment*, 2(2), 179–198.

Couch, L. L., Adams, J. M., & Jones, W. H. (1996). The assessment of trust orientation. *Journal of Personality Assessment*, 67(2), 305–323. https//doi.org/10.1207/s15327752jpa6702_7

Couch, L. L., & Jones, W. H. (1997). Measuring levels of trust. *Journal of Research in Personality*, 31(3), 319–336.

Cropley, A. (2006). In praise of convergent thinking. *Creativity Research Journal, 18*(3), 391–404.

De Bono, E. (2017). *Six Thinking Hats: The Multi-million Bestselling Guide to Running Better Meetings and Making Faster Decisions.* Penguin UK.

De Bono, E. (2014). *Lateral Thinking: An Introduction.* Random House.

Deci, E. L., & Ryan, R. M. (1985). The general causality orientations scale: Self-determination in personality. *Journal of Research in Personality*, 19(2), 109–134.

de Raad, B. (2020). Individuality and personality. In *Psychological Concepts* (pp. 299–323). Psychology Press.

DeRosa, D. M., Smith, C. L., & Hantula, D. A. (2007). The medium matters: Mining the long-promised merit of group interaction in creative idea generation tasks in a meta-analysis of the electronic group brainstorming literature. *Computers in Human Behavior*, 23(3), 1549–1581.

Dietrich, A., & Kanso, R. (2010). A review of EEG, ERP, and neuroimaging studies of creativity and insight. *Psychological Bulletin*, 136(5), 822.

Dow, G. T. (2022). Defining creativity. In *Creativity and Innovation* (pp. 5–21). Routledge.

Fink, A., & Benedek, M. (2014). EEG alpha power and creative ideation. *Neuroscience & Biobehavioral Reviews*, 44, 111–123.

Furnham, A., Batey, M., Anand, K., & Manfield, J. (2008). Personality, hypomania, intelligence and creativity. *Personality and Individual Differences*, 44(5), 1060–1069.

Gadd, K. (2011). *TRIZ for Engineers: Enabling Inventive Problem Solving*. John Wiley & Sons.

Geum, Y., & Park, Y. (2016). How to generate creative ideas for innovation: a hybrid approach of WordNet and morphological analysis. *Technological Forecasting and Social Change*, 111, 176–187.

Gong, Z., Lee, L. H., Soomro, S. A., Nanjappan, V., & Georgiev, G. V. (2022). A systematic review of virtual brainstorming from the perspective of creativity: affordances, framework, and outlook. *Digital Creativity*, 33(2), 96–127.

Gong, Z., Nanjappan, V., Lee, L. H., Soomro, S. A., & Georgiev, G. V. (2023). Exploration of the relationship between culture and experience of creativity at the individual level: a case study based on two design tasks. *International Journal of Design Creativity and Innovation*, 11(3), 185–208.

Green, A. E., Beaty, R. E., Kenett, Y. N., & Kaufman, J. C. (2023). The process definition of creativity. *Creativity Research Journal*, 1–29.

Grech, A., Mehnen, J., & Wodehouse, A. (2023). An extended AI-experience: Industry 5.0 in creative product innovation. *Sensors*, 23(6), 3009.

Guilford, J. P. (1967). Creativity: Yesterday, today and tomorrow. *The Journal of Creative Behavior*, 1(1), 3–14.

Halskov, K., & Lundqvist, C. (2021). Filtering and informing the design space: Towards design-space thinking. *ACM Transactions on Computer-Human Interaction (TOCHI)*, 28(1), 1–28.

Harris, J. A. (2004). Measured intelligence, achievement, openness to experience, and creativity. *Personality and Individual Differences*, 36(4), 913–929.

Holyoak, K. J., & Thagard, P. (1996). *Mental Leaps: Analogy in Creative Thought*. MIT Press.

Hoseinifar, J., Siedkalan, M. M., Zirak, S. R., Nowrozi, M., Shaker, A., Meamar, E., & Ghaderi, E. (2011). An investigation of the relation between creativity and five factors of personality in students. *Procedia-Social and Behavioral Sciences*, 30, 2037–2041.

Isaksen, S. G. (2023). Developing creative potential: the power of process, people, and place. *Journal of Advanced Academics*, 1932202X231156389.

Kaufman, J. C., & Sternberg, R. J. (Eds.). (2010). *The Cambridge Handbook of Creativity*. Cambridge University Press.

Keck-McNulty, C. L. (2004). *Group Leadership Training: What is Learned Using a Fishbowl Method*. Kent State University.

Kilgour, M. (2022). Creative Thinking: Designed for Humans. In *The Handbook of Creativity & Innovation in Business: A Comprehensive Toolkit of Theory and Practice for Developing Creative Thinking Skills* (pp. 179–195). Springer Nature Singapore.

Kleinmintz, O. M., Ivancovsky, T., & Shamay-Tsoory, S. G. (2019). The two-fold model of creativity: the neural underpinnings of the generation and evaluation of creative ideas. *Current Opinion in Behavioral Sciences*, 27, 131–138.

Koutstaal, W., & Binks, J. (2015). *Innovating Minds: Rethinking Creativity to Inspire Change*. Oxford University Press.

Kronfeldner, M. (2014). *Darwinian Creativity and Memetics*. Routledge.

Kuang, C. (2013). The untold story of how the Aeron Chair was born. *Fast Company*, February 5. www.fastcompany.com/1671789/the-untold-history-of-how-the-aeron-chair-came-to-be (Accessed 30/06/2025).

Lebuda, I., & Benedek, M. (2023). A systematic framework of creative metacognition. *Physics of Life Reviews*, *46*, 161–181.

Lewis, P. J. (2000). What is it like to be Schrödinger's cat? *Analysis*, 22–29.

Lewrick, M., Link, P., & Leifer, L. (2020). *The Design Thinking Toolbox: A Guide to Mastering the Most Popular and Valuable Innovation Methods*. John Wiley & Sons.

Luximon, A., & Luximon, Y. (2010). Creativity, inspiration and design exploration through modification of the juicy salif. In *Proceedings of the TMCE 2010* (Eds. I. Horváth, F. Mandorli and Z. Rusák), Ancona, Italy, April 12–16, 2010.

Maaravi, Y., Heller, B., Shoham, Y., Mohar, S., & Deutsch, B. (2021). Ideation in the digital age: literature review and integrative model for electronic brainstorming. *Review of Managerial Science*, 15, 1431–1464.

Maital, S. & Barzani, E. (2021). *Creative Ideation: A Review of the Literature*. Samuel Neaman Institute for National Policy Research (www.neaman.org.il/wp-content/uploads/2024/02/Creativity.pdf)

Malycha, C. P., & Maier, G. W. (2017). The random-map technique: Enhancing mind-mapping with a conceptual combination technique to foster creative potential. *Creativity Research Journal*, 29(2), 114–124.

Manghan, J. P. (2020). Crowd-composition: exploring the qualities of collective creativity (Doctoral dissertation, University of Southampton).

McCauley, P., Strevens, B., Hall, K., Shane, L., & Di Paulo, S. (2016). A human centric approach to research, design and development of an innovative workstation module for the National Aeronautics and Space Administration (NASA). *Journal of Management & Engineering Integration*, 9(1), 52–58.

McFadzean, E., Somersall, L., & Coker, A. (1999). A framework for facilitating group processes. *Strategic Change*, 8(7), 421–431.

Miller, R. L. (2015). Rogers' innovation diffusion theory (1962, 1995). In *Information Seeking Behavior and Technology Adoption: Theories and Trends* (pp. 261–274). IGI Global.

Mumford, M. D., Fichtel, M., England, S., & Newbold, T. R. (2023). Leader thinking, follower thinking: leader impacts on follower creative performance. *Annual Review of Organizational Psychology and Organizational Behavior*, 10, 413–440.

Nguyen, M., & Mougenot, C. (2022). A systematic review of empirical studies on multidisciplinary design collaboration: Findings, methods, and challenges. *Design Studies*, 81, 101120.

Osborn, A. F. (1953). *Applied Imagination*. Charles Scribner.

OECD (2018). *Oslo Manual 2018*. OECD. www.oecd.org/en/publications/oslo-manual-2018_9789264304604-en.html, Guidelines for Collecting, Reporting and Using Data on Innovation, 4th Edition

Paulus, P., & Nijstad, B. (2003). Group creativity: An introduction. *Group Creativity: Innovation Through Collaboration*, 3–11.

Paulus, P. B., Dzindolet, M., & Kohn, N. W. (2012). Collaborative creativity—Group creativity and team innovation. In *Handbook of Organizational Creativity* (pp. 327–357). Academic Press.

Peljko, Ž., & Auer Antončič, J. (2022). Impacts of entrepreneurial openness and creativity on company growth. *Frontiers in Psychology*, 13, 860382.

Peters, D., Loke, L., & Ahmadpour, N. (2021). Toolkits, cards and games–a review of analogue tools for collaborative ideation. *CoDesign*, 17(4), 410–434.

Pinkow, F. (2023). Creative cognition: A multidisciplinary and integrative framework of creative thinking. *Creativity and Innovation Management*, 32(3), 472–492.

Puccio, G. J., Burnett, C., Acar, S., Yudess, J. A., Holinger, M., & Cabra, J. F. (2020). Creative problem solving in small groups: The effects of creativity training on idea generation, solution creativity, and leadership effectiveness. *The Journal of Creative Behavior*, 54(2), 453–471.

Rainone, N. A., Natale, A. N., Alenick, P. R., Kato, A., Patel, K. R., Steele, L. M., & Watts, L. L. (2021). Lost in translation? A review of popular-press claims about organizational creativity and innovation. *Creativity Research Journal*, 33(4), 424–436.

Rane, N., Choudhary, S., & Rane, J. (2023). Enhanced Product Design and Development Using Artificial Intelligence (AI), Virtual Reality (VR), Augmented Reality (AR), 4D/5D/6D Printing, Internet of Things (IoT), and Blockchain: A Review (November 25, 2023). Available at SSRN: https://ssrn.com/abstract=4644059 or http://dx.doi.org/10.2139/ssrn.4644059.

Reiter-Palmon, R., Forthmann, B., & Barbot, B. (2019). Scoring divergent thinking tests: A review and systematic framework. *Psychology of Aesthetics, Creativity, and the Arts*, 13(2), 144.

Ritter, S. M., & Dijksterhuis, A. (2014). Creativity—the unconscious foundations of the incubation period. *Frontiers in Human Neuroscience*, 8, 73722.

Rogers, E. M. (1962). *Diffusion of Innovations* (1st ed.). New York: Free Press of Glencoe.

Ropes, D., van Kleef, H., & Douven, G. (2020). Learning in The World Café: an empirical evaluation. *Journal of Workplace Learning*, 32(4), 303–316.

Ross, V. E. (2006). A model of inventive ideation. *Thinking Skills and Creativity*, 1(2), 120–129.

Runco, M. A., & Jaeger, G. J. (2012). The standard definition of creativity. *Creativity Research Journal*, *24*(1), 92–96.

Rustler, F. (2020). *Thinking Tools for Creativity and Innovation: The Little Handbook of Innovation Methods*. Midas Management Verlag.

Sahin, I. (2006). Detailed review of Rogers' diffusion of innovations theory and educational technology-related studies based on Rogers' theory. *Turkish Online Journal of Educational Technology (TOJET)*, 5(2), 14–23.

Sahoo, P. K., Pany, S., Mohanty, S. P., Dash, K. R., & Rana, S. (2023). Dialogue embedded synectics model of teaching: a hybrid model for promotion of creativity. *Journal of Research in Innovative Teaching & Learning*. https://doi.org/10.1108/JRIT-08-2023-0117

Sanchez-Ruiz, M. J., Santos, M. R., & Jiménez, J. J. (2013). The role of metaphorical thinking in the creativity of scientific discourse. *Creativity Research Journal*, 25(4), 361–368.

Sawyer, R. K. (2003). *Creativity and Development*. Oxford University Press.

Sawyer, R. K. (2021). The iterative and improvisational nature of the creative process. *Journal of Creativity*, 31, 100002.

Shi, Y., Gao, T., Jiao, X., & Cao, N. (2023). Understanding Design Collaboration Between Designers and Artificial Intelligence: A Systematic Literature Review. *Proceedings of the ACM on Human–Computer Interaction*, 7 (CSCW2), 1–35.

Shirazi, M., Modarres, M., Shariati, M., & Dehshiri, A. S. H. (2020). Technical simulation using goldfish bowl method: A medical teaching method for increasing student's creativity. *Archives of Iranian Medicine*, 23(1), 37–43.

Sternberg, R. J., & Lubart, T. I. (1999). The concept of creativity: Prospects and paradigms. *Handbook of Creativity, 1*(3-15).

Stige, Å., Zamani, E. D., Mikalef, P., & Zhu, Y. (2023). *Artificial Intelligence (AI) for User Experience (UX) Design: A Systematic Literature Review and Future Research Agenda.* Information Technology & People.

Suh, S. (2019). A study on the application of SCAMPER techniques for the development of fashion design—Focusing on the development of the SCAMPER questions. *Journal of Fashion Business,* 23(3), 1–9.

Suh, J., & Cho, J. Y. (2020). Linking spatial ability, spatial strategies, and spatial creativity: A step to clarify the fuzzy relationship between spatial ability and creativity. *Thinking Skills and Creativity,* 35, 100628.

Tan, S., & Brown, J. (2005). The world café in Singapore: Creating a learning culture through dialogue. *Journal of Applied Behavioral Science,* 41(1), 83–90.

Torrance, E.P. (1974). *Torrance Tests of Creative Thinking.* Scholastic Testing Service. (www.ststesting.com)

Tzachrista, M., Gkintoni, E., & Halkiopoulos, C. (2023). Neurocognitive profile of creativity in improving academic performance—A scoping review. *Education Sciences,* 13(11), 1127.

Ulloa, M. A., Nesrallah, S., Shafafi, P., Lamar, A., Gonçalves, P. C. R., Gonçalves, M., ... & Bouillon, C. (2023). Designing a youth-led Dialogue Forum tool: The CO-CREATE experience. *Obesity Reviews,* 24, e13611.

Walton, A. P. (2016). Creativity and a human dichotomy: individual or part of a team?. In: Corazza, G., Agnoli, S. (eds) Multidisciplinary Contributions to the Science of Creative Thinking. Creativity in the Twenty First Century. Springer, Singapore. https://doi.org/10.1007/978-981-287-618-8_6.

Wajcman, J. (2017). Automation: is it really different this time? *The British Journal of Sociology,* 68(1), 119–127.

Wang, G., Phan, T. V., Li, S., Wang, J., Peng, Y., Chen, G., ... & Liu, L. (2022). Robots as models of evolving systems. *Proceedings of the National Academy of Sciences,* 119(12), e2120019119.

Weisberg, R. W. (1999). Creativity and knowledge: A challenge to theories. *Handbook of Creativity, 226,* 251.

Xia, T., Kang, M., Chen, M., Ouyang, J., & Hu, F. (2021). Design training and creativity: Students develop stronger divergent but not convergent thinking. *Frontiers in Psychology,* 12, 695002.

Yuan, Y., Humphrey, S. E., & van Knippenberg, D. (2022). From individual creativity to team creativity: A meta-analytic test of task moderators. *Journal of Occupational and Organizational Psychology,* 95(2), 358–404.

7 Prototyping

Move fast and break things.

—Mark Zuckerberg

7.1 INTRODUCTION

Prototyping in design refers to creating a 'visualization' or a touch-and-feel, preliminary version or model of a product, system, or service (Camburn et al., 2017; Hallgrimsson, 2012; Otto & Wood, 2001; Moon, 2022). It is an important step that bridges the gap between ideation and production (Figure 1.5).

Prototyping involves transforming abstract ideas or sketches into tangible or digital representations that can be interacted with, evaluated, and tested (Lang & Mjöberg, 2020). Prototyping aims to bring designs to life in a physical or digital form, allowing designers to visualize and communicate their ideas more effectively, gather feedback, validate design decisions, and uncover potential issues or improvements.

Prototyping can start during concept generation. During the conceptual phase, low-fidelity methods such as sketches, wireframes, or digital mock-ups may be used. In general, prototyping involves planning the scope, objectives, and level of fidelity required. The purpose of prototyping is threefold: to communicate the design concept with stakeholders, to learn how users interact with the prototype as a realization of the design concept, and to lead product development by gaining inspiration from stakeholders and securing investment from investors. It involves selecting the materials, tools, and techniques for creating the prototype. Prototypes can be in a physical or digital form and can be used for user tests to evaluate usability, gather feedback, and make decisions. Based on the user feedback, changes can be made to address usability issues, enhance user experience, and align with user expectations to verify and validate the design at an early stage (Chapter 11). Prototypes can be improved and re-made for further testing. The insights gained from prototyping and user testing guide the final

DOI: 10.1201/9781003476177-7

design. The refined design is then prepared for production, considering manufacturing constraints, cost, and scalability.

7.2 PROTOTYPING AND HUMAN-CENTERED DESIGN

Prototyping is vital in iterative and human-centric design (dos Santos et al., 2021). It helps designers in numerous ways:

1. *Empathy and understanding*: Human-centered design (HCD) begins with gaining a deep understanding of the needs, behaviors, and preferences of the target users (Chapter 4). Prototyping helps designers gather user insights by creating tangible representations of their ideas, allowing users to test and provide feedback. This process enables designers to better understand user needs, challenges, and expectations through direct observation and interactions.
2. *Iterative design*: Human-centered design is an iterative process that involves continuous feedback and improvement. Prototyping facilitates this iterative approach by allowing designers to create and test prototypes quickly, gather user feedback, and make necessary refinements. Prototypes serve as a medium for dialogue and collaboration between designers and users, enabling designers to refine their designs based on user insights.
3. *User testing and validation*: Prototyping allows for user testing and validation, essential components of human-centered design (Chapters 8 and 11). By creating prototypes, designers can simulate a scenario and observe how users interact with the design. User feedback gathered through testing helps identify usability issues, uncover unforeseen challenges, and validate design decisions. This iterative feedback loop ensures that the final product meets the needs and expectations of the users.
4. *Communication and collaboration*: Prototypes serve as a common language among all stakeholders. They provide a tangible representation of the design that can be shared and communicated effectively. Prototypes facilitate collaboration and enable multidisciplinary teams to align their understanding, make informed decisions, and work together toward a human-centered solution.
5. *Risk reduction*: Prototyping allows designers to validate their design concepts, evaluate feasibility, and mitigate risks before investing significant resources in full-scale development. By testing and refining prototypes, designers can uncover design flaws, identify potential technical challenges, and address them early on. This reduces the risk of developing a solution that does not meet user needs or is not technically viable.

7.3 PROTOTYPING TECHNIQUES

Numerous prototyping techniques are available for various products, services, and systems (Camburn et al., 2017). Designers can select different prototyping methods based on their objectives, project constraints, and available resources. Several prototyping techniques are now compared and analyzed.

7.3.1 CHARACTERISTICS OF PROTOTYPING TECHNIQUES

Prototyping methods can be characterized based on factors such as fidelity, materials and medium, and purpose (Warfel, 2009; Coutts et al., 2019; Real et al., 2021) (Table 7.1). The categories provide a framework for classifying prototyping methods based on different criteria. Ultimately, the choice of prototyping method should be based on the project requirements, design goals, available resources, and the design stage. However, these categories are not mutually exclusive, and designers often employ a combination of methods throughout the design process to explore, validate, and communicate their ideas effectively (Hammami et al., 2020). For example, hybrid prototyping, using both physical and digital methods, can provide a comprehensive approach. Designers may create a physical prototype using cardboard or foam and then enhance it with digital elements like 3D-printed components, electronic interfaces, or interactive interfaces. Another example is combining the functional prototype and the visual prototype, thereby integrating both functional performance and aesthetic representation to provide a realistic and interactive prototype. This approach leverages the benefits of combining functional and visual elements, helping designers gain insights into usability and user experience while focusing on the design's intended appearance. Experimentation and adaptation based on the specific project needs and insights gained along the way are key to choosing the most effective prototyping methods.

7.3.1.1 FIDELITY

Fidelity indicates the closeness of the prototype to the final product. Prototypes can be of low-, medium-, or high-fidelity (Zhou & Rau, 2019; Fiorineschi & Rotini, 2019; Starkey et al., 2019; Ranscombe et al., 2020).

1. *Low-fidelity prototypes*: These prototypes are simple, rough, and quick to create (Isa, 2023). They focus on conveying basic concepts and ideas rather than detailed visuals or functionality. Low-fidelity prototypes may lack the visual or functional details necessary for accurate evaluation, and they may not fully capture the user experience or technical feasibility of a design. Examples include sketches, paper prototypes, foam, clay and wireframes.

2. *Medium-fidelity prototypes*: These prototypes balance simplicity and realism. They provide a moderate level of detail and functionality, allowing users to interact with the design to some extent. Examples include digital prototypes with basic interactivity, mockups with visual details, and 3D printing.
3. *High-fidelity prototypes*: These prototypes closely resemble the final design regarding appearance and functionality. They incorporate advanced interactions, visual elements, and sometimes even real data. Examples include interactive digital prototypes with realistic behavior, physical prototypes, or functional prototypes with near-complete functionality (Computer Numerical Control (CNC) machined, injection molded, and other advanced manufacturing technologies).

7.3.1.2 MATERIAL AND MEDIUM

Based on material and medium, prototypes can be broadly grouped into physical or digital prototypes (McElroy, 2016).

1. *Physical prototypes*: These prototypes are tangible, physical representations of a design concept. They enable designers to evaluate physical form, ergonomics, and basic functionality. They are valuable for user testing and validating physical properties. Physical prototyping methods can be time-consuming and expensive and may require specialized equipment or skills. They can be traditionally created using paper, cardboard, foam, modeling clay, natural clay, wood, or plastic. They may also be created using 3D printing or laser cutting and could be scaled models or physical mockups.
2. *Digital prototypes*: These prototypes are created using digital tools and software. They can be interactive, allowing users to navigate screens and interact with elements. They are useful for refining the visual aspects, evaluating user flows, and conducting user testing. This approach reduces initial costs and speeds up the development cycle, enabling more frequent revisions and enhancements based on simulations and virtual testing. However, digital prototypes may not fully capture the physical aspects of a design or the tactile interactions in certain products. They may also require expertise in design software or coding. Examples include interactive digital prototypes created with prototyping software like Figma (www.figma.com), Adobe XD (www.adobexdplatform.com), Miro (www.miro.com), or InVision (www.invisionapp.com).
3. *Hybrid prototype*: Hybrid prototyping combines different methods to leverage the strengths of each approach. It allows for comprehensive

and versatile prototypes that cater to both physical and digital aspects. Hybrid prototyping may require additional resources, expertise, and coordination between different prototyping methods. It may introduce complexity and increase the overall time and cost of the prototyping process.

7.3.1.3 PURPOSE AND FUNCTION

Prototypes can also be grouped based on their function and intended use. These include the following:

1. *Conceptual prototypes*: These prototypes focus on exploring and communicating initial design concepts and ideas (Carfagni et al., 2020). They are often low-fidelity and aim to gather feedback and validate design directions.
2. *Functional prototypes*: These prototypes aim to simulate the behavior and functionality of the final product or system. They focus on testing complex interactions, algorithms, or technical feasibility.
3. *Visual prototypes*: Depending on the product and services, visual prototypes can emphasize the visual aesthetics and branding elements (typography, color schemes, and layout) or CMF (color, material, and finish) (Liu, 2020). CMF prototyping is a process used in product design and development to create physical representations of a product's desired color, material, and finish characteristics. It allows designers and manufacturers to evaluate and refine these aspects before moving into full-scale production.
4. *Usability prototypes*: These prototypes emphasize the user experience and interaction with the design. They test and gather feedback on usability, navigation, and user flows.

7.3.2 PROTOTYPING TECHNIQUES

7.3.2.1 SKETCHING

Although napkin sketches are not a physical prototype, they are a valuable tool for designers: the ability to visually express product ideas without requiring significant time or resources (Suwa & Tversky, 1997). They are quick and easy to create and discard, allowing designers to exercise their creativity and explore the potential functionalities the idea may have. The disadvantage of this method is that it needs to be more precise and effective to depict the functionalities accurately. Figure 7.1 illustrates the initial concepts for an earphone.

FIGURE 7.1 Sketches of earphones.

7.3.2.2 PAPER PROTOTYPING

Paper prototyping is one of the most affordable methods. It is a typical low-fidelity prototyping technique (Snyder, 2003; Rante & Basuki, 2019; Sulaiman et al., 2019). While sketching is used to quickly explore and communicate design ideas, paper prototyping goes a step further by simulating user interaction with the design. Unlike simple sketches, paper prototypes are structured to represent screens, buttons, and flows, allowing for early usability testing without any digital tools. It involves sketching, using sticky notes, or printing design concepts on paper and using them to simulate interactions (Figure 7.2). Paper prototyping is helpful for early-stage ideation and concept exploration (Snyder, 2003). It is commonly used in interaction design.

7.3.2.3 HANDMADE 3D PHYSICAL MODELING

Handmade 3D physical modeling is a traditional method to visualize ideas. Many materials can be used for making a physical model, such as cardboard, foam, clay, plastic, and wood.

Using cardboard or foam materials to create 3D models is cost-effective (Hallgrimsson, 2012; Stickdorn et al., 2018). Cardboard can be easily sourced, and foam boards or foam blocks are relatively inexpensive. These materials can be cut, shaped, and assembled to simulate the form and basic functionality of a design. Similar to paper prototyping, cardboard or foam prototyping does

FIGURE 7.2 Paper prototype of a phone interface.

not require any specific technology. It relies on manual cutting and assembly techniques. Cardboard or foam prototyping is often employed to create physical representations of designs (Figure 7.3). It is suitable for evaluating form, ergonomics, and basic functionality. It is commonly used in product design, industrial design, and architecture.

Clay is a common method for drafting concepts, but it also can create very smooth shapes with fine details. Depending on the details of the model, designers may spend a long-time making clay models with great visual effects. It is common in car prototyping (Figure 7.4).

There are other handmade physical models, including vacuum forming. This method can create a rough shape based on a mold with a plastic sheet using thermoforming (Gruenwald, 1998). For example, a scaled car prototype can be easily created using a Plaster of Paris mold (Figure 7.5). It is a cost-effective method. However, it is generally limited to parts with simple geometries and does not provide as much detail as other forming processes like injection molding.

7.3.2.4 WIREFRAMING

Wireframing is a commonly used prototyping method in user experience (UX) design and web/app development. It helps define the structure and layout of

FIGURE 7.3 Cardboard prototypes of lampshades.

FIGURE 7.4 Clay prototype of a BMW (from museum in Munich).

FIGURE 7.5 Vacuum forming of a scaled car model.

a digital interface, focusing on information hierarchy, navigation, and content placement (Figure 7.6). Wireframing can have low- or high-fidelity and is commonly used at the early stages of design to explore and communicate ideas quickly (Stickdorn et al., 2018; Sutipitakwong & Jamsri, 2020; Gudoniene et al., 2023). Wireframing involves creating basic visual representations of a design, like a blueprint. Digital wireframing uses software tools like Sketch (www.sketch.com/), Adobe XD (www.adobe.com/), Figma (www.figma.com), or Balsamiq (https://balsamiq.com).

7.3.2.5 3D DIGITAL MODELING

Designers often use 3D digital models instead of physical ones because they allow for rapid and easy modifications (Figure 7.7). Creating 3D digital models usually involves the use of specialized software. High-end applications like Autodesk (www.autodesk.com), Blender (www.blender.org), ZBrush (www.maxon.net/en/zbrush), SketchUp (www.sketchup.com), Rhino (www.rhino3d.com/), and SolidWorks (www.solidworks.com/) demand not only strong computing power but also a solid foundation in 3D modeling techniques. Such software is commonly used to design complex mechanical and architectural components. 3D models allow designers to assess dimensions, functionality, assembly processes, and system behavior, while also enabling analysis of physical properties through methods such as finite element analysis (FEA). These tools are widely applied in industries like automotive, aerospace, and engineering. High-fidelity and intricate 3D digital models can also be fabricated using 3D printing technologies (Section 7.3.2.6).

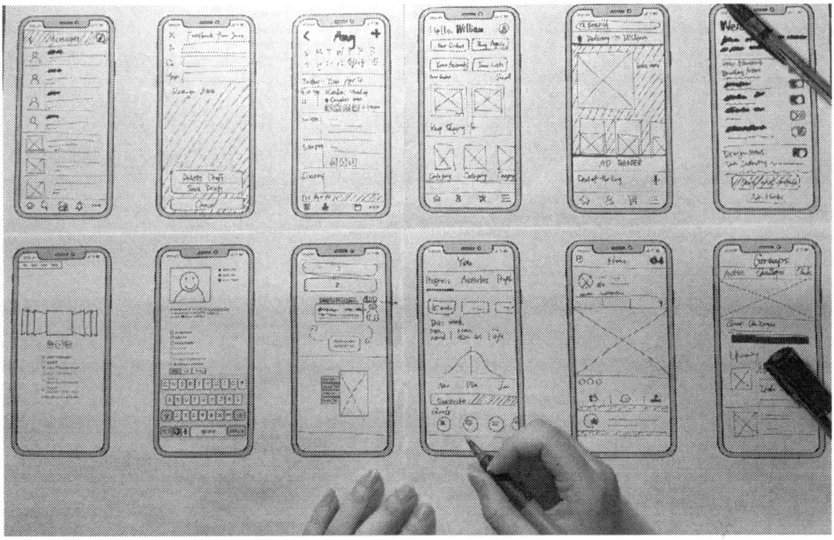

FIGURE 7.6 Wireframes of a phone app.

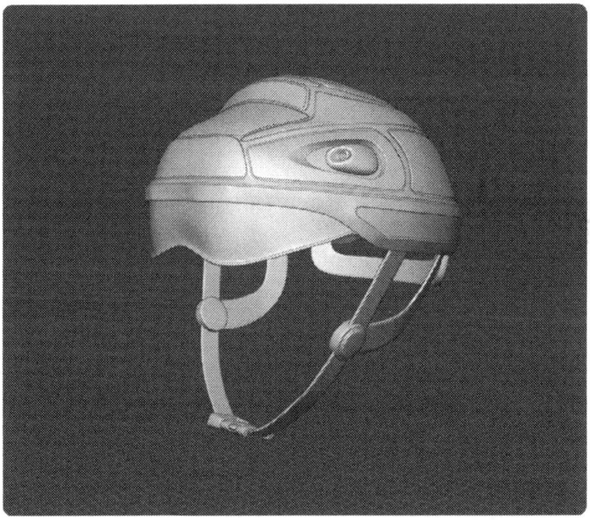

FIGURE 7.7 3D digital model of a helmet using Blender and Rhino software.

7.3.2.6 3D PRINTING

3D printing is an advanced, rapid, and precise prototyping technique based on 3D modeling (Mathias et al., 2019; Parry et al., 2023). 3D printing requires access to a 3D printer, which can be desktop or industrial-grade. 3D printing

FIGURE 7.8 3D printing of a helmet using ABS.

begins with a digital design created using 3D modeling software, which is then exported in a compatible file format such as STL or OBJ. Before printing, the design file is processed by a slicing software, which converts the model into thin horizontal layers and generates the instructions (G-code) the printer will follow. The 3D printer, connected to a computer or network, receives this sliced file and builds the object by depositing material layer by layer. This additive process transforms a virtual model into a physical, three-dimensional prototype. The cost of 3D printing can vary depending on the size, complexity, and material used. Fused deposition modeling (FDM) machines are cost-effective, ranging from a few hundred to several thousand dollars, and can produce durable prototypes with moderate detail and strength. Acrylonitrile butadiene styrene (ABS) and polylactic acid (PLA) are the most used 3D printing thermoplastic materials (Figure 7.8). While the initial investment on a 3D printer can be significant depending on the resolution and the material printed, the cost per print can be relatively low for small to medium-sized prototypes. It may become more expensive for larger or more intricate designs, as they require more time and material. The choice of material, ranging from plastics, resins, and composites to metals, depends on the complexity of the design and the required precision. For example, higher-precision prototypes may require advanced materials such as metal powders or engineering-grade plastics. Metal 3D printing is a specialized form of additive manufacturing that creates strong, durable, and heat-resistant prototypes. It is especially valuable for industries like aerospace, automotive, and healthcare, where functional testing of components is critical. Metal 3D printing typically uses selective laser melting (SLM) or electron beam melting (EBM) to create parts with fine details and complex geometries.

3D printing is valuable for creating physical prototypes with intricate designs, complex geometries, or fine details. It allows designers to test a design's physical form, fit, and function. It is commonly used in product design, fashion, healthcare, and architecture, enabling rapid iteration and innovation.

7.3.2.7 MOCKUP PROTOTYPING

Mockup prototyping refers to the creation of a non-functional, visual representation of a product that shows what it looks like, not necessarily how it works. It is mainly used to evaluate aesthetics, layout, form, and user interface, rather than functionality or performance. Mockups can be low- or high-fidelity. High-fidelity mockups resemble the final prototype, but they too lack any functionality. Mockups can be in physical or digital form and can be in 2D or 3D.

Physical 2D mockups, often created through drawings, are time-consuming and have limited practical use. In contrast, digital 2D mockups are more widely used and require software like Adobe Photoshop (www.adobe.com), Adobe InDesign (www.adobe.com), Blender (www.blender.org), Canva (www.canva.com/), or Figma (www.figma.com). These digital mockups enable designers to craft high-fidelity, interactive, and clickable prototypes, which are essential in user experience (UX) design and web/app development (Baroi & De, 2021; Dave et al., 2021). These mockups can simulate the appearance and basic interactions (Figure 7.9). They are very useful for gathering feedback on the overall user experience and refining the interface design for building functional prototypes.

3D mockups are a widely used technique in product design to showcase the visual appearance and shape of a prototype (Figure 7.10). The production of physical 3D mockups can incur significant costs, and certain designs may be impractical or infeasible to construct manually. Therefore, digital 3D mockups or 3D mockups based on a 3D printed model is quite often used. Digital 3D models use rendering software such as Autodesk (www.autodesk.com), Lumion 3D (https://lumion.com/), SketchUp (www.sketchup.com/en), KeyShot (www.keyshot.com) and Artlantis (https://artlantis.com/en/homepage/).

7.3.2.8 INTERNET OF THINGS (IOT) PROTOTYPING

An IoT prototype is an early model or version of a device or system that demonstrates the basic functionality and concept of an IoT solution (Figure 7.11). It typically includes a mix of hardware components like sensors, actuators, microcontrollers, communication modules (such as Wi-Fi, Bluetooth, or cellular), and software to process and transmit data (Karvinen & Karvinen, 2018).

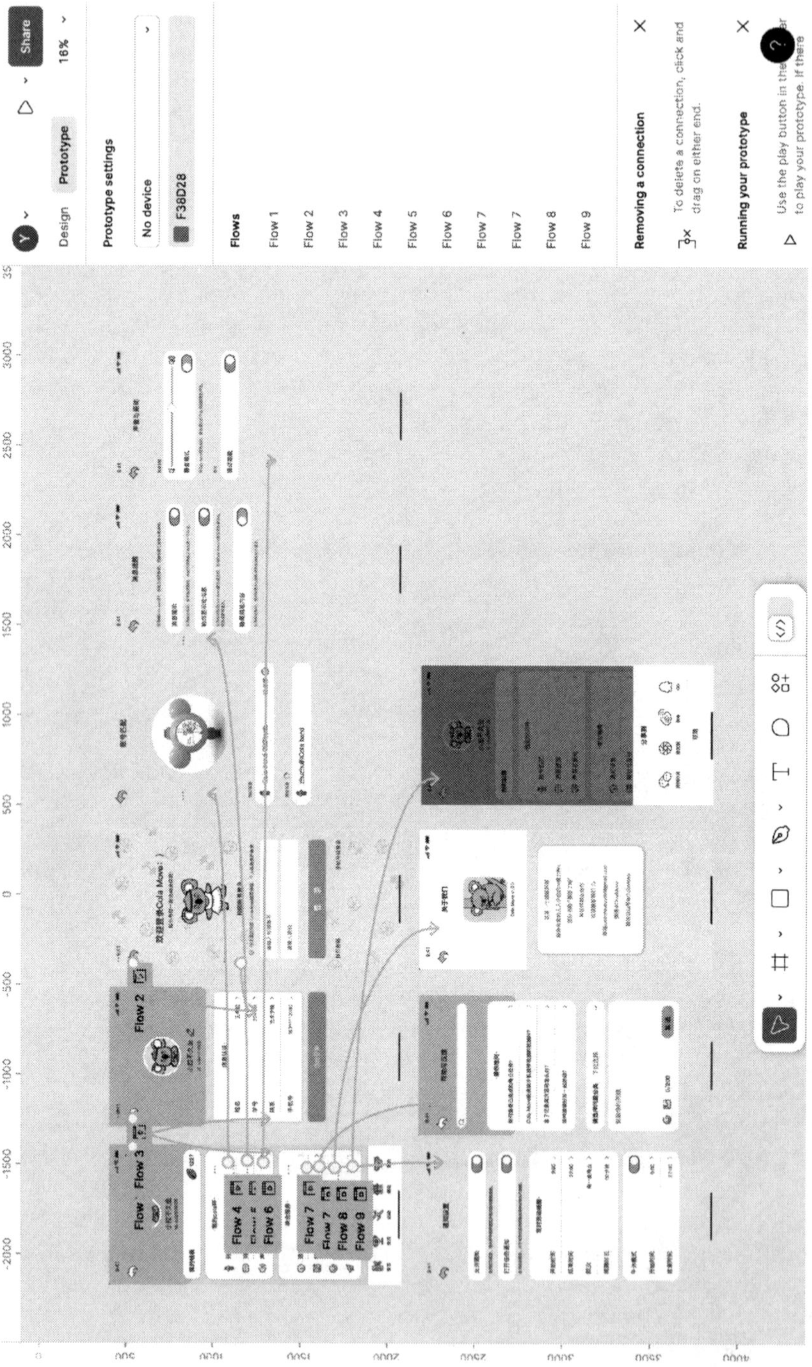

FIGURE 7.9 Digital 2D mockup of a phone app using Figma software.

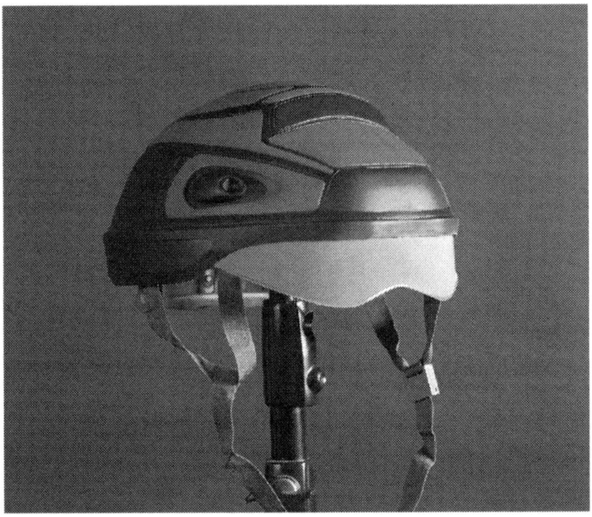

FIGURE 7.10 3D mockup of a helmet based on a 3D printed model.

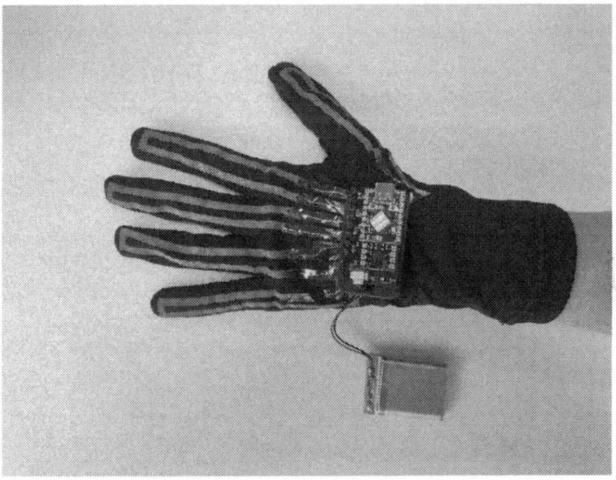

FIGURE 7.11 IoT model of a glove for human–computer interaction.

These models are very rough and the purpose is to demonstrate the core functionalities of the intended IoT device, such as data collection, processing, communication, and user interaction. The performance, reliability, and scalability can be tested and technical challenges and user experience issues may be identified during prototyping.

Various tools and platforms are available to aid in IoT prototyping, such as Arduino, Raspberry Pi, and cloud services like Amazon Web Services (AWS)

IoT or Azure IoT. These platforms provide hardware and software resources to build and test IoT solutions quickly.

7.3.2.9 PRE-PRODUCTION PROTOTYPING

At the late design stage, pre-production prototypes are created to test the functionality of the design and feasibility of the manufacturing process. The types in this stage are casting, injection molding, advanced technologies such as Computer Numerical Control (CNC) machining and laser cutting.

Injection molding is a process used for mass-producing plastic parts and, in some cases, parts made from other materials such as elastomers and metals (Figure 7.12). This process involves creating a mold and injecting molten material, such as plastics, metals, or composites, into it. The choice of material depends on the intended application, desired mechanical properties, and the complexity of the design. For example, engineering plastics are often chosen for their strength and durability, while softer polymers may be used for flexibility or cost efficiency.

Injection molding is generally expensive, as it requires molding machines and the creation of custom molds, which can be time-consuming and costly. It requires expertise in mold design and operation to ensure accuracy and efficiency. This method is typically used in the later stages of prototyping when a high degree of precision, durability, and functionality is required. Prototypes created through injection molding are typically high-fidelity and closely

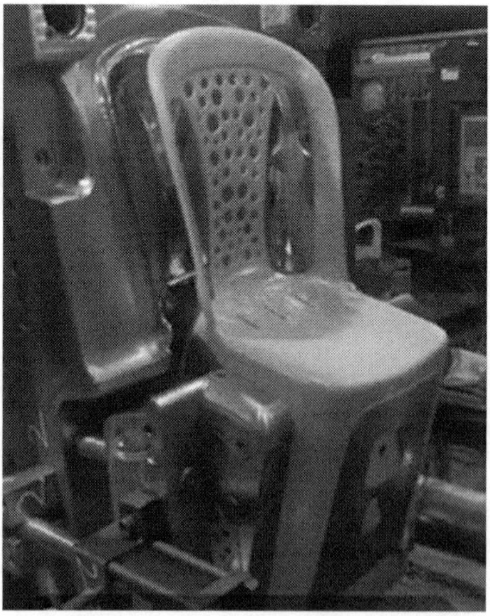

FIGURE 7.12 Injection molding of a plastic chair (from www.youtube.com/watch?v=tUgi2PP3zWc).

FIGURE 7.13 Casting. (From www.dreamstime.com/.)

resemble the final product. It is widely employed in industries like retail goods, electronics, and medical devices.

Casting is a process in which a liquid material, commonly used with metals, is poured into a mold that contains a hollow cavity of the desired shape, and then allowed to solidify (Figure 7.13). The solidified part, known as a casting, is then ejected or broken out of the mold to complete the process. Casting mold can be made from various materials such as sand, wax, plaster, steel, ceramics, and graphite. Casting is used to create complex shapes that would be difficult or uneconomical to make by other methods. It may not have high precision or fine detail.

Advanced technologies like Computer Numerical Control (CNC) machining and laser cutting provide high-fidelity prototyping capabilities (Nafis et al., 2012). They allow for testing manufacturability, performance, and functionality at scale. CNC machining is a subtractive manufacturing method that utilizes computer-controlled machines to cut, drill, and shape materials (Nafis et al., 2012). CNC machining utilizes sophisticated equipment and materials and a longer lead time, contributing to higher costs. It may not be suitable for early iterations or quick design changes. They are often used for complex shapes, intricate details, or when working with materials like metal or acrylic. For example, CNC milling uses computer-controlled machines to remove material and shape the prototype. It requires access to a CNC milling machine, which is connected to a computer that runs software for programming and controlling the milling process. It is commonly used in industries like automotive, aerospace, and mold making. Laser cutting involves using computer-controlled lasers to cut

FIGURE 7.14 CNC laser cutting. (From www.youtube.com/watch?v=OYCRggxN084.)

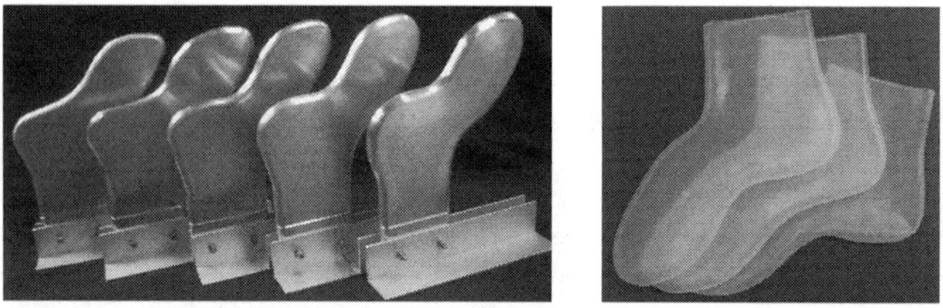

(a) (b)

FIGURE 7.15 Prototypes made from a combination of methods: (a) aluminum molds made using a CNC machine; (b) silicone "socks" made from the aluminum molds (Au, 2008).

and engrave materials. It requires access to a laser cutting machine, which is connected to a computer running software for design and control. Laser cutting is useful for creating precise, intricate, and detailed prototypes from various materials like wood, acrylic, or fabric (Figure 7.14). It is employed in fields like industrial design, fashion, signage, and model making.

In some cases, achieving the desired prototype necessitates the integration of multiple prototyping methods. For example, to assess sensitive regions of the foot and ensure proper footwear fit, silicone insoles (KE-1300T from Shin-Etsu Chemical Co., Ltd. and Elastosil® M 4600 from Wacker Silicones) have been used (Figure 7.15) (Au et al., 2012). Silicone is poured into CNC-cut aluminum molds and slowly rotated to ensure uniform wall thickness. By varying the number of layers and the type of silicone, the sock thickness and the stiffness can be changed, allowing researchers to conduct experiments and determine the necessary allowances for customized footwear, such as ski boots.

When preparing a pre-production prototype, it is crucial to consider design rules set by the manufacturer to ensure the prototype aligns with production constraints. Adhering to these guidelines help streamline the design for manufacturing (DFM) process, reducing the need for major design modifications before full-scale production. By incorporating DFM principles early, designers can avoid costly rework, improve manufacturability, and accelerate the transition to mass production. For a deeper discussion on DFM and other key design considerations, refer to Chapter 9.

7.3.2.10 PROTOTYPING THROUGH EMERGING METHODS

Designs like environmental layouts and specific interactive elements can benefit from using new visualization or simulation tools.

Virtual reality (VR), augmented reality (AR), and mixed reality (MR): VR, AR, and MR prototypes usually require VR/AR hardware and software systems and modeling knowledge and skills (Buruk & Hamari, 2021; De Crescenzio et al., 2019; Freitas et al., 2020). VR prototypes provide immersive, three-dimensional experiences within a virtual environment. The virtual environment is typically created in 3D software and displayed on VR headsets (e.g. Meta Quest and HTC Vive) (Figure 7.16). The common software includes Unity (https://unity.com/)

FIGURE 7.16 VR prototype.

and Unreal Engine (www.unrealengine.com/). VR prototypes are commonly used for testing and refining user experiences, including games, educational tools, training simulations, or other types of experiences. It can create interactive prototypes that mimic the behavior of users, such as swiping, tapping, pinching, and scrolling, to understand how users might interact with the interface. These interactions sometimes require additional equipment like gloves or controller.

For AR prototypes, virtual elements are overlayed onto the real world, allowing users to interact with them in their physical environment. AR glasses and headsets like Microsoft HoloLens and Magic Leap can provide AR experiences by projecting digital content directly into the user's field of view. Common software has Unity (https://unity.com/), Unreal Engine (www.unrealengine.com/), ARKit (https://developer.apple.com/augmented-reality/arkit/), and ARCore (https://developers.google.com/ar). AR prototypes are useful for testing and refining user experiences in augmented reality applications such as games, e-commerce, healthcare, training, and navigation. Mixed reality (MR) is a cutting-edge technology that combines elements of VR and AR. Devices such as the Microsoft HoloLens are expanding opportunities in the gaming and other industries by offering mixed reality experiences. Users can interact with virtual objects, interfaces, and the real world together. For example, you can test the furniture fitting by putting a virtual sofa in a real living room.

Voice user interface (VUI): VUI prototyping involves creating interactive prototypes that simulate voice interactions between users and a digital system or device such as Apple Siri. It uses voice recognition technology and natural language processing (NLP) to interpret and understand the spoken context. This prototype technique allows designers to explore and test the user experience of voice-controlled interfaces and gather feedback to refine the design.

There are other prototyping techniques that are not covered in this chapter. Table 7.1 provides an overview of various prototyping techniques, categorized by their fidelity, cost, technology requirements, purpose and function, medium, and the design stage they are best suited for. It serves as a practical guide for designers to choose the most appropriate prototyping techniques based on project needs, available resources, and the stage of the design process. By comparing key attributes, such as the level of detail (fidelity), the required investment (cost), and specific applications, this summary helps streamline the prototyping approach to maximize efficiency and effectiveness in design exploration and validation.

7.4 HOW TO PLAN AN EFFECTIVE PROTOTYPING PROCESS?

A well-planned prototyping process is essential due to the wide array of available techniques, as it enables designers to navigate through the options and select the most suitable approach for their project. A step-by-step process is described

TABLE 7.1 An Overview of Various Prototyping Techniques

	Design Stage			Purpose and Function				Fidelity			Technology Requirements			Cost		
Prototyping techniques	**Early**	**Middle**	**Late**	**Concep-tual**	**Visual**	**Func-tional**	**Usability**	**Low**	**Medium**	**High**	**Low**	**Medium**	**High**	**Low**	**Medium**	**High**
Sketching	✓			✓				✓			✓			✓		
Paper prototyping	✓			✓				✓			✓			✓		
Handmade 3D physical modeling	✓	✓		✓	✓	✓	✓	✓	✓		✓	✓		✓	✓	
Wireframing	✓	✓		✓		✓		✓	✓			✓		✓	✓	
3D digital modeling		✓			✓	✓	✓		✓	✓		✓	✓		✓	✓
3D printing		✓	✓		✓	✓	✓		✓	✓			✓		✓	✓
Mockup prototyping		✓			✓	✓	✓		✓	✓		✓	✓		✓	✓
IoT prototyping		✓	✓			✓	✓		✓	✓		✓	✓		✓	✓
Pre-production prototyping			✓		✓	✓	✓			✓			✓			✓
Prototyping through emerging methods (VR, AR, VUI)	✓	✓	✓		✓	✓	✓	✓	✓	✓		✓	✓		✓	✓

here to help plan the most effective prototyping process. Table 7.1 could be used as a reference for selection.

Step 1: Assess the stage of the design
Consider the stage of your design process, as different prototyping methods are better suited for various phases. Choose the one that best aligns with the current stage of your design. In the early stages, you may focus on exploring and refining ideas; in later stages, you may be more concerned with validating specific aspects. Low-fidelity techniques like sketching or paper prototyping are suitable for early ideation and concept exploration. In contrast, high-fidelity techniques like 3D printing or pre-production prototypes provide more accurate representations of the final product. They are more suitable for later design stages.

Step 2: Define prototyping goals
Clearly defining what you want to achieve with your prototypes. Think of whether you are aiming to explore the concepts, test functionality and technical feasibility, evaluate user interactions, validate the physical form, or gather feedback on aesthetics? For example, a mockup can help with the evaluation of appearance. Understanding your prototyping goals will guide your decision-making process.

Step 3: Consider resource constraints
Evaluate the resources available to you, including budget, time, materials, and expertise. Some prototyping methods may require specialized equipment, skills, or expensive materials, while others may be more accessible and cost-effective. Choose methods that align with your available resources.

Step 4: User involvement
Determine the level of user involvement required. If user feedback is crucial, opt for prototyping methods allowing user testing and observation such as digital 2D mockups and pre-production prototypes. Usability tests can help validate design decisions and ensure the final product meets user needs.

Step 5: Combine methods
Be open to using multiple prototyping methods. Prototyping combining different methods to capture various aspects of your design is often beneficial. Start with easy and low-fidelity methods to explore ideas, then progress to higher fidelity as your design evolves.

Step 6: Test and refine
Remember that prototyping is meant to gather feedback and refine your design. Be ready to iterate and make changes based on the insights gained from evaluating and testing your prototypes. The chosen prototyping methods should allow for flexibility and easy modifications.

This six-step prototype planning framework, as shown in Figure 7.17, provides a structured approach to selecting and adapting prototyping

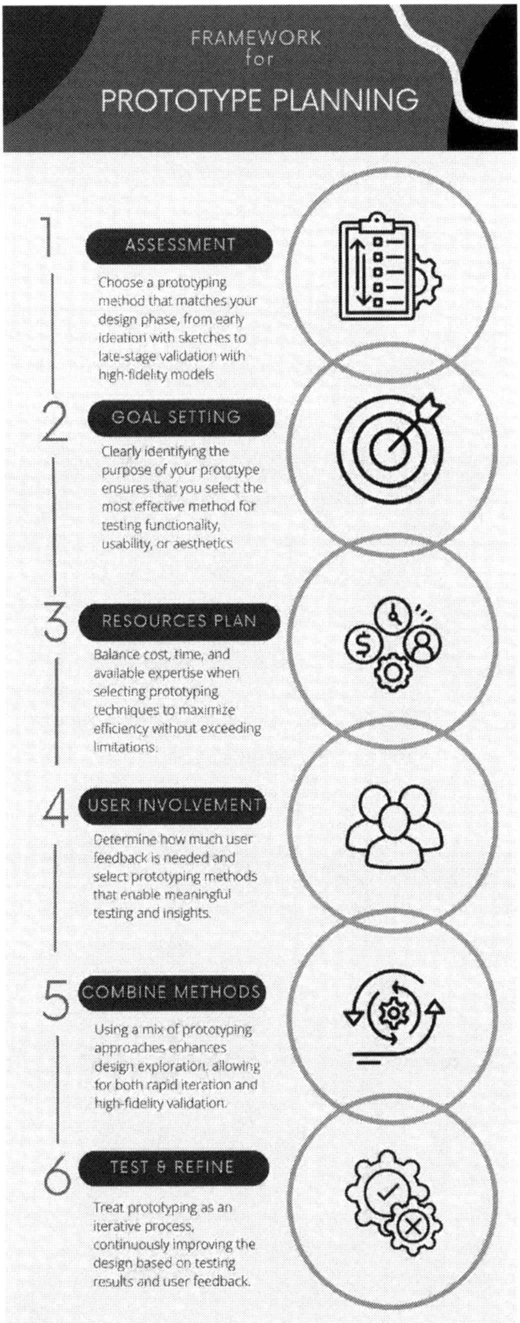

FIGURE 7.17 Framework for prototype planning.

methods. It emphasizes effective communication among stakeholders, enabling shared understanding and feedback. By aligning prototyping methods with project goals, available resources, the design stage, and the level of user involvement, designers can make informed choices. Prototyping is a dynamic and iterative process, and you can adapt your prototyping methods as your design evolves.

7.5 CONCLUSION AND CHALLENGES IN PROTOTYPING

This chapter provides a practical and easy-to-understand overview of prototyping in human-centered design. By grasping the concepts and insights presented here, designers can improve their ability to create innovative solutions that truly meet the needs of their users.

Prototyping is an essential part of the design process because it offers several important benefits. By incorporating prototyping into the design process, designers can visualize and evaluate their ideas, refine designs based on user feedback, reduce risks, and create more human-centered and effective solutions. Prototyping enables designers to validate design decisions, address potential issues early on, and increase the chances of success when the final product is developed.

While prototyping offers numerous benefits, it also has certain challenges including time and cost, limited functionality, scale and manufacturing constraints, interpretation and communication, limited user feedback, material and physical constraints. It is important to note that these limitations are not absolute constraints, and designers often find ways to address them creatively. Despite its limitations, prototyping remains an invaluable tool in the design process before committing to the final product.

> Design is not just what it looks like and feels like. Design is how it works.
>
> —Steve Jobs

REFERENCES

Au, Y. L. (2008). Psychophysics of footwear fit and a methodology for needs elicitation. PhD thesis. HKUST Electronic Theses.

Au, E. Y. L., Goonetilleke, R. S., Witana, C. P., & Xiong, S. (2012). A methodology for determining the allowances for fitting footwear. *International Journal of Human Factors Modelling and Simulation*, 2(4), 341–366. https://doi.org/10.1504/IJHFMS.2011.045003.

Baroi, I., & De, S. (2021). A novel application of neuromarketing for designing user interface mockups to enhance user experience in software development. In 2021 IEEE 2nd International Conference on Technology, Engineering, Management for Societal Impact Using Marketing, Entrepreneurship and Talent (TEMSMET) (pp. 1–6). IEEE.

Buruk, O. T., & Hamari, J. (2021). Immersive video sketching: Low-fidelity extended reality prototyping for everyone. In Proceedings of the 24th International Academic Mindtrek Conference (pp. 165–175).

Camburn, B., Viswanathan, V., Linsey, J., Anderson, D., Jensen, D., Crawford, R., Otto, K., & Wood, K. (2017). Design prototyping methods: state of the art in strategies, techniques, and guidelines. *Design Science*. 3, e13. https://doi.org/10.1017/dsj.2017.10

Carfagni, M., Fiorineschi, L., Furferi, R., Governi, L., & Rotini, F. (2020). Usefulness of prototypes in conceptual design: students' view. *International Journal on Interactive Design and Manufacturing (IJIDeM)*, 14, 1305–1319.

Coutts, E. R., Wodehouse, A., & Robertson, J. (2019). A comparison of contemporary prototyping methods. *In Proceedings of the Design Society: International Conference on Engineering Design* (Vol. 1, No. 1, pp. 1313–1322). Cambridge University Press.

Dave, H., Sonje, S., Pardeshi, J., Chaudhari, S., & Raundale, P. (2021). A survey on Artificial Intelligence based techniques to convert User Interface design mock-ups to code. In 2021 International Conference on Artificial Intelligence and Smart Systems (ICAIS) (pp. 28–33). IEEE.

De Crescenzio, F., Bagassi, S., Asfaux, S., & Lawson, N. (2019). Human centred design and evaluation of cabin interiors for business jet aircraft in virtual reality. *International Journal on Interactive Design and Manufacturing (IJIDeM)*, 13, 761–772.

dos Santos, T. B., Campese, C., Marcacini, R. M., Sinoara, R. A., Rezende, S. O., & Mascarenhas, J. (2021). Prototyping for user involvement activities: How to achieve major benefits. *CIRP Journal of Manufacturing Science and Technology*, 33, 465–472.

Fiorineschi, L., & Rotini, F. (2019). Unveiling the multiple and complex faces of fidelity. In *Proceedings of the Design Society: International Conference on Engineering Design* (Vol. 1, No. 1, pp. 1723–1732). Cambridge University Press.

Freitas, G., Pinho, M. S., Silveira, M. S., & Maurer, F. (2020). A systematic review of rapid prototyping tools for augmented reality. In 2020 22nd Symposium on Virtual and Augmented Reality (SVR) (pp. 199–209). IEEE.

Gruenwald, G. (1998). *Thermoforming: A Plastics Processing Guide*, 2nd edition. Routledge, Taylor & Francis Group.

Gudoniene, D., Staneviciene, E., Buksnaitis, V., & Daley, N. (2023). The scenarios of artificial intelligence and wireframes implementation in engineering education. *Sustainability*, 15(8), 6850.

Hallgrimsson, B. (2012). *Prototyping and Modelmaking for Product Design*. Hachette UK.

Hammami, H., Camara, F. G., Calvary, G., Riahi, M., & Moussa, F. (2020). The benefits of combining paper-and video-based prototypes for user interface evaluation. ACHI 2020: The Thirteenth International Conference on Advances in Computer-Human Interactions.

Isa, S. S. B. (2023). *Enhancing Creativity through Low-Fidelity Prototyping: Strategies and Approaches for Idea Generation and Concept Formation*. PhD thesis, Norwegian University of Science and Technology (NTNU), Norway.

Karvinen, K., & Karvinen, T. (2018). IoT rapid prototyping laboratory setup. *International Journal of Engineering Education*, 34(1), 263–272.

Lang, F., & Mjöberg, A. (2020). Prototyping as a requirements engineering technique. LU-CS-EX. Master thesis, Lund University, Sweden.
Liu, Y. (2020). Research on the value of CMF design in industrial products. DRS Biennial Conference Series. DRS 2020, Synergy, Australia.
Mathias, D., Snider, C., Hicks, B., & Ranscombe, C. (2019). Accelerating product prototyping through hybrid methods: Coupling 3D printing and LEGO. *Design Studies*, 62, 68–99.
McElroy, K. (2016). *Prototyping for Designers: Developing the Best Digital and Physical Products*. O'Reilly Media.
Moon, A. (2022). Creating prototypes: products of the design thinking process (Master's thesis, SUNY Polytechnic Institute).
Nafis, O.Z.M., Nafrizuan, M.Y., Munira, M.A., & Kartina, J. (2012). Review on CNC-rapid prototyping. *IOP Conference Series: Material Science Engineering* 36, 012032.
Otto, K., & Wood, K. (2001). *Product Design: Techniques in Reverse Engineering and New Product Design*. Prentice-Hall.
Parry, G.R., Felton, H.J., Ballantyne, R., Su, S., & Hicks, B. (2023). Reducing prototype fabrication time through enhanced material extrusion process capability. *Proceedings of the Design Society* 3, 3025–3034. https://doi.org/10.1017/pds.2023.303
Real, R., Snider, C., Goudswaard, M., & Hicks, B. (2021). Dimensions of knowledge in prototyping: a review and characterisation of prototyping methods and their contributions to design knowledge. *Proceedings of the Design Society*, 1, 1303–1312.
Rante, H., & Basuki, A. (2019). The role of paper prototyping in designing visual novel game as learning media for children. In 2019 5th International Conference on Education and Technology (ICET) (pp. 24–28). IEEE.
Ranscombe, C., Bissett-Johnson, K., Mathias, D., Eisenbart, B., & Hicks, B. (2020). Designing with LEGO: exploring low fidelity visualization as a trigger for student behavior change toward idea fluency. *International Journal of Technology and Design Education*, 30, 367–388.
Snyder, C. (2003). *Paper Prototyping: The Fast and Easy Way to Design and Refine User Interfaces*. Morgan Kaufmann, Elsevier Science.
Starkey, E. M., Menold, J., & Miller, S. R. (2019). When are designers willing to take risks? How concept creativity and prototype fidelity influence perceived risk. *Journal of Mechanical Design*, 141(3), 031104.
Stickdorn, M., Hormess, M., Lawrence, A., & Schneider, J. (2018). *This Is Service Design Doing: Applying Service Design Thinking in the Real World*. 1st Edition. O'Reilly Media.
Sulaiman, S., Musthafa, N. H., & Ab Aziz, N. (2019). Customizing interface design for users with impairment using a paper prototyping. *Platform: A Journal of Science and Technology*, 2(2), 33–39.
Sutipitakwong, S., & Jamsri, P. (2020). Pros and cons of tangible and digital wireframes. In 2020 IEEE Frontiers in Education Conference (FIE) (pp. 1–5). IEEE.
Suwa, M., & Tversky, B. (1997). What do architects and students perceive in their design sketches? A protocol analysis. *Design Studies*, 18(4), 385–403. https://doi.org/10.1016/S0142-694X(97)00008-2
Warfel, T. Z. (2009). *Prototyping: A Practitioner's Guide*. Rosenfeld Media.
Zhou, X., & Rau, P. L. P. (2019). Determining fidelity of mixed prototypes: Effect of media and physical interaction. *Applied Ergonomics*, 80, 111–118.

Prototype Selection and Testing

8

> The best design isn't necessarily an object, a space, or a structure: it is a process—dynamic and adaptable.
>
> —Don Norman

8.1 INTRODUCTION

Prototype selection and testing are pivotal in the design process, bridging initial prototypes and final products. Prototype selection is a vital step that helps designers assess the feasibility of the different prototype designs and identify flaws in the early design stage. It also reduces costs and fast-tracks the design process by allowing quick iterations of various prototypes.

Prototype testing can take various forms, from low-fidelity sketches to high-fidelity interactive models. Engaging real users in testing provides critical insights into usability and functionality, allowing designers to identify potential problems and areas for improvement. This feedback loop is crucial for refining prototypes, minimizing risks, and reducing the likelihood of costly changes during later development phases.

Testing with diverse participants guarantees that the designs cater to a wide range of users' and stakeholders' needs and enhance the product's overall effectiveness. Ultimately, the integration of prototype selection and testing advances a people-centered design approach, increasing the likelihood of delivering successful and impactful products that satisfy the needs and wants of target audiences. This iterative process allows designers to create solutions that meet functional requirements and enhance user satisfaction.

DOI: 10.1201/9781003476177-8

8.2 PROTOTYPE SELECTION METHODS

Effective prototype selection involves evaluating multiple design alternatives and determining which best addresses user needs, expectations, and project goals (Honkala et al., 2007). This process requires careful consideration of context, evaluation strategies, and decision criteria. The nature of this process enhances the quality and success of the final product. The prototype should appeal and add value to the users. Prototype selection also aids in eliminating implausible ideas and refining promising ones to maximize performance and minimize future problems with the product. It ensures the selected prototype can be built using the available technologies and procedures, and also helps funnel the resources, like time, cost, and effort, to the most promising directions (Figure 8.1).

It is imperative to establish clear selection criteria, consider different evaluation methods, and iteratively refine the prototypes to optimize performance. Table 6.10 used novelty, usefulness, feasibility, impact, originality, adaptability, practicality, marketability, sustainability, and alignment with objectives as the selection criteria for evaluating design ideas. Prototypes can be selected using

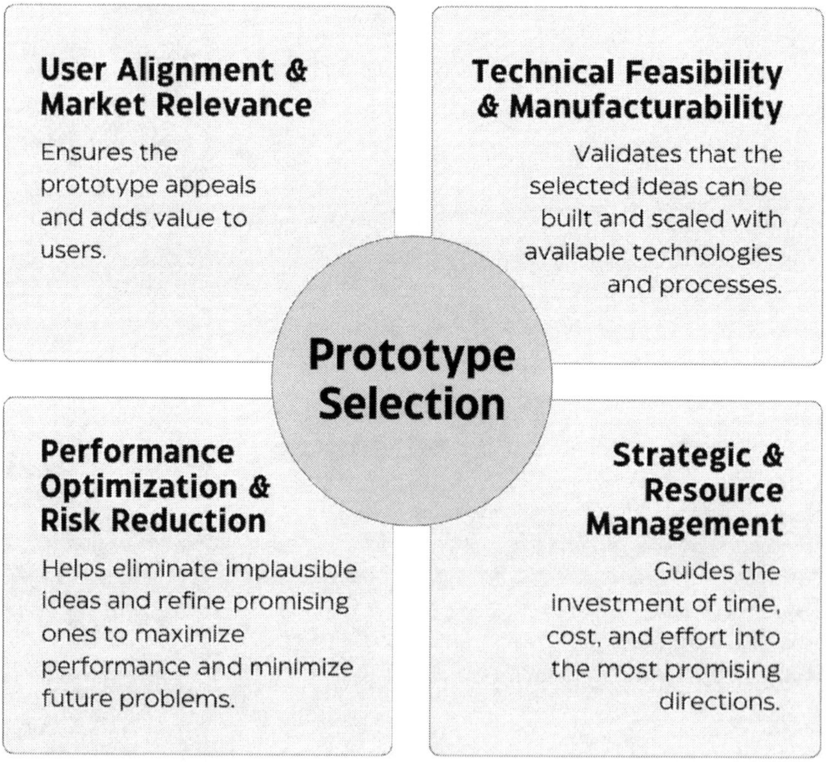

FIGURE 8.1 Reasons for prototype selection.

similar criteria with different weights. For example, depending on the context and the product, the weighting can be functionality 20 percent, usability 20 percent, sustainability 40 percent, and feasibility 20 percent.

8.2.1 METHODS FOR PROTOTYPE SELECTION

There is more than one way to choose a winning prototype; we give you five (see Table 8.1). Each method represents a unique mindset, ranging from quick gut-checks to data-heavy analysis. Whether you are going with simple prioritization for speed, criteria-based scoring for balance, strategic and risk-oriented selection for future-proofing, people-centered evaluation for empathy and usability, or AI/data-driven selection for computational insight, each pathway offers a distinct lens through which to evaluate your ideas and make smart, context-sensitive decisions.

Simple prioritization methods include dot voting, MoSCoW, consensus building, and prototyping workshops, to effectively filter ideas and foster team alignment during early design stages. Dot voting (IDEO, 2015) allows team members to express their preferences visually, facilitating quick consensus on the most promising prototypes. Consensus building (Kaner, 2014) promotes collaborative discussions, ensuring all voices are heard and considered in decision-making. The MoSCoW method (Clegg & Barker, 1994) is a prioritization technique with four categories of "must have", "should have", "could have", and "won't have". Prototyping workshops (Brown, 2009) encourage hands-on collaboration, allowing participants to quickly iterate and refine prototypes collectively, enhancing engagement and creativity.

Criteria-based scoring methods like the weighted scoring method, Pugh method, modified Pugh, Kepner–Tregoe decision analysis (KTDA), cost–benefit analysis, and using standards offer more structured, quantitative ways to compare alternatives based on predefined evaluation criteria. The weighted scoring model (Hassan et al. 2016; Project Management Institute, 2017) allows teams to assign importance to various criteria, facilitating importance scoring. The Pugh method (Pugh, 1991) helps identify the best options by comparing alternatives against a baseline, while the modified Pugh (Fung et al., 2007) enhances sensitivity through weight adjustments. Kepner–Tregoe decision analysis (KTDA) (Kepner & Tregoe, 1965) simplifies complex decisions by breaking them into manageable components. Cost–benefit analysis (Boardman et al., 2018) assesses financial feasibility, and with standards (example: ISO 9241; FDA (www.fda.gov/)) ensuring compliance with industry benchmarks, collectively enhancing informed decision-making and collaboration within design teams.

Strategic and risk-oriented methods, including SWOT (strengths–weaknesses–opportunities–threats) analysis, analytic hierarchy process (AHP), and the fuzzy Delphi method, help align design decisions with long-term objectives and market positioning. SWOT analysis (Humphrey, 2005) allows teams to identify strengths, weaknesses, opportunities, and threats related to their

TABLE 8.1 Prototype Selection Methods.

Category	Method	Method Description	Administration	
			Effort	References
Simple prioritization	Dot voting	Team votes using dots or stickers. Ideal for early-stage or low-stakes decisions.	*	IDEO, 2015; Gibbons, 2019
	MoSCoW method	Categorize items as Must have, Should have, Could have, or Won't have.	**	Clegg & Barker, 1994; Ahmad et al., 2017
	Consensus building	Group discussion to reach mutual agreement, supporting team alignment.	**	Kaner, 2014; Muller & Hiete, 2021
	Prototyping workshops	Collaborative sessions for rapid prototyping and team feedback.	***	Brown, 2009; IDEO Toolkit; Wegener et al., 2019
Criteria-based scoring method	Pugh method	Compare alternatives using + (positive), – (negative), and S (same). Useful for early elimination.	*	Pugh, 1991; Guler, 2021
	Modified Pugh method	Add weighting to the criteria in the original Pugh matrix, improving sensitivity.	**	Fung et al., 2007; Gutierrez et al., 2021
	Weighted Scoring Model	Assign weights to evaluation criteria and scores each prototype numerically. (similar to Chapter 6, Table 6.10)	**	Project Management Institute 2017; Hassan et al., 2022
	Kepner–Tregoe decision analysis (KTDA)	Separate 'musts' and 'wants,' then ranks and scores each. Useful for multi-factor choices.	**	Kepner & Tregoe, 1965; Parker & Moseley, 2008
	Cost–benefit analysis	Compare prototypes in terms of financial return vs. cost.	***	Boardman et al., 2018; Shen et al., 2019
	Using standards	Use regulatory or industry standards as benchmarks for compliance and evaluation.	***	ISO 9241; FDA; Gutierrez et al., 2014

Strategic and risk-oriented	SWOT analysis	Analyze strengths, weaknesses, opportunities, and threats of each prototype.	***	Humphrey, 2005; Niekurzak et al., 2023
	Analytic hierarchy process (AHP)	Use pairwise comparisons to prioritize criteria and alternatives.	*****	Saaty, 1980; Sudin et al., 2017
	Fuzzy Delphi method	Combine expert consensus (Delphi) with fuzzy logic to reduce ambiguity.	*****	Murry & Hammons, 1995; Hsu & Sandford, 2007; Yao et al., 2022
People-centered evaluation and selection	Kano model	Classify features as basic, performance, or excitement for user satisfaction.	****	Kano et al., 1984; Rotar & Kozar, 2017
	User-centric evaluation	Test prototypes with real users to uncover usability and emotional response.	****	Norman, 2013; Nielsen, 1994; Sreejith & Mathirajan, 2024
	Conjoint analysis	Analyze user trade-offs among feature sets using choice-based surveys.	*****	Green & Srinivasan, 1990; Salhieh & Al-Harris, 2014
AI/Data-driven selection	AI-Based concept evaluation	Use AI to simulate, score, and predict prototype performance and appeal.	**	Luo, 2025

Note: The effort required is indicated with "*" denoting easy and "*****" denoting more difficult.

designs, providing a comprehensive view of the competitive landscape. The analytic hierarchy process (AHP) (Saaty, 1980; Sudin et al., 2017) facilitates structured decision-making by breaking down complex problems into simpler components, allowing for prioritization based on multiple criteria. Meanwhile, the fuzzy Delphi method (Murry & Hammons, 1995; Hsu & Sandford, 2007) gathers expert opinions to achieve consensus on uncertain factors, enhancing the quality of decision-making. Together, these methods enable prototype selection based on informed and strategic information.

People-centered evaluation methods, rooted in human-centered design, emphasize understanding stakeholders' preferences, behaviors, and perceptions to create products that improve the lives of the targeted population. The Kano model (Kano et al., 1984) categorizes features based on how they influence user satisfaction, distinguishing between basic needs and those that delight users. User-centric evaluation (Norman, 2013; Nielsen, 1994) involves real users interacting with prototypes, providing invaluable feedback on usability and functionality. This direct engagement helps identify usability problems and rank and select different prototypes. Conjoint analysis (Green & Srinivasan, 1990; Srinivasan, 2000) explores user trade-offs among other features, revealing how preferences change based on varying combinations. These methods enable design selection, categorization, and ranking of different options.

Finally, AI/data-driven methods (Luo, 2025) utilize advanced predictive analytics and large-scale data processing to enhance selection. These methods analyze vast amounts of data from various sources, allowing teams to identify patterns and trends that might not be immediately apparent through traditional evaluation techniques. By leveraging machine learning algorithms, AI can predict user preferences and potential market success, providing knowledge that supports informed decision-making. This evidence-based approach enhances the reliability of selections, ensuring that the chosen prototypes align closely with user needs and expectations.

8.2.2 THE SELECTION PRINCIPLES

Prototype selection is not just a checkbox in the design process but a strategic moment of truth. It demands more than a quick pick; it requires a thoughtful blend of context awareness, sharp evaluation strategies, and clear decision criteria. The methods you choose or the creative mix you devise can shape the entire trajectory of your project, influencing not just what you build, but how well you understand its potential. In this section, we will unpack how to choose the right selection methods, when to combine them, how to make sense of what the evaluation tells you, and, perhaps most importantly, how to know when to stop selecting and start testing.

Choosing the right method: The prototype selection method should be aligned with the specific context and needs of the design project. Informal techniques like dot voting, MoSCoW, or consensus building are effective when the goal is quick consensus or prioritizing among numerous early-stage ideas. These methods are simple to administer and work well in collaborative settings. For more structured analysis, especially when comparing multiple alternatives based on diverse criteria, the Pugh method, weighted scoring model, or Kepner–Tregoe decision analysis provides a more rigorous framework. When emphasizing user insights and emotional response, people-centric evaluation, the Kano model, or conjoint analysis are ideal for validating desirability and user value. In contrast, projects requiring strategic decision-making or risk awareness may benefit from SWOT analysis, the fuzzy Delphi method, or the analytic hierarchy process (AHP), which are suited for high-level assessment and long-term planning. Finally, AI-based concept evaluation offers predictive analytics to support concept refinement in data-rich environments.

Using a single method vs. combining methods: While a single method can offer clarity and focus, most prototype selection processes benefit from combining multiple methods. For instance, a team might begin with dot voting to capture quick input and initial preferences from stakeholders, followed by a weighted scoring model to compare the shortlisted concepts using quantitative criteria. This can be complemented with a people-centric evaluation of the final few candidates to ensure the prototype meets user needs and expectations. Combining methods in this way enhances the robustness of the decision-making process by blending qualitative intuition, analytical structure, and real-world validation.

Evaluating the selection process: To ensure that the prototype selection process is effective, teams should periodically reflect on the clarity, alignment, and validity of their decisions. Key questions to ask include: Are the criteria well-defined and consistently applied? Do the selected concepts address the problem space and user needs? Is the process transparent if reviewed by external stakeholders? It is also essential to keep a record for knowledge management and future reference for the selection procedure, and to be used as a benchmark.

Knowing when to stop: Finally, has the team reached a point of diminishing returns, where further evaluation yields little to no new insight? When these questions can be confidently answered, the team can proceed knowing their process has produced a well-justified decision. Prototype selection should not become an endless loop of reconsideration. A clear stopping point is reached when the team has narrowed the field to one or two strong, viable prototypes that meet technical and user-defined requirements. When consensus is strong, the design aligns with project goals, and further iteration does not yield substantial new information, it is time to conclude the selection phase. Moving forward too early risks overlooking important alternatives, but delaying unnecessarily may waste valuable time and resources.

Transitioning from selection to testing: The process shifts to prototype testing once an "optimal" prototype has been selected. This transition begins with setting well-defined testing goals, whether to assess usability, functionality, emotional appeal, or technical performance. The team should then choose appropriate testing methods. Testing aims to improve the prototype and validate the selection decision to confirm that the selected prototype is ready for further development into commercial products.

8.3 PROTOTYPE TESTING

Before a product ever hits the shelves, it must prove itself in the real world, and that means testing. Prototype testing is not just a technical formality; it is a vital checkpoint to ensure your design works as intended, and more importantly, works for people. This stage examines everything from functionality and safety to aesthetics, usability, and the emotions your prototype stirs in users. Great design does not happen in isolation. That is why many teams bring lead users into the process, those passionate early adopters who push boundaries and give honest feedback. Their insights help spot flaws, refine features, and uncover hidden opportunities. By listening carefully and iterating early, design teams avoid costly missteps and move one step closer to creating a product that performs and succeeds.

This collaborative methodology engages the users in the design of the prototype and can build trust with the users. The fine-tuning and refinement of the prototype is valuable in making design decisions to guarantee the product meets the user's needs and requirements. User needs and requirements may sound similar but play different roles in great design. Think of user needs as the human side of the equation, what people hope to feel or experience. When shopping for shoes, for instance, we instinctively look for comfort, the right fit, maybe even a sense of style that matches who we are. We want shoes that do not pinch our toes or make us dread walking. User requirements, on the other hand, speak the language of functionality and performance. They are the specs behind the scenes, the shoe's durability, material quality, grip, or water resistance. In short, needs connect to emotions and expectations, while requirements translate those into measurable design targets. The magic happens when a product meets both. Prototyping and testing complement each other to create the best solution to address the defined problem (Figure 1.5).

8.3.1 PRINCIPLES RELATED TO PROTOTYPE TESTING

The first principle is that prototype testing should be goal-oriented. The goals set by the design team determine the appropriate measures to be collected during the user testing. What do you want to test, or what are you trying to test and discover? Knowing the end goal confirms that the designer gets the results needed and takes the necessary actions after the prototype testing.

The second principle is that prototype testing is iterative, as you will continuously test and gather feedback throughout the design process. Table 8.2 provides a guide depending on the prototype fidelity level, the testing objectives, characteristics, and who will test your prototype.

8.3.2 PROTOTYPE MEDIUM AND TEST FORMATS

By carefully planning and implementing prototype testing, designers can gain valuable insights into the designs, ultimately developing a delightful product.

TABLE 8.2 Fidelity Level and Testing Objectives.

Fidelity Levels	Examples	Test Objectives	Characteristics	Testers
Low-fidelity	Paper, sketches, clay, and wireframes.	Communicate and show the layout or outline of the interface (website or app) or the look of the product.	Involves imagination on the participants.	Members of the design team and target customers.
Medium-fidelity	Digital prototypes with basic interactivity, mockups with visual details and 3D printing.	Functionality and usability of the prototype.	Provides more details compared to low-fidelity prototypes and allows users to interact with the prototype.	Stakeholders and target customers.
High-fidelity	Interactive digital prototypes with realistic behavior, physical prototypes, or functional prototypes with near completed functionality (CNC) machined, injection molded, and other advanced manufacturing technologies.	Functionality, usability, safety and other technical performance.	High-fidelity prototypes are incredibly detailed, almost like the final product, and can be interactive. It involves more time to create and is more costly to build.	Stakeholders and target customers.

TABLE 8.3 Test Methods and Their Use When Testing Physical and Digital Prototypes.

Medium (McElroy, 2016)	Context	Test
Physical prototype	■ Evaluate the overall design and test the functionality, usability and other technical performance.	■ Usability testing ■ A/B testing ■ Breadboarding testing ■ Force testing ■ Decibel testing ■ Airflow testing ■ Dielectric testing ■ Compatibility testing (Biocompatibility testing and sterilization testing) ■ Packaging testing
Digital prototypes	Assess the user interface or navigation menu. Examine interactive UI elements such as drop-down menus and pop-ups. Review visual components like image quality, font size, and typography. Observe user tasks and interactions within a digital interface.	Usability testing A/B testing AR/VR testing Simulation testing

Table 8.3 summarizes key characteristics of physical and digital prototypes (McElroy, 2016) and the possible testing approaches.

8.3.3 USABILITY TESTING

Usability testing and user feedback are essential in people-centered design, offering valuable insights into a product or service's usability, satisfaction, and effectiveness (Nielsen, 2012a). Usability testing focuses on users' conscious interactions with a product. This includes evaluating ease of use, intuitiveness, and overall user experience (Nielsen, 1993; Albert & Tullis, 2022).

8.3.3.1 QUANTITATIVE VS. QUALITATIVE USABILITY TESTING

Quantitative testing provides numerical and quantifiable data such as task completion time, success rates, error counts, and user satisfaction ratings. It looks into trends and patterns in the data for further investigation. An example is where a mobile phone interface is evaluated. The participants are given different tasks, and their completion times are recorded.

Qualitative testing provides illustrative information from think-aloud protocols or open-ended feedback, like guerilla interviews. A guerrilla interview is used for low-fidelity prototypes, as you would want to get perceptions from users regarding the designs quickly and iteratively (Diederichs et al., 2020).

One that is qualitative and quantitative is behavior mapping. It is a structured approach combining different methods for documenting, mapping, and counting activities performed by people passing and occupying a defined space at a given moment (Dahlkvist et al., 2020).

8.3.3.2 IN-PERSON VS. REMOTE TESTING

In-person testing occurs in a controlled environment, such as a laboratory or the participant's work area. A researcher (facilitator or designer) asks participants to perform tasks using one or more specific user interfaces or products. While the participant completes each task, the researcher observes the participant's behavior and listens to feedback (Hornbaek, 2006). This helps the designer understand if the product meets their needs and if they find it modified.

Remote usability testing involves conducting the usability testing via video conferencing or other internet-based software (Hill et al., 2021; Thompson et al., 2004). This is applicable for both moderated and unmoderated usability testing. Software such as Figma (www.figma.com) makes remote testing cost-efficient and more accessible.

8.3.3.3 NEUROSCIENTIFIC APPROACHES

Neuroscientific approaches include techniques such as functional magnetic resonance imaging (fMRI) (Hay et al., 2022), electroencephalography (EEG) (Deng & Wang, 2019), and eye-tracking (Khalighy et al., 2015), to investigate users' cognitive and emotional responses during their interactions with a product or service (Zhang et al., 2024; Tholen et al., 2020; Lencastre et al., 2022; Wever et al., 2022). These methods offer insights into the underlying neural mechanisms that shape users' perceptions, preferences, and decision-making processes. By monitoring brain activity or eye movements, designers can unveil implicit reactions and preferences that may not be readily expressed through conventional self-report scales or behavioral measures. For example, a study by Wan and their team (2018), while studying cardboard products, found that consumers spent less than average fixation duration on cardboard prototypes that had familiar form and functionality. Such neuroscientific approaches provide a unique and objective perspective on user perception, enabling designers to make

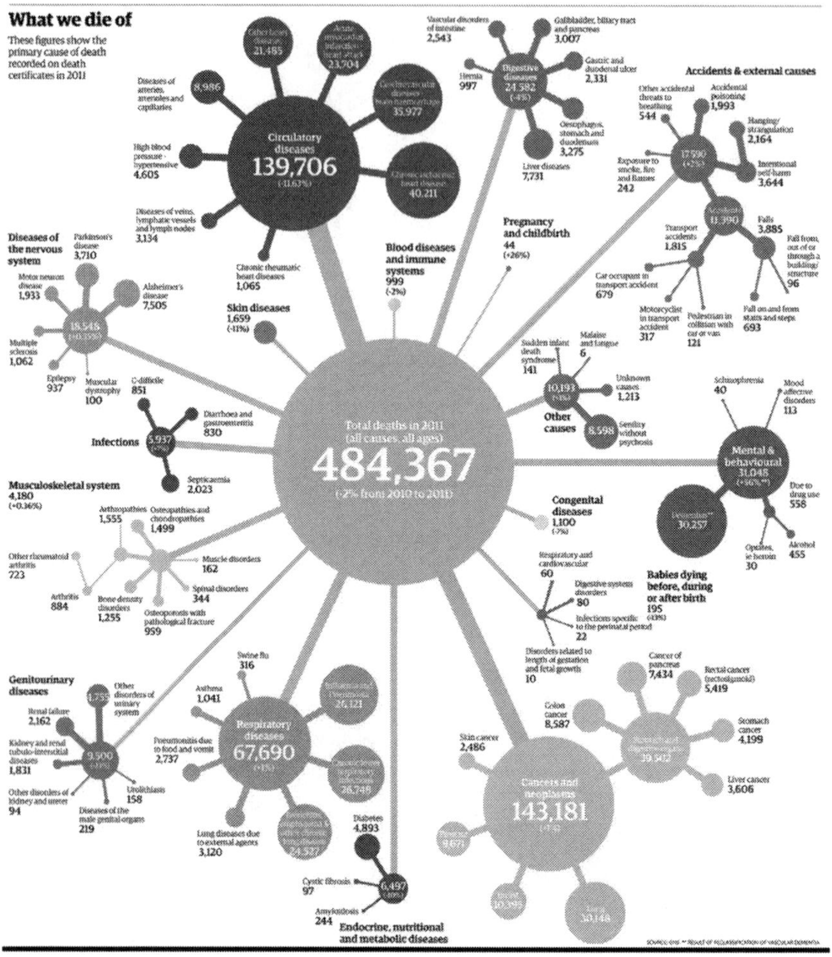

FIGURE 8.2 Infographic on causes of deaths in 2011. (*Source*: The Guardian, 2012)

informed decisions grounded in neurophysiological data. Figure 8.3 presents an eye-tracking analysis of the infographic shown in Figure 8.2. The color-coded heat map, using red, yellow, and green to represent high, moderate, and low fixation counts, effectively illustrates how hues, contrast, and color combinations influence visual attention. For example, the dark magenta-colored “139,706” attracts more visual attention than the light green “143,181,” highlighting the importance of carefully selecting hues and contrast to guide the viewer’s focus when designing.

Neuroscientific approaches offer a powerful means of understanding users’ cognitive and emotional responses, leveraging advanced neuroimaging techniques to uncover implicit reactions and preferences.

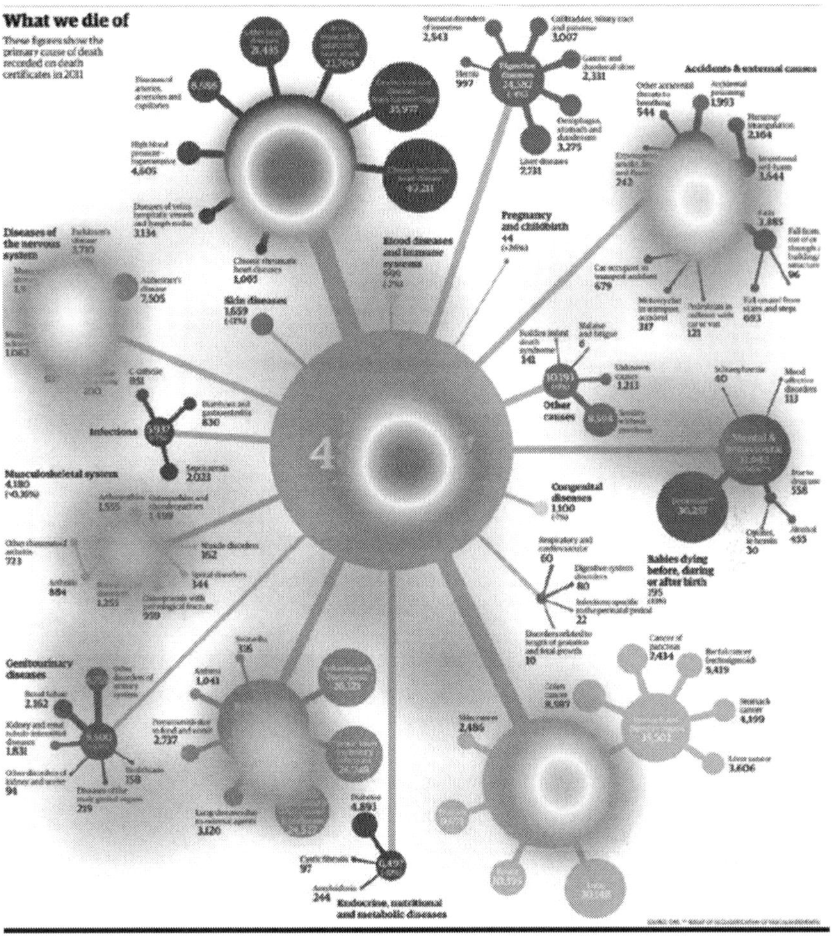

FIGURE 8.3 The eye gaze heatmap on the infographic shows how color overpowers the size of the circle. The higher the number of fixations, the more red you will see in the heatmap. (*Source*: The Guardian, 2012)

8.3.3.4 EXPERT-BASED USABILITY EVALUATION METHODS

A. Heuristic Evaluation

The main goal of heuristic evaluation is to find problems with the design interfaces. The most popular are the ten usability heuristics (Nielsen, 2023). These heuristics can be used as a guide to identify areas for improvement while conducting usability testing. It is helpful as it can discover usability problems quickly. Using the correct set of heuristics for each use context is essential (Lodhi, 2010). Some common heuristics include keeping users informed about what is happening, using language that is easy to understand, providing user control, maintaining uniform design elements, and preventing errors, to name a few (Table 8.4).

TABLE 8.4 Useful Heuristics, Including Nielsen's First Ten Usability Heuristics (Nielsen, 2023).

1. System Status: Keep users informed about what's happening. (Example: When downloading a file, it shows an icon.)

2. Familiarity: Use familiar language and concepts. (Example: A kitchen scale that displays the numerical value and unit of measurement.)

3. User Control: User freedom to select their actions. (Example: A digital camera with a mode dial for quick switching between photo and video modes.)

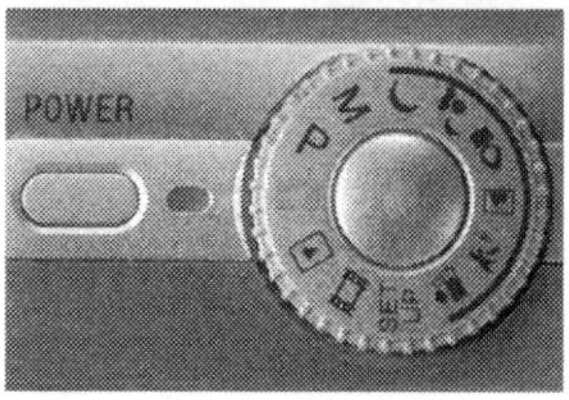

4. Consistency: Maintain uniform design elements. (Example: Similarity of Microsoft Word and Microsoft Excel to show consistency.)

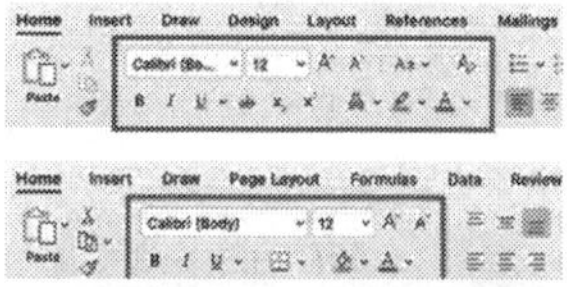

5. Error Prevention: Minimize chances of user errors. (Example: Error message feedback is provided.)

6. Recognition: Use visible options to reduce memory load. (Example: Phone interface used consistent icons with labels to minimize memory load.)

7. Flexibility: Cater to different needs of users. (Example: A multi-tool with interchangeable heads for different tasks.)

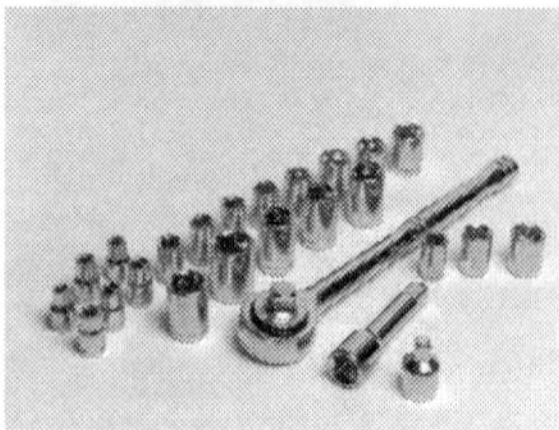

TABLE 8.4 (Continued)

8. Minimalist design: Avoid unnecessary clutter. (Example: Google search bar engine provides a minimalist and understandable design.)

9. Error Recovery: Provide clear error message and solutions. (Example: When using Word document, it provides an undo icon.)

10. Guidance: Offer accessible guides and instructions. (Example: A washing machine with icons illustrating wash cycles, and type of fabric.)

11. Usage clarity: Design should suggest how to use the product. (Example: A handle for pulling and a plate for pushing. Designed by Jeon Hwan Soo.)

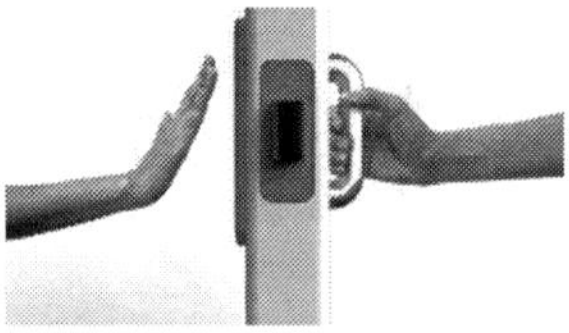

12. Feedback: Provide feedback on user actions. (Example: A whistling kettle that emits a sound when water is boiling.)

13. Comfort: Ensure products are comfortable and easy to use. (Example: A handheld vacuum cleaner designed with a comfortable grip for prolonged use.)

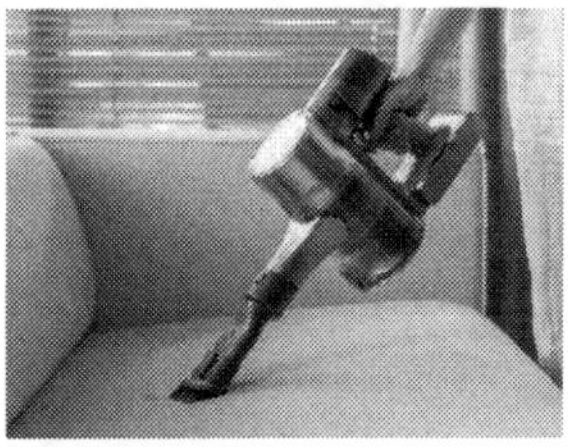

14. Accessibility: Make products usable for people with disabilities. (Example: An elevator with a braille button for those with visual impairment.)

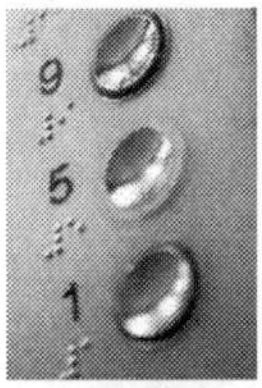

15. Durability: Ensure products withstand wear and tear. (Example: A kitchen knife made from high-quality materials for longevity)

(continued)

TABLE 8.4 (Continued)

16. Logical layout: Organize controls and features logically. (Example: A smart home control panel.)	
17. Sustainability: Use eco-friendly materials for long-term use. (Example: A reusable eco-bag made from sustainable materials.)	
Cultural awareness: Design to consider various cultural practices. (Example: Color to be location dependent. Green in Hong Kong and red in the United States.)	

B. Cognitive Walkthrough

A cognitive walkthrough is a method used to assess how easily a user can complete tasks in a system. It identifies the user's goals and actions to accomplish each step. This method is fast, cost-effective, and valuable for detecting design issues early. A group of evaluators simulates the experience of a first-time user, walking through each task step-by-step. At every step, they ask four key questions:

- Will the user do the right thing?
- Will the user know what to do?
- Will the user know that they are doing the right thing?
- Will the user know that they are making progress?

The evaluators then note the answers to these questions and identify the problems.

Enhanced cognitive walkthrough (ECW) is an analytical method to systematically identify usability issues (Bligard & Osvalder, 2013). Such an approach was used to assess the user interfaces of various medical devices, including home care ventilators, infusion pumps, dialysis machines, and insulin pumps. The ECW provides a structured framework for evaluating how effectively users can interact with complex healthcare equipment, focusing on enhancing safety and usability in critical care contexts (Khajouei et al., 2017).

8.3.3.5 USER QUESTIONNAIRE FOR USABILITY TESTING OF SYSTEM

Questionnaires are standard tools in usability testing, allowing users to provide feedback during or after task completion. However, they can be time-consuming for participants (Andreasen et al., 2007). Some usability questionnaires are:

1. The System Usability Scale (SUS), created by Brooke in 1996, is a "quick and dirty" tool. With just ten questions on a Likert scale, SUS offers a fast, accessible way to gauge how usable a system feels to users. Despite its brevity, research by Sauro (2011), and Tullis and Stetson (2004) has confirmed that SUS delivers highly reliable and valid results, making it a go-to method in usability testing across industries.
2. The Questionnaire for User Interaction Satisfaction (QUIS) gathers user opinions and evaluates their acceptance (Harper & Norman, 1993). The questionnaire assesses key aspects such as ease of use, consistency, system capability, and learning.
3. Software Usability Measurement Inventory Method (SUMI) is a licensed questionnaire with 50 questions (Kirakowski & Corbett, 2006). It is used to measure the perceived quality of software end-user experience.
4. The Post-Study System Usability Questionnaire (PSSUQ), developed by IBM in 1992, is a heavyweight in the world of usability testing. With its 16 carefully crafted questions and seven-point rating scale, it dives deep into how users feel about a system after using it. Whether it is a website, software application, or physical product, the PSSUQ (Lewis, 2002) captures users' perceived satisfaction, offering rich insights into what worked and what did not.

8.3.4 A/B TESTING

A/B testing (King et al., 2017) is a structured approach to hypothesis testing crucial in evaluating prototypes. This method involves comparing two alternative designs or prototypes, designated as Design A and Design B, to assess which performs better regarding user interaction, satisfaction, or performance (Quin et al., 2024). One of the key advantages of A/B testing is its methodical approach, which relies on random assignment of users to either design. A/B testing allows designers to gather empirical data about user preferences and behaviors, making informed decisions based on user feedback.

While A/B testing is widely recognized for its applications in digital products (Firmenich et al., 2019), its principles can also be applied to physical prototypes (McElroy, 2016). For instance, Lou (2023) used A/B testing to test the effectiveness and suitability of a surf shoe prototype. The thoughts gained from A/B testing can guide teams in refining digital interfaces and physical products, guaranteeing that the final offerings align closely with user expectations.

Incorporating A/B testing into the prototype development process can significantly improve the precision of decision-making in innovation activities (Zou & Ertug, 2024). By systematically testing different prototype iterations, teams can foster a culture of experimentation and continuous improvement. This iterative process helps optimize products and ensures that the final designs effectively solve the original problems set in Chapter 5.

8.3.5 AUGMENTED REALITY (AR) /VIRTUAL REALITY (VR) TESTING

AR is a promising method for building products and enables stakeholder communication (Kang et al., 2022; Li & Fessenden, 2016; Zampelis et al., 2012). The vast flexibility of AR in visualizing information and creating virtual representations of products or features is one of the relevant research projects. These visualizations can expedite communication and increase the efficiency of the design process by reducing the need for internal prototype modifications (Zampelis et al., 2012; Zorriassatine et al., 2003). AR can be an effective alternative for products that are difficult to create using traditional methods.

8.3.6 COMPATIBILITY

When medical devices touch the human body, the stakes are high, and safety is not optional but essential. That is why evaluating biocompatibility and sterilization is not just a regulatory checkbox but a critical part of the design process. Biocompatibility testing ensures that materials do not provoke harmful reactions when in contact with tissues or fluids, helping detect risks like toxicity, allergies, or inflammation (Huzum et al., 2021). On the other hand, sterilization must go beyond surface-level cleaning; it needs rigorous validation to eliminate all pathogens, reducing the risk of infection (Diep et al., 2019). These two pillars form the foundation of safe, effective medical devices. When done right, they protect patients and boost device performance and trust, raising the bar for healthcare innovation.

8.3.7 PROTOTYPE TESTING TO TEST THE FUNCTIONALITY AND OTHER TECHNICAL FEATURES

Some types of prototype testing do not involve user testing, as the focus is on evaluating functionality, safety, feasibility, and other technical aspects. The following discussion outlines the various testing methods used for these prototypes.

8.3.7.1 SIMULATION

Simulation is increasingly important in various fields, particularly in developing and evaluating electronic products and systems (Bot & Segal, 2024; Sierra et al., 2019). It entails building a model that precisely represents the system or process you are testing. This involves using software, hardware, or a combination of both. It is also vital to confirm that the simulation captures the essential aspects of the real-world system. Computer-aided programs such as SolidWorks (https://my.solidworks.com/) and CAD models using finite element analysis (FEA) are used to predict real-world design behavior. Computer simulation models test how parts and assemblies work under stress. Simulation testing is applicable for high-fidelity prototypes and is a crucial tool across different domains, enabling the evaluation and refinement of products and systems before physical implementation, thereby enhancing overall reliability and efficiency. However, it relies heavily on commercially available simulation software (Li et al., 2019).

Khutal et al. (2020) used the finite element method for bicycle frames to minimize the development time and testing using linear static analysis to determine the test rig boundary conditions and harmonic fatigue analysis. Both methods were compared with the laboratory test rig findings. The stress amplitude computed using harmonic fatigue analysis was one-third of the test rig's experimental stress amplitude. On the other hand, the finite element static analysis computed was the same as the experimental results. They recommended conducting fatigue harmonic stress before prototype testing using a "fast vibration test rig" for reliability.

8.3.7.2 OTHER TESTING METHODS

Other testing methods include breadboard testing, force testing, decibel testing, airflow testing, dielectric testing, and packaging testing. Breadboard testing (Peppler et al., 2023) is a vital step in electronics development, allowing engineers to assemble and modify prototypes quickly to validate circuit designs. Force testing (Zhu et al., 2021) evaluates a product's durability under various loads, ensuring it can withstand operational stress without failure. Decibel testing (Murphy & King, 2016) measures sound levels to assess compliance

TABLE 8.5 Other Prototype Testing Methods.

Testing Method	Description	Sources
Breadboard testing	Breadboard or proof of concept prototypes are cost cost-effective method for weighing the risks and benefits of different design concepts. This test the feasibility of an idea before developing it further. A breadboard is a plastic board with several holes that allows you to insert the components to prototype an electronic circuit. The connection is temporary and allows you to remove a component. You can build a circuit without soldering it.	Peppler et al., 2023
Force testing	Force testing determines the force needed to operate a prototype and claims the components of the prototype are strong enough to support those forces.	Zhu et al., 2021
Decibel testing	Decibel testing uses a sound level meter to measure the sound for prototype that emits noise such as fans, motors or alarms. It confirms that these devices is within the standard noise requirements.	Murphy & King, 2016
Airflow testing	Airflow testing identifies design flaws or assembly problems that obstruct the flow of air such as fan or blower.	Kelemenova, 2019
Dielectric testing	Dielectric testing also called "hipot" testing. This test is important to validate that no risk to user of electrocution or shock when using the prototype.	Liu et al., 2011
Packaging testing	Packaging testing involves several methods such as force and drop testing to certify that the product reach the customer in good condition. It should protect the product from physical and chemical damage. It should also include ensuring the product shelf life and can withstand different temperatures.	Deshwal et al., 2019

with noise regulations and enhance user comfort, while Airflow testing (Kelemenova, 2019) examines the efficiency of ventilation systems, ensuring optimal performance and energy efficiency. Dielectric testing (Liu et al., 2011) checks the insulation properties of electrical components, preventing failures that could lead to safety hazards. Finally, packaging testing (Deshwal et al., 2019) ensures that product packaging is robust and functional, protecting items during transit and providing ease of use for consumers. These prototype testing methods are crucial for ensuring product safety, reliability, and user satisfaction.

TABLE 8.6 Description and Benefits of Metrics.

Metrics	Description	Example	Benefits	References
Usability	Focus on the ease of use and effectiveness of a product or service in achieving users' goals.	Task success rate, time on task, error rates, and completion rates.	By analyzing these metrics, designers can identify usability issues and evaluate the efficiency of design solutions.	Han & Moghaddam, 2021
Satisfaction	Measure users' subjective feelings and attitudes toward a product or service.	Satisfaction ratings, Net Promoter Score (NPS), and customer satisfaction surveys.	By collecting feedback on users' overall satisfaction levels, designers can gauge the success of design interventions and identify areas for improvement.	Tyo & Desroches, 2023
Engagement	Assess users' level of involvement and interaction with a product or service.	Time spent on the platform, frequency of visits, and user engagement scores.	By tracking user engagement metrics, designers can understand how effectively their designs capture users' attention and encourage continued usage.	Han & Moghaddam, 2021
Effectiveness	Measure the extent to which a product or service achieves its intended goals and delivers value to users.	Conversion rates, task completion rates, and key performance indicators (KPIs) relevant to the specific domain or industry.	By evaluating effectiveness metrics, designers can assess the impact of design interventions on users' ability to accomplish tasks and achieve desired outcomes.	Miao et al., 2023
Retention and churn	Measure users' loyalty and attrition rates over time.	Retention rate, churn rate, and customer lifetime value (CLV)	By monitoring retention and churn metrics, designers can assess the long-term sustainability of their designs and identify opportunities to enhance user loyalty and retention.	Stagg et al., 2023
Accessibility	Evaluate the degree to which a product or service is accessible to users with disabilities or impairments.	Compliance with accessibility standards (e.g., WCAG), screen reader compatibility, and keyboard navigation support.	By prioritizing accessibility metrics, designers can confirm that their designs are inclusive and accessible to all users.	Han & Moghaddam, 2021

8.3.8 PROCEDURE FOR PROTOTYPE TESTING WITH USERS

8.3.8.1 START WITH A SCRIPT

Think of it as your usability test's blueprint, the test plan. This written document maps out exactly how your testing session will unfold. It captures every detail, from how the designer interacts with participants to the step-by-step testing procedure and the tasks and questions users will tackle. It is your guide to staying consistent, organized, and focused, ensuring the test runs smoothly and delivers meaningful, actionable insights.

8.3.8.2 RECRUITING PARTICIPANTS

The target users should be recruited and compensated for their time. The participants will also be asked to maintain product anonymity and to sign a consent form for the test. The number of participants may vary depending on the level of desired conclusiveness(Rogers, 2011; Schmettow et al., 2012). For qualitative usability studies, a minimum of five participants is often sufficient to identify most usability issues. For quantitative usability studies, particularly when collecting performance metrics such as task completion time or error rates, a minimum sample size of 20 participants is recommended (Nielsen, 2012b). It is essential to document participant demographics, including age, gender, education level, profession, and other relevant details, as these factors may influence performance and interpretation of results. At the same time, ethical considerations must be rigorously observed to ensure responsible research practices. Researchers and designers must uphold user privacy principles, obtain informed consent, and take measures to minimize any potential discomfort experienced during testing (Britt et al., 2020).

8.3.8.3 BRIEFING PARTICIPANTS

Before diving into testing, it is essential to brief your participants thoroughly. Start by clearly explaining the purpose of the evaluation, what you are trying to learn, and not to judge their skills or intelligence. Reassure them that this is not a test of their abilities, but a chance to explore the design as naturally as possible. Encourage them to perform tasks just as they would in real life. To set the stage, provide a quick, hands-on demonstration of the prototype, whether a physical device or a digital interface, so they know exactly what to expect.

8.3.8.4 THE TEST

Participants are guided into the testing area, which may feature equipment, computers, cameras, and, in some cases, an environmentally controlled room.

Here, they prepare to embark on a hands-on journey with the prototype. They receive a clear list of tasks to complete, setting the stage for focused exploration. As they work through each task, they are encouraged to think aloud, sharing their immediate thoughts, reactions, and frustrations, giving the designer a window into their experience and mindset (Ishaq et al., 2021). Meanwhile, the entire session is preferably recorded, capturing every detail and nuance for deeper analysis later. This combination of action and insight turns testing into a powerful tool for uncovering hidden usability gems.

8.3.8.5 DEBRIEF PARTICIPANTS

After completing all the tasks, the participants should be debriefed. During the debriefing, the participants should be encouraged to express their thoughts about the whole experience. This session is essential to establish rapport with the participant and encourage them to give their views about the prototype (physical or interface) they used.

8.3.8.6 EVALUATE RESULTS AND REPORT

The results of the qualitative and quantitative measures used to assess user experience, including usability, satisfaction, engagement, and effectiveness, are essential for stakeholders. These metrics provide designers with valuable insights into how users perceive and interact with a prototype (physical or interface), helping them identify areas for improvement and track the success of design interventions (see Table 8.6).

User experience metrics provide designers with objective measures to assess their designs' usability, satisfaction, engagement, and effectiveness. By tracking these metrics, designers can identify areas for improvement, evaluate design interventions' success, and confirm that their designs meet users' needs and expectations. Once the overall metrics (satisfaction, completion error, completion time) are obtained, the next step is to prepare the report for the stakeholders.

8.3.9 IMPROVING THE PROTOTYPE

Based on the evaluation results and the issues uncovered, the next step is to refine the prototype by weaving in user feedback, guided by the insights from relevant chapters in this book. Once the improvements are made, the updated prototype is returned to users for another round of testing and fresh feedback. This iterative loop of testing, learning, and refining continues until users are delighted and the design reaches its polished, final form with no more tweaks needed (Figure 1.5).

8.3.10 FINAL VERSION OF HIGH-FIDELITY PROTOTYPE

Once users give their nod of approval to the prototype, the journey is not over yet. The next critical step is to verify that every specification, standard, and technical requirement is fully met, a crucial checkpoint before moving into full-scale manufacturing. This ensures that users love the product and that it is built to rigorous quality and safety benchmarks. For an in-depth look at navigating specifications and compliance, turn to Chapter 10, where these essential steps are explored in detail.

8.4 CASE STUDY: TRAIN CONTROL LEVER

This case aims to evaluate a prototype train control lever that is ergonomically sound and minimizes driver wrist and arm fatigue.

Background: This case will demonstrate the process for user testing for different prototypes. Most trains rely on a vital guardian known as the "deadman switch." This clever safety feature keeps the driver alert and in control by requiring a constant, deliberate action: the driver must push the lever forward and hold it down with steady pressure of about 20 newtons throughout the journey. It is a simple but powerful way to ensure vigilance. If the driver's grip falters from fatigue, distraction, or an emergency, the lever releases, and the train automatically grinds to a halt. In that moment, the deadman switch is a silent protector, preventing accidents and safeguarding lives.

Drawing from the observational studies, contextual inquiries, and ethnographic interviews detailed in Chapter 4, it was observed that drivers approach the lever either from the top or the side, depending mainly on their seated posture and driver's height. The postures adopted while operating the lever are illustrated in Figure 8.4. According to the ergonomic principles discussed in Chapter 2, most drivers do not maintain a neutral posture on the wrist during this task (Figure 8.4a). Two distinct strategies can be observed: some drivers reach for the lever from above, while others use a handshake grip (Figure 8.4). The handshake posture (Figure 8.4b) keeps the wrist in a more neutral, ergonomically desirable alignment. In contrast, reaching from above causes the wrist to flex excessively, a posture associated with increased risk of wrist strain and injury.

To mitigate hand stress and reduce the likelihood of musculoskeletal disorders, several strategies can be considered:

(a) redesigning the lever to promote better wrist alignment,
(b) modifying the task to encourage safer hand positions,
(c) providing ergonomic training to promote correct posture,
(d) implementing exercise programs aimed at improving the strength, flexibility, and endurance of the hand and finger muscles, and
(e) developing personal protective equipment (PPE) to minimize both awkward motions and the force transmitted through the wrist.

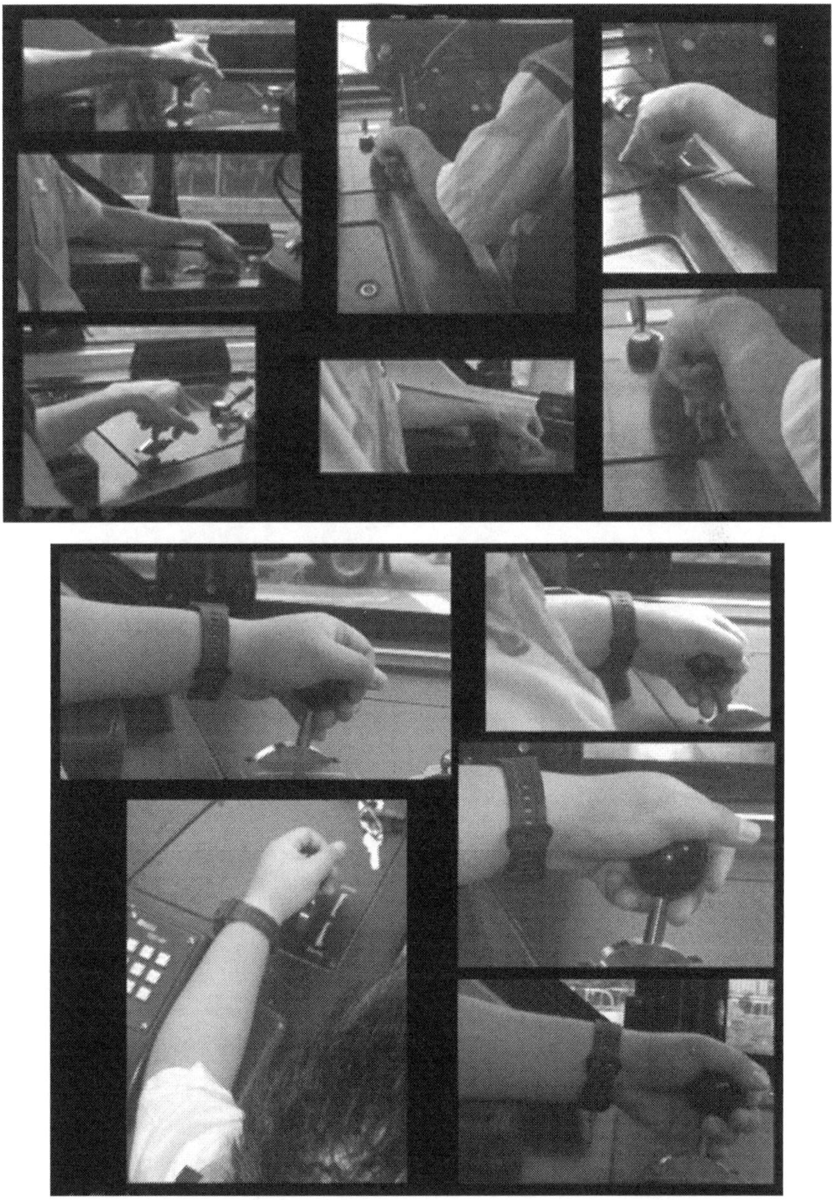

FIGURE 8.4 Driver wrist postures (a) Hand/wrist posture of different drivers. (b) Required hand posture for least stress on wrist.

Even though all of the above are important, the focus was on redesigning the lever (option a). Inspired by the ergonomic design of gear handles in sedans and sports cars, a series of alternative lever designs was prototyped. Two new designs are shown in Figure 8.5 to cater to different hand sizes. A key feature of the new design is a spherical top paired with a contoured power grip handle, which encourages a neutral wrist posture. The handle also includes a flat bottom surface, enabling drivers to apply the required downward force to operate the deadman switch using the edge of the palm.

To evaluate the performance of the three control designs shown in Figure 8.5, a custom jig replicating the deadman switch unit was fabricated as shown in Figure 8.6.

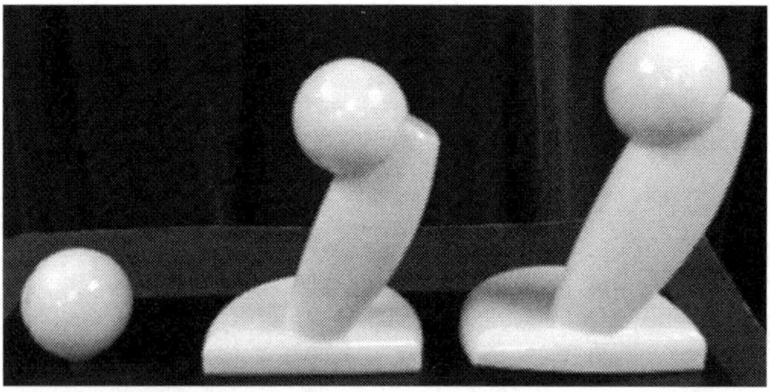

FIGURE 8.5 Test handles (a) sphere, (b) small handle, (c) large handle.

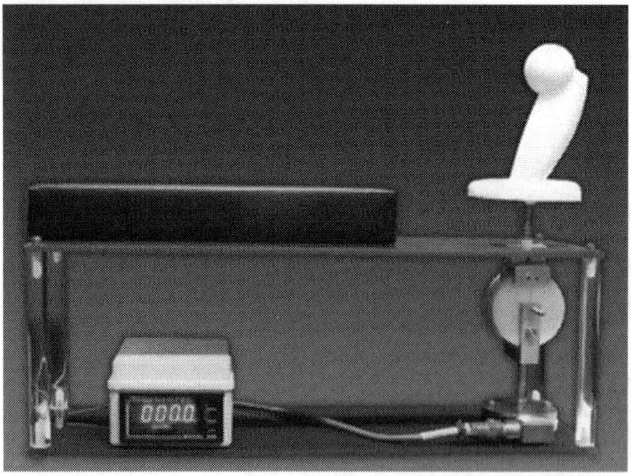

FIGURE 8.6 The special jig was fabricated for testing. A pre-calibrated digital force gauge measured the downward force and the holding time.

The prototypes were tested using the methodology described in Section 8.3.8:

- Test script: A detailed test protocol was developed outlining the procedures for evaluating each of the three handle designs. These included the order in which the three handles were tested with each participant, so there was no fatigue or bias in the results.
- Participants: The participants for the test were train operators.
- The test: Each participant tested all three handle designs in a controlled test environment. They were instructed to apply their maximum downward force on each handle and maintain this force for approximately five minutes or until the force stabilized. The exerted forces were recorded for all three handles.
- Debrief participants: Following the test, participants were debriefed and asked to provide qualitative feedback. Most operators preferred the larger handle, citing the ergonomic advantage of being able to alternate between a power grip and a "hand-on-top-of-sphere" posture. This flexibility in grip style enhances comfort and control during operation.
- Evaluate the results: The maximum force rapidly drops within the first minute, illustrating the concept of fatigue discussed in Chapter 2. The force stabilizes after around 4 minutes. This force is around 16 newtons for the sphere and the small handle, while the larger handle can exert a maximum of around 24 newtons. The video recording showed minimal wrist deviations with small and large handles.
- Prepare the report: A report was prepared with the complete methodology and the results, concluding with the preferred control lever from qualitative and quantitative measures.
- Refine and iterate: Based on the evaluation and feedback from train operators, the downward force was lowered to 15 newtons to minimize arm and wrist fatigue using the larger handle.

Testing revealed that designs (b) and (c) allowed users to exert greater force over longer durations compared to the existing design (a), with significantly less fatigue. Even in a static posture, users could consistently maintain a force of at least 15 newtons using the new designs. As a result, the risk of musculoskeletal injury associated with the driving control task is significantly reduced, if not eliminated. Adopting a power grip in these designs supports a neutral wrist posture, effectively lowering strain and improving comfort during operation.

8.5 CONCLUSION

Prototype selection includes evaluating and choosing the most favorable design solutions from several alternatives, prioritizing those that best meet customer needs, functionality, feasibility, safety, reliability, and other crucial factors. The method or combination of methods a team chooses can significantly influence the quality and clarity of the design decisions made. This chapter outlined how to select appropriate strategies and procedures, when to combine them, how to evaluate the process, and when to stop and transition to testing. For a product to succeed, testing your prototype to determine if it works as intended before releasing it to the market is imperative. The prototype testing involves evaluating the functionality, usability, safety, aesthetics, and other technical factors that will lead to the product's success. Most prototype testing involves real users in the design process. These tests are usability, A/B testing, AR/VR testing, and testing for human compatibility, since we want to ensure that the product is safe and will not cause harm or illnesses to the users. Other prototype testing methods that test the safety and functionality of the prototype and do not involve users are simulation testing, breadboarding, decibel testing, force testing, airflow testing, dielectric testing, and packaging testing.

A prototype testing procedure that engages users is presented, and these steps include starting with a script, recruiting participants, briefing participants on the task to accomplish, conducting the user testing, debriefing the participants, and evaluating the results and sharing the report with stakeholders. The results of the qualitative and quantitative measures used to assess user experience, including usability, satisfaction, engagement, and effectiveness, are important for stakeholders. These metrics provide designers with valuable insights into how users perceive and interact with a prototype (physical or interface), helping them identify areas for improvement and track the success of design interventions. Lastly, a case study was presented to show how to evaluate three 3D prototypes of lever design for train operators and observe their gripping and wrist deviations while using the prototypes to determine the best design prototype.

"Design is really an act of communication, which means having a deep understanding of the person whom the designer is communicating" - Donald Norman

REFERENCES

Ahmad, K.S., Ahmad, N., Tahir, H., & Khan, S. (2017). Fuzzy_MoSCoW: A fuzzy-based MoSCoW method for the prioritization of software requirements. International Conference on Intelligent Computing, Instrumentation and Control Technologies (ICICICT), Kerala, India, 2017, pp. 433–437, https://doi.org/10.1109/ICITCICT1.2017.8342602.

Albert, W., & Tullis, T. (2022). Measuring the User Experience: Collecting, Analyzing, and Presenting UX Metrics. 3rd ed. Morgan Kauffmann.

Almashani, Y., Alamri, A., Al-Ghassani, M., & Kulkarni, M.(2023). Stress analysis using SolidWorks simulation. *EPRA International Journal of Multidisciplinary Research (IJMR)*, 9(9), 2455–3662.

Andreasen, M.S., Nielsen, H.V., Schrøder, S.O., & Stage, J. (2007). What happened to remote usability testing? an empirical study of three methods. In Proceedings of the SIGCHI Conference on Human Factors in Computing Systems (CHI '07). Association for Computing Machinery, New York, NY, USA, 1405–1414. https://doi.org/10.1145/1240624.1240838.

Bligard, L.O., & Osvalder, A.L. (2013). Enhanced cognitive walkthrough: development of the cognitive walkthrough method to better predict, identify, and present usability problems. *Advances in Human–Computer Interaction*, 931698. https://doi.org/10.1155/2013/931698.

Boardman, A. E., Greenberg, D. H., Vining, A. R., & Weimer, D. L. (2018). *Cost-Benefit Analysis: Concepts and Practice* (5th ed.). Cambridge University Press.

Britt, V., Santiago, C., & Lytle, A. L. (2020). Ethical considerations in usability testing: A framework for decision-making. *Journal of Business Ethics*, 166(3), 585–600.

Brown, T. (2009). *Change by Design: How Design Thinking Creates New Alternatives for Business and Society*. Harvard Business Press.

Bot, Y., & Segal, A. (2024). New trends in testing electronic products by simulation during design using programmable rules, circuit analysis and AI. Annual Reliability and Maintainability Symposium (RAMS), Albuquerque, NM, USA, 2024, pp. 1–7, doi: 10.1109/RAMS51492.2024.10457651.

Brooke, J. (1996). SUS-A quick and dirty usability scale. *Usability Eval. Ind.*, 189, 4–7.

Clegg, D., & Barker, R. (1994). *Case Method Fast-Track: A RAD Approach*. Addison-Wesley.

Dahlkvist E, Engström M., &.Nilsson A. (2020). Residents' use and perceptions of residential care facility gardens: A behavior mapping and conversation study. *International Journal of Older People Nursing*, 15(1), 1–10. https://doi.org/10.1111/opn.12283.

Deng, L., & Wang, G. (2019). Application of EEG and interactive evolutionary design method in cultural and creative product design. *Computational Intelligence and Neuroscience*, 2019(1), 1860921.

Deshwal, G.K., Panjagari, N.R., & Alam T. (2019). An overview of paper and paper-based food packaging materials: health, safety, and environmental concerns. *Journal of Food Science and Technology*, 56(10), 4391–4403. https://doi.org/10.1007/s13197-019-03950-z.

Diederichs, F., Niehaus, F., & Hees, L. (2020). Guerilla evaluation of truck HMI with VR. In: Chen, J.Y.C., Fragomeni, G. (eds), *Virtual, Augmented and Mixed Reality. Design and Interaction*. HCII 2020. Lecture Notes in Computer Science, vol 12190. Springer. https://doi.org/10.1007/978-3-030-49695-1_1.

Diep, B., Moulin, J., Bastic-Schmid, V., Putallaz, T., Gimonet, J., Valles, A., & Klijn, A. (2019). Validation protocol for commercial sterility testing methods. *Food Control* 103, 1–8.

Firmenich, S., Garrido, A., Grigera, J., Rivero, J. M., & Rossi, G. (2019). Usability improvement through A/B testing and refactoring. *Software Quality Journal*, 27, 203–240.

Fung, R.Y.F., Chen, Y., & Tang, J. (2007). A quality engineering based approach for conceptual product design, *International Journal of Advanced Manufacturing Technology*, Vol. 32, pp.1064–1073.

Gibbons, S. (2019). Dot Voting: A Simple Decision-Making and Prioritizing Technique in UX. www.nngroup.com/articles/dot-voting/

Green, P. E., & Srinivasan, V. (1990). Conjoint analysis in marketing: New developments with implications for research and practice. *Journal of Marketing*, 54(4), 3–19.

The Guardian (2012). What do people die of? Mortality rates and data for every cause of death in 2011 are visualised. www.theguardian.com/news/datablog/2012/nov/06/deaths-mortality-rates-cause-death-2011.

Guler, K., & Petrisor, D.M. (2021). A Pugh Matrix-based product development model for increased small design team efficiency. *Cogent Engineering*, 8(1), 1–16. https://doi.org/10.1080/23311916.2021.1923383.

Gutierrez, A.M.J., Galang, M., Seva, R. Lu, M., & Ty, D. (2014). Designing an improved respirator for automotive painters. *International Journal of Industrial Ergonomics*, 44, 131–139.

Gutierrez, M. D. C., Alcayde, A., Salcedo, R., & Salazar, R. E. (2021). Using a modified Pugh matrix approach, designing trash bins for improving waste segregation behavior on a university campus. *International Journal of Environment and Waste Management*, 27(3), 237–255. https://doi.org/10.1504/IJEWM.2021.114952.

Han, S., & Moghaddam, A. (2021). A framework for evaluating digital user experience using behavioral metrics. *Journal of Design and Technology Education*, 26(2), 120–136.

Harper, B.D. & Norman, K.L. (1993, February). Improving user satifaction: The questionnaire for user interaction satisfaction version 5.5. In Proceedings of the 1st annual mid-Atlantic Human Factors Conference Vo.224, p. 228.

Hassan, M.F., Saman, M.Z.M., Sharif, S., & Omar, B. (2016). Sustainability evaluation of Alternative part configurations in product design: weighted decision matrix and artificial neural network approach. *Clean Technologies and Environmental Policy* 18 (1), 63–79. https://doi.org/10.1007/s10098-015-0990-1.

Hay, L., Duffy, A. H. B., Gilbert, S. J., & Grealy, M. A. (2022). Functional magnetic resonance imaging (fMRI) in design studies: Methodological considerations, challenges, and recommendations. *Design Studies*, 78, 101078.

Hill, J.R., Brown, J.C., Campbell, N.L., & Holden, R.J. (2021). Usability-in-place-remote usability testing methods for homebound older adults: rapid literature review. *JMIR Formative Research*, 5, 1–9.

Hornbæk, K. (2006). Current practice in measuring usability: Challenges to usability studies and research, I*nternational Journal of Human–Computer Studies*, 64(2), 79–102, https://doi.org/10.1016/j.ijhcs.2005.06.002.

Honkala, S., Hamalainen, M. & Salonen, M. (2007). Comparison of four existing concept selection methods. International Conference on Engineering Design, ICED '07 (28-31 August 2007).

Humphrey, A.S. (2005). History Corner: Swot Analysis for Management Consulting. SRI Alumni Association.

Huzum, B., Puha, B., Necoara, R.M., Gheorghevici, S., Puha, G., Filip, A., Sirbu, P.D., & Alexa, O. (2021). Biocompatibility assessment of biomaterials used in orthopedic devices: An overview (Review). *Experimental and Therapeutic Medicine* 22(5), 1315. doi: 10.3892/etm.2021.10750.

Hsu, Y.-L., & Sandford, B. A. (2007). The Delphi technique: making sense of consensus. *Practical Assessment, Research, and Evaluation*, 12(10), 1–8.

IDEO. (2015). Design Kit. The Field Guide to Human-Centered Design, https://www.designkit.org/

Ishaq, K., Rosdi, F., Zin, N. & Abid, A. (2021). Heuristic and think aloud method to evaluate the low fidelity prototype of game-based language learning application. 2021 International Conference on Innovative Computing (ICIC), Lahore, Pakistan, pp. 1–10, http://doi.org/10.1109/ICIC53490.2021.9693022.

International Organization for Standardization. (2018). Ergonomics of Human-System Interaction- Part 11:Usability: Definitions and Concepts (ISO Standard No. 9241-11:2018). Retrieved from www.iso.org/standard/63500.html

Kaner, S. (2014). *Facilitator's Guide to Participatory Decision-Making* (3rd ed.). Jossey-Bass.

Kang, B., Crilly, N., Ning, W., & Kristensson, P.O. (2022. Prototyping to elicit user requirements for product development: using head-mounted augmented reality when designing interactive devices. *Design Studies*, 84, 1–33.

Kano, N., Seraku, N., Takahashi, F., & Tsuji, S. (1984). Attractive quality and must-be quality. *Hinshitsu (Quality, The Journal of Japanese Society for Quality Control)*, 14(2), 39–48.

Kepner, C. H., & Tregoe, B. B. (1965). *The Rational Manager: A Systematic Approach to Problem Solving and Decision-Making*. McGraw-Hill.

Kelemenova, T. (2019). Airflow measurement test device for airflow sensors. *Acta Mechatronica—International Scientific Journal about Mechantronics*.4(4), 17–21.

Khajouei, R., Esfahani, M.Z., & Jahani, Y. (2017). Comparison of heuristic and cognitive walkthrough usability evaluation methods for evaluating health information systems. *Journal of the American Medical Informatics Association*, 24(e1), e55–e60, https://doi.org/10.1093/jamia/ocw100.

Khalighy, S., Green, G., Scheepers, C., & Whittet, C. (2015). Quantifying the qualities of aesthetics in product design using eye-tracking technology. *International Journal of Industrial Ergonomics*, 49, 31–43.

Khutal, K., Kathiresan, G., Ashok, K., Simhachalam, B., & Jebaseelan, D. D. (2020). Design validation methodology for bicycle frames using finite element analysis. *Materials Today: Proceedings*, 22, 1861–1869.

Kirakowski, J., & Corbett, M. (2006). SUMI: the Software Usability Measurement Inventory. British Journal of Educational Technology. 24. 210 - 212. https://doi.org/10.1111/j.1467-8535.1993.tb00076.x.

Lencastre, J. A., Monteiro, L., & Ferreira, J. (2022). Neurodesign insights: EEG-based assessment of emotional UX. *Journal of Usability Studies*, 17(4), 234–249.

Lewis, J. R. (2002). Psychometric evaluation of the PSSUQ using data from five years of usability studies. *International Journal of Human–Computer Interaction*, 14(3–4), 463–488.

Li, Y., Deng, H., Xu, X., & Wang, W. (2019). Modeling and testing of in-wheel motor drive intelligent electric vehicles based on co-simulation with CARSIM/SIMULINK. *IET Intelligent Transport System*, 13(1), 115–123.

Li, A., & Fessenden, T. (2016). Augmented reality: What does it mean for UX?. https://www.nngroup.com/articles/augmented-reality-ux.

Liu, X. Pu, J. Jiang, H. Ma, H., & Wang, Z. (2011). Research on dielectric loss testing device for large-capacity and high-voltage electrical equipment. 2011 4th International Conference on Electric Utility Deregulation and Restructuring and Power Technologies (DRPT), Weihai, China, pp. 649–652, https://doi.org/10.1109/DRPT.2011.5993973.

Lodhi, A. (2010). Usability heuristics as an assessment parameter: for performing usability testing. 2nd International Conference on Software Technology and Engineering, San Juan, PR, USA, pp. V2-256-V2-259, https://doi.org/10.1109/ICSTE.2010.5608809.

Lou, H. (2023). A Proposed Shoe Design to Improve Surfing Performance, Master's thesis, Pratt Institute.

Luo, J. (2025). Development of an Artificial-Intelligence-Driven Product Design Evaluation Model Using Multi-model Data. PhD Dissertation. School of Design, The Hong Kong Polytechnic University.

McElroy, K. (2016). *Prototyping for Designers: Developing the Best Digital and Physical Products*. O'Reilly Media.

Miao, X., Zhang, W., & Lee, C. (2023). Data-driven UX optimization using KPIs and user analytics. *International Journal of Human–Computer Interaction*, 39(1), 1–18.

Muller, D.P., & Hiete, M. (2021). Visualization supported corporate decision making for life cycle sustainability assessment—illustrated using a case study for selecting a sustainable packaging system for self-leveling compounds. *Journal of Cleaner Production*, 313.

Murphy, E., & King, E. (2016). Testing the accuracy of smartphones and sound level meter applications for measuring environmental noise. *Applied Acoustics* 106, 16–22.

Murry, J. W., & Hammons, J. O. (1995). Delphi: A versatile methodology for conducting qualitative research. *The Review of Higher Education*, 18(4), 423–436.

Niekurzak, M., Lewicki,W., Coban, H., & Brelik, A. (2023). Conceptual design of a semi-automatic process line for recycling photovoltaic panels as a way to ecologically sustainable production. *Sustainability*, 15(3), 2822.

Nielsen, J. (1993). *Usability Engineering*. Morgan Kaufmann.

Nielsen, J. (1994). *Usability Engineering*. Academic Press.

Nielsen, J. (2012a). *Usability 101: Introduction to Usability*. https://www.nngroup.com/articles/usability-101-introduction-to-usability/ Date Retrieved: June 3,2025.

Nielsen, J. (June 2012b). How Many Test User in Usability Study? Retrieved from www.nngroup.com/articles/how-many-test-users. Date Retrieved: May 25, 2025.

Nielsen, J. (July 2023). The ten heuristics reimagined. Retrieved from www.uxtigers.com/post/10-heuristics-reimagined. Date Retrieved: March 3, 2025.

Norman, D. A. (2013). *The Design of Everyday Things* (Revised and Expanded Edition). Basic Books.

Parker, J., & Moseley, J. (2008). Kepner Tregoe decision analysis as a tool to aid route selection part 1. *Organic Process Research & Development*, 12(6), 1041–1043.

Peppler, K., Sedas, M., & Thompson, N. (2023). Paper circuits vs. breadboards: materializing learners' powerful ideas around circuitry and layout design. *Journal of Science Education and Technology*, 32(4), 1–24.

Project Management Institute. (2017). *A Guide to the Project Management Body of Knowledge* (PMBOK® Guide) (6th ed.). PMI.

Pugh, S. (1991). *Total Design: Integrated Methods for Successful Product Engineering*. Addison-Wesley.

Quin, F., Weyns, D., Galster, M., & Costa Silva, C. (2024). A/B Testing: A systematic literature review. *Journal of Systems & Software*, 211, pp. 1–28.

Rogers, Y. (2011). Interaction design gone wild: Striving for wild theory. *Interactions*, 18(4), 58–62.

Rotar, L., & Kozar, M. (2017). The use of the kano model to enhance customer satisfaction. *Organizacija*, 50(4), 339–351.

Saaty, T. L. (1980). *The Analytic Hierarchy Process: Planning, Priority Setting, Resource Allocation*. McGraw-Hill.

Salhieh, S., & Al-Harris, M. (2014). New product concept selection: an integrated approach using data envelopment analysis (DEA) and conjoint analysis (CA). *International Journal of Engineering & Technology*. 3(1), 44–55.

Sarker, I., Hoque, M., Uddin, M., & Alsanoosy, T. (2021). Mobile data science and intelligent apps: concepts, AI-based modeling and research directions, *Mobile Networks and Applications Journal*, 26(1), 1–19.

Sauro, J. (2011). *A Practical Guide to the System Usability Scale*. Measuring Usability.

Schmettow, M., Bach, C., & Scapin, D. (2014). Optimizing usability studies by complementary evaluation methods. Proceedings of the 28th British HCI Conference (BSC-HCI 2014). Southport, UK.

Schmettow, M. (2012). Sample size in usability studies. *Communications of the ACM*, 55(4), 64–70.

Shen, K., Cheng, C., Li, X., & Zhang, Z. (2019). Environmental cost–benefit analysis of prefabricated public housing in Beijing. *Sustainability* 11, 207; https://doi.org/10.3390/su11010207.

Sierra, A., Gercek, C., Übermasser, S., & Reinders, A. (2019). Simulation-supported testing of smart energy product prototypes. *Applied Sciences*, 9(10), 1–13. https://doi.org/10.3390/app9102030.

Sreejith, S.S., & Mathirajan, M. (2022). A modified Pugh matrix method for continuous performance evaluation of employees. *South Asian Journal of Human Resources Management*, 9(1), 149–163.

Srinivasan, E. (2000). The predictive power of internet-based product concept testing using visual depiction and animation. *Journal of Product Innovation Management*, 17(2), 99–109.

Stagg, S., Greaves, A., & Kerr, A. (2023). Post-launch feedback integration for iterative product development. *Design Management Journal*, 18(2), 89–102.

Sudin, M., Salim,M., Alkahari,M., Abdullah, M. Mansor, M., Akop, M., & Ramil, F. (2017). AHP method and application example for the robust multi-criteria design concept selection. *ARPN Journal of Engineering and Applied Sciences* 12(7), 2107–2114.

Takai, S., & Ishii, K. (2004). Modifying Pugh's design concept evaluation methods. Proceedings of the ASME 2004 International Design Engineering Technical Conferences and Computers and Information in Engineering Conference. https://doi.org/10.1115/DETC2004-57512.

Tholen, A., Plank, M., & Weber, A. (2020). Empathy in the brain: Investigating emotional responses using fMRI and EEG in design. *Design Studies*, 68, 45–61.

Thompson, K.E., Rozanski E.P., & Haake, A.R. (Oct., 2004). Here, there, anywhere: Remote usability testing that works. Presented at SIGITE04: ACM Special Interest Group for Information Technology Education Conference. Salt Lake City, Utah, USA, p.132–137. https//doi.org/10.1145/1029522.1029567.

Tullis, T. S., & Stetson, J. N. (2004). A comparison of questionnaires for assessing website usability. In Usability Professional Association Conference (Vol. 1). https://citeseerx.ist.psu.edu/viewdoc/download?doi=10.1.1.396.3677&rep=rep1&type=pdf

Tyo, J., & Desroches, R. (2023). UX Measurement Framework for Inclusive Digital Products. *User Experience Magazine*, 23(1), 17–28.

Wan, Q., Song, S. S., Li, X. H., Zhang, Q., Yang, X., Zhang, Y. C., & Design, A. (2018). The visual perception of the cardboard product using eye-tracking technology. *Wood Research*, 63(1), 165–178.

Wever, R., Desmet, P. M. A., & Groenesteijn, L. (2022). Emotion and cognition in product experience: Applications of neuroscientific methods in design research. *Design Journal,* 25(1), 22–40.

Wegener, F.E., Guerreiro Goncalves, & M., Dankfort, Z. (Aug.2019). Reflection- in-Action When Designing Organizational Processes: Prototyping Workshops for Collective Reflection-in-Action. International Conference on Engineering Design, ICED 19, 5-8 August 2019, Delft, The Netherlands.

Yao, K, Lai, J., Huang, W., & Tu, J. (2022). Utilize fuzzy Delphi and analytic process to construct consumer product design evaluation indicators. *Mathematics Journal,* 10(3), 1–25.

Zampelis, D., Gill, S., Loudon, G., & Walker, D. (2012). Augmented reality-centered rapid prototyping—the 26th BCS conference on human-computer interaction. The 26th BCS Conference on Human Computer Interaction (HCI). Sept. 12–14, 2012 1e6.

Zhang, Y., Zhang, Z., & Xu, M. (2024). Neuroimaging-based evaluation of human-centered design. *Cognitive Technologies Journal,* 38(2), 145–162.

Zhu, R., Takamura, Y., Yamamoto, I., Matsumoto, K., & Nagayasu, T. (2021). Research and development of testing device for evaluating force transmission, friction, and contact pressure distribution of surgical spring forceps (tweezers). *Sensors and Materials,* 33(3), 919–932.

Zorriassatine, F., Wykes, C., Parkin, R., & Gindy, N. (2003). A survey of virtual prototyping techniques for mechanical product development. *Proceedings of the Institution of Mechanical Engineers, Part B: Journal of Engineering Manufacture,* 217(4), 513e530.

Zou, T., & Ertug, G. (2024). Unwitting participants at our expense: A/B testing and digital exploitation. *Business & Society,* 63(7), 1513–1517.

Design for X and Product Architecture

9

> Good design is actually a lot harder to notice than poor design, in part because good designs fit our needs so well that the design is invisible.
>
> —Don Norman

9.1 DESIGN FOR X

The trends in design have come a long way (Cross, 2021). We have moved from a narrow focus on products, where the main concerns were manufacturing and production quality. Then, the emphasis was on the consumer and user experience, which is directed at better product design and improved business performance. But now, we are at a turning point. There is a growing awareness of the importance of societal impact, environmental responsibility, and sustainability (Clark et al., 2009). Businesses are starting to recognize that their actions have real consequences, and they cannot just chase profits without considering the effects on the community and the planet. With the rise of AI and digital technologies, this shift has only accelerated, pushing us toward a knowledge-based society where technology plays a central role.

As we adapt to these changes, the concept of Design for X (DfX) is also evolving. It is no longer enough to focus solely on traditional metrics like quality; we need to embrace a holistic view that considers societal needs and future trends. Let us explore several key factors that are essential for effective design (see Figure 9.1) and what this new approach means for us all: (1) design for people and ethics (DfP)is about ensuring that designs consider people, is inclusive and uphold ethical standards; (2) design for fabrication and making (DfF) is how products and services will be fabricated and manufactured; (3) Design for sustainability and environment (DfS) is considering the

DOI: 10.1201/9781003476177-9

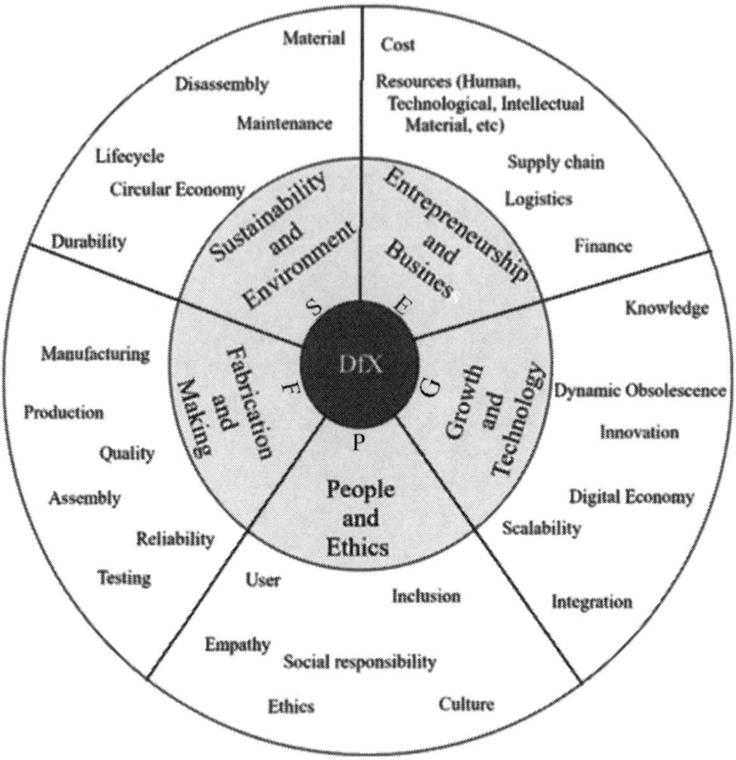

FIGURE 9.1 Factors involved in today's design for X.

environmental impact of our designs and striving for sustainability; (4) design for entrepreneurship and business (DfE) focuses on the business and financial factors that will guide our design decisions; (5) design for growth and technology (DfG) is incorporating relevant technologies to support future growth and advancement.

9.1.1 PEOPLE AND ETHICS (P)

In product development, integrating people and ethics is essential for ensuring that designs align with social responsibility and universal design principles (McAdam, & Kostovich, 2011; Null, 2013). Designers should consider the broader social implications, such as privacy, data security, and labor practices, throughout the design process. We have explored the capabilities and limitations of people (Chapter 2), the importance of inclusive design for diverse populations (Chapter 3), and the role of empathy (Chapter 4) in creating meaningful products. By merging these elements within the DfP framework, companies can

TABLE 9.1 Considerations When Designing for People and Ethics.

Element	Principle	Example
User (Chapter 2)	Minimize stress and error.	Designing banking mobile apps that are user-friendly, error-free, and minimize mental workload.
Inclusion (Chapter 3)	Accommodate all within reasonable bounds.	Designing interfaces that are accessible to people with disabilities.
Empathy (Chapter 4)	Look at aspects from the point of view of the user.	Creating personas based on real user experiences to inform design decisions.
Social responsibility	Focus on user needs and behaviors throughout the design process.	Conducting usability tests with different user groups.
Ethics	Ensure the responsible use of user data, prioritizing privacy and consent.	A fitness tracker that anonymizes data before sharing them with third parties.
Culture	Respect and incorporate various cultural values and traditions in design.	Designing a mobile app that accommodates multiple languages.

develop products that are not only technologically advanced but also contribute to a more equitable and just world. Table 9.1 shows some of the principles to consider for DfP.

Social responsibility (Bissett-Johnson & Radcliffe, 2021; Boughzala, 2021) is a cornerstone of the DfP approach. It encourages organizations to align their product development with ethical practices. This involves prioritizing the well-being of all stakeholders. For instance, companies need to consider worker safety and fair labor practices when evaluating manufacturing processes. Moreover, it is crucial to consider how products can enable social good, such as improving access to essential services or promoting sustainable lifestyles. By embedding social responsibility into the DfP framework, organizations can create products that satisfy technical and commercial demands while positively impacting the society and the planet.

Ethical considerations (Taebi, 2021) are also foundational to the DfP approach. They ensure that product development reflects responsible practices. Organizations should prioritize the rights and well-being of all stakeholders. This means assessing potential harm, the impact on safety, privacy and data security, and the social consequences throughout the product lifecycle. Additionally, companies should explore ways their products can encourage ethical behavior, whether that's through enhancing transparency or empowering users to make informed decisions. By incorporating ethical principles into the DfP framework, organizations can create products that meet commercial needs and uphold high moral and social standards.

Cultural considerations (De Mul, 2015; Edensor, 2020) are increasingly vital in the DfP approach, as they promote respect for and celebration of the diversity of human experiences. When designing with cultural sensitivity, organizations should strive to understand and incorporate their target user groups' unique values, traditions, and preferences. This may involve adapting product features and marketing strategies to align with the cultural norms of different communities. Moreover, companies should consider how their products can facilitate cultural exchange and understanding, contributing to a more inclusive global society. By embedding cultural considerations in the DfP framework, organizations can create products that fit with diverse user groups, creating a sense of belonging and cultural pride.

9.1.2 FABRICATION AND MAKING (F)

Fabrication usually refers to the industrial process of constructing products, while Making is often associated with craftsmanship and prioritizes artistic expression and uniqueness. DfF is about taking a comprehensive approach that connects product design with the entire manufacturing and production ecosystem (Table 9.2), such as manufacturing, production, quality, assembly, testing, and reliability. This means looking beyond just the aesthetics or functionality of a product; it is about ensuring that everything can be produced efficiently and

TABLE 9.2 Considerations in Design for Fabrication and Making.

Process	Principle	Example
Manufacturing	Optimize designs for efficient and cost-effective manufacturing processes.	Simplifying parts to reduce complexity and manufacturing costs.
Production	Consider the entire production process, including volumes, layout, and material flow.	Designing a production line layout that minimizes material handling.
Quality	Integrate quality principles into the design to ensure product or service meets standards.	Using failure mode and effects analysis (FMEA) to identify potential failure modes early in design.
Assembly	Simplify assembly processes by reducing part count and assembly steps.	Creating snap-fit components that allow tool-less assembly.
Testing	Incorporate testability into the product design for effective quality assurance.	Adding built-in test points for easy diagnostics during production.
Reliability	Ensure the product consistently performs over its expected lifespan under varied conditions.	Designing with redundant components to mitigate failure risks.

reliably while meeting quality and performance standards. This holistic strategy encompasses various principles that guide the designer in creating successful products.

Design for manufacturing (Boothroyd et al., 2010; Poli & Poli, 2001; ElMaraghy et al., 2012; El Wakil, 2019; Anderson, 2020), which focuses on optimizing designs for efficient and cost-effective manufacturing, is very important. Its goal is to ensure that a product can be made easily without sacrificing quality. This involves simplifying parts, selecting the right materials, and managing tolerances. For instance, designers might aim to reduce the number of components and use standardized parts, while considering how different materials affect costs and machinability.

Design for production (Tolio, 2008; Nahmias & Cheng, 2009; Yin et al., 2018) takes a broader look at the entire production process. It is about optimizing the entire system for efficiency, scalability, and cost-effectiveness. Factors like production volumes, material flow, and automation play significant roles here. Different production systems like job shops or mass production might be combined to meet various manufacturing needs, depending on the product's characteristics and the manufacturer's goals.

Design for quality (El-Haik, 2005; Dale, et al., 2007; Park & Antony, 2008; Ma et al, 2020; Adjoul, 2021) integrates quality principles into the design phase to ensure that a product consistently meets customer expectations throughout its lifecycle. Designers use advanced techniques like failure mode and effects analysis (FMEA) to identify potential issues early on. By collaborating closely with manufacturing teams, they can optimize product features and tolerances to minimize quality-related costs.

Design for assembly (Boothroyd et al., 2010; Molloy, et al., 2012; Battaïa et al, 2018) is another essential principle that simplifies the assembly process. Designers aim to reduce the number of parts and assembly steps, which can significantly cut down on labor costs and time. Techniques like modular design or snap-fit features can make assembly easier and more efficient. It is all about making sure that the product can be assembled quickly.

Design for testing (Turino, 2012; Huhn & Drechsler, 2021) focuses on incorporating testability into the product design, ensuring it can be thoroughly tested throughout the manufacturing process. This includes adding built-in test points and designing for automation in testing. By doing this, companies can streamline quality assurance and deliver higher-quality products.

Design for reliability (Raheja, 2012; Kapur & Pecht, 2014; Crowe & Feinberg, 2017) verifies that a product performs reliably over its expected lifetime under various conditions. This involves designing for environmental resilience and incorporating redundancy and fault tolerance. By thinking ahead about potential failures and integrating predictive maintenance capabilities, designers can enhance the product's overall reliability, ultimately building customer trust.

Incorporating these principles into the design process isn't just about improving efficiency or cutting costs; it is about creating products that truly match users' expectations while maintaining high standards of quality and reliability. When these holistic strategies are embraced, we enhance manufacturability and contribute to a sustainable and responsible approach.

9.1.3 SUSTAINABILITY AND ENVIRONMENT (S)

DfS is of utmost importance these days. Incorporating DfS principles involves using renewable, recyclable, or biodegradable materials, optimizing manufacturing processes, integrating renewable energy sources, and planning for how a product will be treated at the end of its life, think disassembly, reuse, and recycling. It aims to minimize environmental impact while maximizing sustainability throughout a product's entire lifecycle. This means considering a wide range of factors and principles such as material, disassembly, maintenance, lifecycle, circular economy, and durability (Table 9.3).

Within the domain of DfS, there are several important subsets. For example, design for disassembly (Battaïa et al, 2018; Rios et al., 2015; Soh et al., 2014) focuses on making it easy to take products apart at the end of their useful life. This allows for the recovery and reuse of components and materials, ultimately reducing waste. Key strategies for this include using standardized, easily separable fasteners and providing clear disassembly instructions.

Design for maintenance (Gits, 1992; Neeley & Robbins, 2024) ensures that products can be easily maintained and repaired, extending their lifespan and reducing waste. This involves designing for accessibility, using standardized components, and simplifying repair tasks. Resource utilization can be maximized

TABLE 9.3 Principles for Design for Environment and Sustainability.

Element	Principle	Example
Material	Choose sustainable materials that minimize environmental impact.	Using bamboo or recycled plastics in product design.
Disassembly	Ensure products can be easily taken apart for repair, reuse, or recycling.	Furniture designed with removable parts for easy recycling.
Maintenance	Facilitate easy maintenance and repair to extend product lifespan.	Modular electronics that allow for easy replacement of broken components.
Lifecycle	Evaluate the environmental impact of a product throughout its entire lifecycle.	Conducting a life cycle assessment (LCA) for a new product to identify hotspots for improvement.
Circular Economy	Create products that support a circular economy through reuse and recycling.	Clothing made from recycled fibers that can be returned for recycling.
Durability	Design products to last longer and withstand wear and tear, reducing the need for replacements.	High-quality tools that are built to last for years.

and environmental impact can be minimized over time by making maintenance efficient.

Design for lifecycle management (Saaksvuori & Immonen, 2008; Liu et al., 2020) takes a holistic view of a product's journey from conception to end-of-life. The goal is to minimize environmental impact while maximizing economic and social value. This includes considerations like modular design, end-of-life planning, and lifecycle assessments. When manufacturers adopt a lifecycle perspective, they can make smarter decisions that enhance sustainability and support a circular economy.

Design for circular economy (Bocken et al., 2016; Den Hollander, et al., 2017; Moraga et al., 2019) is a strategic approach aimed at creating products that can thrive in a circular economy. Unlike traditional linear models, design for circular economy encourages the continuous cycling of materials, helping to minimize waste and environmental impact. This involves designing modular components, selecting materials that can be easily recovered, and developing product-service systems that promote longer lifespans.

Design for durability (Mesa, et al., 2022) is essential in sustainability and environmental design, as it helps create products that last longer and require fewer resources over their lifecycle. By prioritizing high-quality materials and repairable components, durable design minimizes waste and reduces environmental impact. This approach encourages consumers to value longevity over disposability, enabling a shift toward sustainable consumption patterns. Additionally, durable products contribute to a circular economy by allowing for upgrades and repairs, which helps keep materials in use longer. Ultimately, focusing on durability not only benefits the environment but also promotes responsible production and consumption practices.

By incorporating these principles into product design, we can contribute to a more sustainable and regenerative economic model that meets user needs and supports the health of our planet!

9.1.4 ENTREPRENEURSHIP AND BUSINESS (E)

DfE plays a crucial role in product design, directly influencing the viability and profitability of what we create (Bocken et al., 2016; Amit & Zott, 2020). In a fast-paced market, it is essential for organizations to assess various factors like costs and pricing, resources, supply chain, and logistics (Table 9.4). Design impacts operational efficiency, inventory management, and after-sales service. Companies that incorporate entrepreneurial and business perspectives into their design processes can create products that not only meet technical and environmental standards but also deliver substantial financial benefits and align with their broader business goals (Bridge & O'Neill, 2017; Meluso et al., 2022).

Cost is a critical element in the design strategy (Agard & Bassetto, 2013; Mörtl & Schmied, 2015). Evaluating manufacturing costs, material choices, process optimization, and supply chain logistics is vital. For instance, analyzing how

TABLE 9.4 Principles of Design for Entrepreneurship and Business.

Element	Principle	Example
Cost	Design products that minimize production and operational costs.	Using automation in manufacturing to reduce labor costs.
Resources	*Human resources*: Invest in training employees to enhance skills relevant to product development and customer service.	Offering workshops on new technologies and design techniques.
	Technological resources: Leverage advanced technologies to improve product design and development processes.	Utilizing 3D printing for rapid prototyping and testing.
	Intellectual resources: Protect innovations through patents, trademarks, or copyrights to secure competitive advantage.	Filing patents for unique product designs or technologies.
	Financial resources: Ensure financial resources are allocated effectively to support product development and sustainability initiatives.	Creating a budget that prioritizes investment in eco-friendly technologies.
	Material resources: Select sustainable materials that reduce environmental impact and enhance product life.	Sourcing recycled materials for packaging and product components.
Supply Chain	Optimize the supply chain to ensure timely delivery and cost-effectiveness.	Implementing just-in-time inventory systems to reduce storage costs.
Logistics	Streamline logistics to improve efficiency and reduce transportation costs.	Using route optimization software to minimize delivery times.

design decisions impact raw material procurement and production workflows can lead to more cost-effective solutions. Additionally, considering lifecycle costs helps create products that are financially viable over time. By taking a comprehensive view of cost drivers, companies can design products that are both competitive and sustainable.

Designing with resources and assets in mind is vital for any business (St-Pierre & Audet, 2011; Raddats et al., 2015; Hasan & Starly, 2020). Whether it is manufacturing or service-oriented, firms need a mix of physical, intellectual, and financial assets. For manufacturers, this means having the right equipment and skilled workers. On the other hand, service firms rely heavily on human capital and intangible assets, like brand reputation. Strategically investing in these resources is crucial for achieving competitive advantage and long-term success. Understanding the capabilities and limitations of available manufacturing infrastructure helps safeguard efficient production. This includes assessing equipment capabilities and optimizing workflows to enhance safety and productivity (Sirkin

& ten Houten, 1994; Ulrich & Eppinger, 2016; Qi & Tao, 2018). Effective human resources management is also key to attracting and retaining talent (Jabbour et al., 2013). When it comes to design, focusing on employee well-being and professional development can lead to better outcomes (Yong et al., 2020). By minimizing physical strain in manufacturing and providing clear documentation and training, organizations can encourage their workforce, ultimately improving overall efficiency. Leveraging collective knowledge and expertise is important for intellectual resource (Gómez & Vargas, 2012; Kamasak, 2017). Bringing together cross-functional teams encourages creativity and leads to innovative solutions. Establishing platforms for knowledge sharing creates continuous learning and improvement. Addressing material resources is crucial for minimizing environmental impacts (Oberle et al, 2019; Södersten et al., 2020; Black & Kohser, 2020). Selecting sustainable or recycled materials and optimizing their usage are key considerations. By carefully evaluating material properties and lifecycle impacts, companies can create products that are responsible, cost-effective, and aligned with sustainability goals. Financial resources (Tang et al., 2002) focuses on optimizing resources to support product development and innovation. This involves creating budgets that prioritize investments in technologies and processes, ensuring that financial decisions align with design goals. Effective financial planning allows for exploring alternative funding sources, such as grants or partnerships, to support development initiatives. Ultimately, integrating financial considerations into the design process enhances profitability and contributes to long-term success.

Supply chain considerations are equally important (Pashaei & Olhager, 2015; Ebikake, et al., 2018). Efficient management of the supply network directly impacts on a product's cost, quality, and environmental footprint. Addressing factors like supplier selection, material sourcing, and end-of-life logistics is essential. Strategies like modular design and standardized components can enhance supply chain flexibility and responsiveness. Integrating sustainable practices, such as using local materials, can further improve environmental performance.

Designing for logistics stresses the importance of efficiently moving, storing, and distributing products (Shamout, et al., 2022; Bielecki & Hadas, 2018). Key factors include packaging design, transportation modes, and warehousing requirements. Strategies such as modular packaging and optimizing product dimensions can enhance distribution efficiency. Moreover, planning for reverse logistics supports circular economy principles by facilitating product recovery and reuse.

By integrating these different resources and considerations into the design process, organizations can create products that are innovative, efficient, environmentally supportive, and aligned with business objectives.

9.1.5 GROWTH AND TECHNOLOGY (G)

DfG considerations are crucial for creating innovative and scalable products that can keep up with the ever-changing people's demands and technological

TABLE 9.5 Principles of Design for Growth and Technology.

Element	Principle	Example
Dynamic obsolescence	Design products that can adapt to changing technologies and consumer preferences over time.	A smartphone with upgradeable components to keep up with tech advancements.
Innovation	A culture of continuous improvement and creativity in product design.	Developing a new wearable device that incorporates cutting-edge health monitoring features.
Knowledge	Leverage industry perceptions and user feedback to inform design decisions.	Conducting user research to guide the development of a new software application.
Scalability	Check if products can be scaled up or down when technology changes without significant redesign.	A modular furniture design that allows for easy adjustments in size and functionality.
Digital economy	Utilize digital tools and platforms to enhance product reach and customer experience.	An e-commerce platform that uses data analytics to personalize shopping experiences.
Integration	Incorporate IoT and AI features to enhance product functionality and user engagement.	Smart home devices that learn user preferences for automation.

changes (Liedtka & Ogilvie, 2011; Tassey, 2013). Some of the considerations in DfG are dynamic obsolescence, innovation, knowledge, scalability, digital economy, and integration (Table 9.5).

Dynamic obsolescence (Alzaydi, 2024) refers to how products, services, or technologies can become outdated not just from wear and tear, but due to advancements or shifts in consumer preferences. It highlights how innovation and evolving market trends can make existing offerings less desirable, even though they function well. Designing for dynamic obsolescence means creating products that adapt to the changing technologies and consumer needs. Strategies like modular design or platform-based architecture can help ensure that products remain relevant and viable. Moreover, companies should consider how their products can drive new industries or transform existing ones, contributing to economic and societal progress. Organizations can set themselves up for sustained success in a rapidly evolving marketplace by aligning product design with technological advancements and growth potential.

Innovation allows for the development of groundbreaking products that solve complex problems (Bruton, 2011; Cantamessa & Montagna, 2016). When designing for innovation, it is important to create a culture of creativity and encourage collaboration across teams. Engaging with various stakeholders and adopting design thinking methodologies can play a significant role

(Chapters 4–6). Embracing a "fail-fast, learn-fast" mindset can lead to valuable understandings and breakthroughs through fast prototyping (Chapter 7). Companies should also consider how their products can disrupt existing industries and create new market opportunities while contributing to sustainable development. By making innovation a core principle in their design framework, organizations can push the limits of what's possible, delivering exceptional value to customers and driving positive change.

Knowledge management is vital in the design process, as it helps organizations leverage the collective expertise of their teams and partners (Chouikha, 2016; Secundo et al., 2020). When focusing on knowledge, companies should prioritize sharing and applying both explicit and tacit knowledge throughout product development. Establishing knowledge management systems and enabling collaboration can create a rich environment for innovation. By continuously learning from experiences and capturing lessons learned, organizations can incorporate this knowledge into future designs. This approach enables more informed decision-making and helps create products that better meet the evolving needs of customers and the market.

For scalability, it is essential for organizations to create products that can thrive in dynamic market conditions (Putnik et al., 2013; Jogalekar & Woodside, 2000). This involves evaluating opportunities for expansion and diversification while being mindful of potential risks. Incorporating modular or platform-based architecture and exploring new distribution channels can significantly contribute to a product's adaptability and success. By embedding growth considerations into the design framework, companies can develop products and services that are not only innovative but also capable of sustained, profitable expansion over time.

Considering the digital economy is critical for designing products and services that are optimized for a rapidly evolving technological landscape (Bukht & Heeks, 2017; Xue et al., 2022). Companies should evaluate how emerging technologies—like cloud computing, AI, and the Internet of Things—can enhance product functionality and user experiences (Tao et al., 2014). Integrating smart features and exploring data-driven knowledge can help create seamless interactions between users and products. Additionally, organizations should think about how their offerings can contribute to new digital ecosystems and support a more connected and sustainable economic environment. By aligning product design with the principles of the digital economy, companies can position themselves for success in today's fast-paced marketplace.

Design for integration (Kuebler et al., 2024) focuses on seamlessly incorporating emerging technologies like IoT and AI into products to enhance functionality and user experience. By designing systems that communicate and adapt, products can deliver personalized services and improve efficiency. This approach enables innovation, allowing businesses to respond quickly to market changes and consumer demands. For instance, smart appliances can learn user habits and automate tasks, creating a more intuitive environment. Emphasizing integration not only drives growth by attracting tech-savvy consumers but also positions companies for future advancements, ensuring they remain competitive in a rapidly evolving technological landscape.

9.2 PRODUCT ARCHITECTURE

Product architecture and DfX are closely related concepts in the field of product development. Product architecture is the fundamental structural design of a product, defining how its various components and subsystems are organized and integrated to deliver the desired functionality and performance (Marion et al., 2015; Pashaei & Olhager, 2015; Mesa et al., 2019; Ulrich, 1995). It lays the foundation for the product's development, manufacturing, life cycle management, and cost. Product architecture encompasses the physical arrangement of components, the interfaces between them, and the flow of energy, materials, and information within the product. Designers must carefully consider factors such as modularity, standardization and customization, scalability, and adaptability when defining the product architecture. The decision on these factors will be influenced by company policies (Section 9.6).

9.2.1 TYPES

Product architecture can be broadly classified into modular, integrated, and hybrid. The choice of product architecture depends on various factors, including DfX methodologies, product complexity, the pace of technological change, the need for customization, and the overall business strategy of the organization (Yin et al., 2014; Bonvoisin et al., 2016; Yassine & Naoum-Sawaya, 2017; Ulrich, 1994).

In modular architecture, the product is divided into distinct, self-contained subsystems or modules that can be designed, manufactured, and tested independently (Kamrani & Sa'ed, 2013). Each module has well-defined interfaces and standardized connections, allowing them to be easily interchanged or upgraded without affecting the overall product. The modular design promotes flexibility, customization, and ease of maintenance, as individual modules can be replaced or updated without disrupting the entire system. This approach is often used in products with rapidly evolving technologies where adapting to changing customer needs is crucial.

In an integrated architecture, the product is designed as a cohesive, tightly coupled system with closely interdependent components and subsystems (Yin et al., 2014). The interfaces between the components are often proprietary or specialized, making it difficult to modify or replace individual parts without impacting the overall product performance. Integrated architecture typically results in a more compact, streamlined, and optimized design, as the components are tailored to work seamlessly together. This approach is commonly used in products where performance, efficiency, and a high degree of integration are the primary design objectives, such as in high-end electronics or aerospace applications.

A hybrid architecture combines elements of both modular and integrated design approaches. It features a mix of standardized, modular subsystems and customized, tightly integrated components (Lin et al., 2010). This approach allows for the benefits of both modularity (e.g., flexibility, customization) and integration (e.g., optimized performance, efficiency). Hybrid architecture is often used in complex products, where a balance between adaptability and optimization is required, such as in automotive or industrial equipment design.

9.2.2 STANDARDIZATION

Product architecture involves the balance between standardization (Wang et al., 2016) and customization (Manavis et al., 2024). Companies must carefully consider DfX methodologies, analyze their product portfolio, target market, and manufacturing capabilities to determine the optimal level of standardization and customization that aligns with their strategic objectives.

Standardization refers to the use of common, interchangeable components, interfaces, and processes across a product or product family (Baud-Lavigne et al., 2012). Some benefits of standardization in product architecture include cost optimization, quality and reliability, ease of maintenance and servicing, and knowledge and skill transfer.

Standardized components and interfaces enable the replacement of individual parts. Standardization enables economies of scale in component sourcing, manufacturing, inventory management, and maintenance, leading to reduced overall costs. Standardized components and processes are typically well-tested and proven, ensuring consistent quality and performance across the product line. Standardized components and interfaces simplify the maintenance, repair, and servicing of products, as technicians only need to be familiar with a limited set of parts and procedures. Standardization enables the transfer of knowledge, skills, and best practices across product development, manufacturing, and service teams.

9.2.3 SCALABILITY

Scalability refers to the ability of a product or system to handle increasing or decreasing demands over time, without significant changes to its underlying structure or performance (Matt, 2013; Grijota et al., 2021). In the context of product architecture, scalability can be achieved through various design approaches. A modular product architecture enables scalability by allowing for the easy addition, removal, or replacement of components. Hence, products can be easily scaled up or down in size, capacity, or functionality to meet changing market demands.

Other scalability approaches are flexible interfaces, scalable components, and distributed or cloud-based architecture. Flexible interfaces include well-defined interfaces between the product's components and subsystems that facilitate scalability by seamlessly integrating new or updated parts. Flexible interfaces make adapting the product's architecture easier to accommodate changes in technology, customer requirements, or market conditions. Designing system infrastructure, such as power supply, processors or storage units, to be scalable in capacity and performance allows the overall product to be easily scaled without major architectural changes. This approach ensures that the product can grow or shrink in capability as needed without compromising its core functionality. Adopting a distributed or cloud-based architecture, where certain product functions or resources are hosted on remote servers or in the cloud, can enhance the product's scalability. This approach allows the product to dynamically allocate and scale computing power, storage, or other resources as demand fluctuates, without requiring significant changes to the on-device product architecture. Scalable product architecture is essential for companies to maintain a competitive edge, enabling them to adapt their products to changing market conditions, technological advancements, and customer needs. By designing for scalability, companies can future-proof their products, maximize their return on investment, and ensure long-term relevance in the marketplace.

9.2.4 ADAPTABILITY

Adaptability refers to the ability of a product to adapt to changing requirements, technologies, and market conditions over its lifetime (Uckun et al., 2014). Adaptable product architecture allows for the seamless integration of new features, the incorporation of technological advancements, and customization to meet the evolving customer needs. Modularity and scalability can enable the development of adaptable product architecture. In addition, consideration for extensibility, upgradability, and user configurability can enable the adaptability of product architecture. Designing the product architecture with extensibility enables the easy addition of new features, capabilities, or integration points. Extensibility can be achieved through open APIs, plugin-based architectures, or the incorporation of expansion slots or modular components. Incorporating upgradability into the product architecture allows customers to easily update or enhance the product over time without needing a complete replacement. Upgradability can be accomplished through software updates, hardware component swaps, or integrating new technologies. Providing users with the ability to configure and customize the product to their specific needs enhances its adaptability. User configurability can include adjusting settings, preferences, or selecting various features or components.

9.3 OTHER CONSIDERATIONS

9.3.1 DIGITAL VS PHYSICAL

The balance between digital and physical considerations (Goebeler et al., 2024; Weichel et al., 2015) is a complex and nuanced aspect of the DfX approach. As the world becomes more digitized with advancements in digital twins, organizations must thoughtfully assess how to effectively blend the digital and physical domains to develop products and services that align with customers' changing needs and preferences. When designing for this balance, companies should strive to leverage the strengths of each domain to enhance user experiences, improve operational efficiency, and enable seamless integration between the online and offline environments.

In the digital domain, organizations can explore the integration of smart technologies, data-driven knowledge, and virtual/augmented reality to enhance product functionality, personalization, and accessibility. This may involve developing robust digital platforms, implementing intuitive user interfaces, and leveraging the power of the cloud to deliver innovative, scalable, and on-demand solutions. Conversely, in the physical domain, companies should consider the tactile and sensory qualities of their products, the ergonomics and aesthetics of their designs, and the importance of tangible interactions that enable emotional connections with users. This may involve the use of proper materials, thoughtful ergonomic design, and the incorporation of physical touchpoints that complement the digital experience.

By striking the right balance between digital and physical considerations, organizations can create products that are not only technologically advanced but also deeply rooted in human-centric design principles, delivering a cohesive and holistic experience that matches customer expectations and drives long-term success in the evolving marketplace.

9.3.2 REAL VS VIRTUAL

The balance between real and virtual considerations (Pontonnier et al., 2014) comes into play as organizations navigate the blurring boundaries between the physical and digital domains. In an era where virtual and augmented reality technologies are rapidly advancing, and the metaverse is emerging as a new frontier for human interaction and commerce, designers must carefully evaluate how to integrate the best of both the real and virtual worlds to create products and services that seamlessly cater to user needs and preferences.

When designing for this balance, companies should strive to leverage the unique strengths and capabilities of each domain to enhance user experiences, create deeper connections, and drive innovation. In the real world, the tactile, sensory, and emotional qualities of physical products and environments remain deeply important, offering opportunities for immersive, embodied interactions that tap into our innate human desires for tangible engagement. Conversely, the virtual sphere presents new possibilities for exploration, experimentation, and the creation of experiences that exceed the limitations of the physical world, enabling personalization, scalability, and the integration of cutting-edge technologies.

By carefully considering the interplay between real and virtual elements, organizations can develop value that provide the best of both worlds, delivering seamless and enriching experiences that fit with users across different touchpoints and platforms. This balance can unlock new avenues for creativity, productivity, and the continuous evolution of products and services, ultimately positioning companies for success in the increasingly blended landscape of real and virtual domains.

9.3.3 PRODUCT VS SERVICE

Organizations seek to create holistic offerings that deliver exceptional value to customers. In an era where the business offering between physical products and digital services is becoming increasingly unclear, designers must carefully evaluate how to integrate these two fundamental elements to develop solutions that cater to evolving consumer needs and preferences (Pezzotta et al., 2014).

When designing for the balance between product and service, companies should strive to leverage the unique strengths and capabilities of both to create compelling and differentiated offerings. On the product side, the tangible nature of physical goods can provide a sense of ownership, tactile engagement, and emotional connection with users, while also offering opportunities for customization and personalization. Services, on the other hand, can introduce greater flexibility, scalability, and the ability to adapt to changing market conditions, often through the integration of digital technologies and data-driven knowledge.

By carefully considering the interplay between product and service components, organizations can develop holistic solutions that seamlessly combine the best of both worlds. This may involve the creation of smart, connected products that are supported by comprehensive service offerings, or the development of service-centric platforms that are enriched by physical touchpoints and experiences. Striking the right balance between product and service considerations can unlock new avenues for value creation, customer loyalty, and sustainable competitive advantage in the evolving marketplace.

9.4 CHALLENGES IN IMPLEMENTING DFX

It will be nice to talk about the conflicting nature of the various DfX and implementing will require tradeoffs. Not all can be achieved at the same time. How can one trade off to improve humanity? As the world progresses, there may be ways to incorporate new aspects while foregoing others. Sometimes it happens in reverse. In ancient times, people used sustainable materials for roofing (e.g., clay, bamboo roof tiles). But then it changed to materials like asbestos that were toxic and carcinogenic. Now eco-friendly roofing is emerging again. Use of trade-off analysis. Decision matrices, lifecycle thinking, total cost of ownership etc. Use digital twins and AI optimization to simulate trade-offs in a virtual environment before committing to a design.

9.5 UNITED NATIONS' (UN) SUSTAINABILITY GOALS AND DFX

The alignment of the United Nations' (UN) Sustainable Development Goals (SDGs) (https://sdgs.un.org/goals) with the DfX frameworks of DfP, DfF, DfS, DfE, and DfG is represented in Table 9.6. Each contributes differently to the realization of the 17 SDGs, marked by asterisks indicating the strength of alignment, with * signifying the weakest connection and ***** indicating the strongest. For example, for Zero Hunger, sustainable agriculture practices (DfS) contribute significantly to reducing hunger (*****); innovative fabrication techniques (DfF) can improve food production methods (****); ethical and business considerations (DfP and DfE) in food production and distribution are important but less direct. (**). New development in technology (DfG) might have future implications, but for the current situation, new technology can be expensive and will have a less direct impact on hunger (*).

9.6 BENCHMARKING AND PRODUCT POSITIONING

Product benchmarking using DfX involves a comprehensive evaluation of products against established best practices and recognized standards in various areas (Table 9.7). This includes metrics such as efficiency, sustainability, and resource management. For instance, Efficiency metrics might assess production cycle times, material usage, and waste reduction strategies, while standards like LEED (Leadership in Energy and Environmental Design) (www.usgbc.org/leed) and ISO 14001 (Environmental Management Systems) (www.iso.org) provide frameworks for evaluating environmental performance and sustainability in product design.

When comparing two or more similar products, assessing their adherence to each DfX principle is essential. This evaluation can be performed using both objective and subjective ratings (see Table 9.7). Due to the inherent subjectivity in these assessments, it is crucial to involve a multidisciplinary team comprising higher management, design, and technical staff. This collaborative effort will ensure that the DfX ratings align with the product requirements, user needs, the company's corporate identity, brand positioning, and competing products and services.

TABLE 9.6 Alignment of UN Sustainability Goals with DfX.

UN sustainability goals	People and Ethics (DfP)	Fabrication and Making (DfF)	Sustainability and Environment (DfS)	Entrepreneurship and Business (DfE)	Growth and Technology (DfG)
1. No Poverty	***	***	**	****	*
2. Zero Hunger	**	****	*****	**	*
3. Good Health and Well-being	***	***	*****	**	*
4. Quality Education	*****	**	***	**	*
5. Gender Equality	*****	**	**	**	*
6. Clean Water and Sanitation	**	***	*****	**	*
7. Affordable and Clean Energy	**	****	*****	****	*****
8. Decent Work and Economic Growth	***	*****	***	*****	****
9. Industry, Innovation, and Infrastructure	**	*****	****	*****	*****
10. Reduced Inequalities	*****	**	**	**	*
11. Sustainable Cities and Communities	**	****	*****	***	****
12. Responsible Consumption and Production	**	*****	*****	****	****
13. Climate Action	**	****	*****	****	*****
14. Life Below Water	**	**	*****	*	*
15. Life on Land	**	**	*****	*	*
16. Peace, Justice, and Strong Institutions	*****	**	**	**	*
17. Partnerships for the Goals	**	***	****	****	****

Notes: The scores reflect perceived contributions of each design framework to achieving the respective SDGs, based on typical practices and research. A higher number of asterisks indicates a more significant alignment and potential impact of the design frameworks on achieving the goals.

TABLE 9.7 Rating of the DfX.

	Level	Notes
Design for people and ethics		
User		
Inclusion		
Empathy		
Social responsibility		
Ethics		
Culture		
Design fabrication and making		
Manufacturing		
Production		
Quality		
Assembly		
Testing		
Reliability		
Design for environment and sustainability		
Material		
Disassembly		
Maintenance		
Lifecycle		
Circular economy		
Durability		
Design for entrepreneurship and business		
Cost		
Resources		
Supply chain		
Logistics		
Design for growth and technology		
Dynamic obsolescence		
Innovation		
Knowledge		
Scalability		
Digital economy		
Integration		

The rating level can be defined as follows:

Very high (*****): The product excels in this area, demonstrating outstanding performance and adherence to best practices and standards.

High (****): The product performs well, with only minor improvements needed to reach the highest standards.

Normal (***): The product meets basic expectations but has room for improvement in several areas.
Low (**): The product falls short of accepted standards and requires significant enhancements to become competitive.
Very low (*): The product shows minimal or no consideration for the principle, indicating a need for major redesign or overhaul.

The above is a structured approach for benchmarking products or even prototypes.

9.7 CONCLUSION

The design for X principles presented emphasize the critical importance of integrating various design considerations throughout the product development lifecycle. By examining distinct aspects, we can appreciate the multifaceted nature of effective design. Each section highlights specific strategies and methodologies that enhance product performance, sustainability, and market viability. For instance, DfF emphasizes the need for efficient making processes, while DfE focuses on minimizing environmental impact through sustainable practices. Additionally, the discussions on resources and finance reveal how prudent management can lead to cost-effective innovations. The principles of inclusion and ethics further ensure that products are designed with social responsibility in mind, creating a culture of accessibility and equity. Meanwhile, DfG illustrates the role of technology in driving growth and innovation, emphasizing the importance of adapting to a rapidly changing digital landscape.

In conclusion, embracing DfX principles permits organizations to create products that are not only high-performing and cost-effective but also sustainable, ethical, and aligned with societal needs. As we move forward, integrating these principles into design processes will be crucial for enabling innovation and achieving long-term success in an increasingly complex and interconnected world. Based on the DfX, a clear product specification will make sure that every aspect of the product aligns with stakeholders and regulatory requirements.

> Design is intelligence made visible.
>
> —Alina Wheeler

REFERENCES

Adjoul, O., Benfriha, K., & Aoussat, A. (2021). Design for maintenance: new algorithmic approach. *Journal of Quality in Maintenance Engineering*, 27(1), 129–143.
Agard, B., & Bassetto, S. (2013). Modular design of product families for quality and cost. *International Journal of Production Research*, 51(6), 1648–1667.

Alzaydi, A. (2024). Balancing creativity and longevity: The ambiguous role of obsolescence in product design. *Journal of Cleaner Production*, 445, 141239.

Amit, R., & Zott, C. (2020). *Business Model Innovation Strategy: Transformational Concepts and Tools for Entrepreneurial Leaders*. Wiley.

Anderson, D. M. (2020). *Design for Manufacturability: How to Use Concurrent Engineering to Rapidly Develop Low-cost, High-quality Products for Lean Production*. Productivity Press.

Battaïa, O., Dolgui, A., Heragu, S. S., Meerkov, S. M., & Tiwari, M. K. (2018). Design for manufacturing and assembly/disassembly: joint design of products and production systems. *International Journal of Production Research*, 56(24), 7181–7189.

Baud-Lavigne, B., Agard, B., & Penz, B. (2012). Mutual impacts of product standardization and supply chain design. *International Journal of Production Economics*, 135(1), 50–60.

Bielecki, M., & Hadas, L. (2018). Selected aspects of product logistic efficiency as part of the design for logistics concept-case study. *Business Logistics in Modern Management*. https://ojs.srce.hr/index.php/plusm/article/view/7887

Bissett-Johnson, K., & Radcliffe, D. F. (2021). Engaging engineering students in socially responsible design using global projects. *European Journal of Engineering Education*, 46(1), 4–26.

Black, J. T., & Kohser, R. A. (2020). *DeGarmo's Materials and Processes in Manufacturing*. Wiley.

Bocken, N. M., De Pauw, I., Bakker, C., & Van Der Grinten, B. (2016). Product design and business model strategies for a circular economy. *Journal of Industrial and Production Engineering*, 33(5), 308–320.

Bonvoisin, J., Halstenberg, F., Buchert, T., & Stark, R. (2016). A systematic literature review on modular product design. *Journal of Engineering Design*, 27(7), 488–514.

Boothroyd, G., Dewhurst, P., & Knight, W.A. (2010). *Product Design for Manufacture and Assembly*. Taylor & Francis.

Boughzala, Y. (2021). The role of design thinking in corporate social responsibility (CSR) strategy and its influence on innovation. *Journal of Innovation Economics & Management*, 34(1), 169–195.

Bridge, S., & O'Neill, K. (2017). *Understanding Enterprise: Entrepreneurs and Small Business*. Bloomsbury Publishing.

Bruton, D. (2011). Learning creativity and design for innovation. *International Journal of Technology and Design Education*, 21, 321–333.

Bukht, R., & Heeks, R. (2017). Defining, conceptualising and measuring the digital economy. Development Informatics working paper, 68. Available at SSRN:https://ssrn.com/abstract=3431732 or http://dx.doi.org/10.2139/ssrn.3431732

Cantamessa, M., & Montagna, F. (2016). *Management of Innovation and Product Development* (p. 377). Springer.

Chouikha, M. B. (2016). *Organizational Design for Knowledge Management*. Wiley.

Clark, G., Kosoris, J., Hong, L. N., & Crul, M. (2009). Design for sustainability: current trends in sustainable product design and development. *Sustainability*, 1(3), 409–424.

Cross, N. (2021). *Engineering Design Methods: Strategies for Product Design*. Wiley.

Crowe, D., & Feinberg, A. (2017). *Design for Reliability*. CRC Press.

Dale, B. G., Van Der Wiele, T., & Van Iwaarden, J. (2007). *Managing Quality*. Wiley.

De Mul, J. (2015). The game of life: Narrative and ludic identity formation in computer games. In *Representations of Internarrative Identity* (pp. 159–187). Palgrave Macmillan UK.

Den Hollander, M. C., Bakker, C. A., & Hultink, E. J. (2017). Product design in a circular economy: Development of a typology of key concepts and terms. *Journal of Industrial Ecology*, 21(3), 517–525.

Ebikake, O. E., Sassanelli, C., & Terzi, S. (2018). PSS design through Design for Supply Chain: State of the art review. *Procedia CIRP*, 70, 198–203.

El-Haik, B. (2005). *Axiomatic Quality: Integrating Axiomatic Design with Six-sigma, Reliability, and Quality Engineering*. Wiley.

ElMaraghy, W., ElMaraghy, H., Tomiyama, T., & Monostori, L. (2012). Complexity in engineering design and manufacturing. *CIRP Annals*, 61(2), 793–814.

El Wakil, S. D. (2019). *Processes and Design for Manufacturing*. CRC Press.

Edensor, T. (2020). *National Identity, Popular Culture and Everyday Life*. Routledge.

Goebeler, L., Hukal, P., & Xiao, X. (2024). Four roles of physicality in digital innovation: A theoretical review. *Journal of Strategic Information Systems*, 33(4), 101862. https://doi.org/10.1016/j.jsis.2024.101862

Gits, C. W. (1992). Design of maintenance concepts. *International Journal of Production Economics*, 24(3), 217–226. https://doi.org/10.1016/0925-5273(92)90133-R.

Gómez, J., & Vargas, P. (2012). Intangible resources and technology adoption in manufacturing firms. *Research Policy*, 41(9), 1607–1619.

Grijota, C. G., Acero, R., & Yagüe-Fabra, J. A. (2021). Product development methodology" scalability". *Procedia CIRP*, 100, 571–576.

Hasan, M., & Starly, B. (2020). Decentralized cloud manufacturing-as-a-service (CMaaS) platform architecture with configurable digital assets. *Journal of Manufacturing Systems*, 56, 157–174.

Huhn, S., & Drechsler, R. (2021). *Design for Testability, Debug and Reliability*. Springer International Publishing.

Jabbour, C. J. C., de Sousa Jabbour, A. B. L., Govindan, K., Teixeira, A. A., & de Souza Freitas, W. R. (2013). Environmental management and operational performance in automotive companies in Brazil: the role of human resource management and lean manufacturing. *Journal of Cleaner Production*, 47, 129–140.

Jogalekar, P., & Woodside, M. (2000). Evaluating the scalability of distributed systems. *IEEE Transactions on Parallel and Distributed Systems*, 11(6), 589–603.

Kamrani, A. K., & Sa'ed, M. S. (2013). *Product Design for Modularity*. Springer Science & Business Media.

Kamasak, R. (2017). The contribution of tangible and intangible resources, and capabilities to a firm's profitability and market performance. *European Journal of Management and Business Economics*, 26(2), 252–275.

Kapur, K. C., & Pecht, M. (2014). *Reliability Engineering* (Vol. 86). Wiley.

Kuebler, M., Lange, C., Glasmacher, B., Düser, T., & Albers, A. (2024). Future-robust product design–validating influencing factors on upgradeable mechatronic systems. *Proceedings of the Design Society*, 4, 633–642.

Liedtka, J., & Ogilvie, T. (2011). *Designing for Growth: A Design Thinking Tool Kit for Managers*. Columbia University Press.

Lin, S., Narayan, R. J., & Lee, Y. S. (2010). Hybrid client–server architecture and control techniques for collaborative product development using haptic interfaces. *Computers in Industry*, 61(1), 83–96.

Liu, Y., Zhang, Y., Ren, S., Yang, M., Wang, Y., & Huisingh, D. (2020). How can smart technologies contribute to sustainable product lifecycle management?. *Journal of Cleaner Production*, 249, 119423.

Ma, X., Yan, X., & Guo, Y. (2020). Robust design for quality characteristics of mechanical processing products based on algorithm decision-making. *Journal of Intelligent & Fuzzy Systems*, 38(6), 6917–6926.

Manavis, A., Minaoglou, P., Efkolidis, N., & Kyratsis, P. (2024). Digital Customization for Product Design and Manufacturing: A Case Study within the Furniture Industry. *Electronics*, 13(13), 2483.

Marion, T. J., Meyer, M. H., & Barczak, G. (2015). The influence of digital design and IT on modular product architecture. *Journal of Product Innovation Management*, 32(1), 98–110.

Matt, D. T. (2013). Design of a scalable assembly system for product variety: a case study. *Assembly Automation*, 33(2), 117–126.

McAdams, D. A., & Kostovich, V. (2011). A framework and representation for universal product design. *International Journal of Design*, 5(1), 29–42.

Meluso, J., Johnson, S., & Bagrow, J. (2022). Flexible environments for hybrid collaboration: Redesigning virtual work through the four orders of design. *Design Issues*, 38(1), 55–69.

Mesa, J. A., Gonzalez-Quiroga, A., Aguiar, M. F., & Jugend, D. (2022). Linking product design and durability: A review and research agenda. *Heliyon*, 8(9), e10734.

Mesa, J. A., Esparragoza, I., & Maury, H. (2019). Trends and perspectives of sustainable product design for open architecture products: Facing the circular economy model. *International Journal of Precision Engineering and Manufacturing-Green Technology*, 6, 377–391.

Molloy, O., Warman, E. A., & Tilley, S. (2012). *Design for Manufacturing and Assembly: Concepts, Architectures and Implementation*. Springer Science & Business Media.

Moraga, G., Huysveld, S., Mathieux, F., Blengini, G.A., Alaerts, L., Van Acker, K., de Meester, S., & Dewulf, J. (2019) Circular economy indicators: What do they measure? *Resource Conservation and Recyclying*, 146, 452–461. https://doi.org/10.1016/j.resconrec.2019.03.045.

Mörtl, M., & Schmied, C. (2015). Design for cost—a review of methods, tools and research directions. *Journal of the Indian Institute of Science*, 95(4), 379–404.

Nahmias, S., & Cheng, Y. (2009). *Production and Operations Analysis* (Vol. 6). McGraw-Hill.

Neeley, J. R., & Robbins, W. W. (2024). Design for maintenance. In Department of Defense Human Factors Engineering Technical Advisory Group (DOD HFE TAG).

Null, R. (Ed.). (2013). *Universal Design: Principles and Models*. CRC Press.

Oberle, B., Bringezu, S., Hatfield-Dodds, S., Hellweg, S., Schandl, H., & Clement, J. (2019). *Global Resources Outlook: 2019*. International Resource Panel, United Nations Envio, Paris, France.

Park, S. H., & Antony, J. (2008). *Robust Design for Quality Engineering and Six Sigma*. World Scientific.

Pashaei, S., & Olhager, J. (2015). Product architecture and supply chain design: a systematic review and research agenda. *Supply Chain Management: An International Journal*, 20(1), 98–112.

Pezzotta, G., Pinto, R., Pirola, F., & Ouertani, M. Z. (2014). Balancing product-service provider's performance and customer's value: The service engineering methodology (SEEM). *Procedia CIRP*, 16, 50–55.

Poli, C., & Poli, C. (2001). *Design for Manufacturing: A Structured Approach*. Butterworth-Heinemann.

Pontonnier, C., Dumont, G., Samani, A., Madeleine, P., & Badawi, M. (2014). Designing and evaluating a workstation in real and virtual environment: toward virtual reality based ergonomic design sessions. *Journal on Multimodal User Interfaces*, 8, 199–208.

Putnik, G., Sluga, A., ElMaraghy, H., Teti, R., Koren, Y., Tolio, T., & Hon, B. (2013). Scalability in manufacturing systems design and operation: State-of-the-art and future developments roadmap. *CIRP Annals*, 62(2), 751–774.

Qi, Q., & Tao, F. (2018). Digital twin and big data towards smart manufacturing and industry 4.0: 360 degree comparison. *IEEE Access*, 6, 3585–3593.

Raddats, C., Burton, J., & Ashman, R. (2015). Resource configurations for services success in manufacturing companies. *Journal of Service Management*, 26(1), 97–116.

Raheja, D. (2012). Design for reliability paradigms. *Design for Reliability*, 1–13.

Rios, F. C., Chong, W. K., & Grau, D. (2015). Design for disassembly and deconstruction – challenges and opportunities. *Procedia Engineering*, 118, 1296–1304. https://doi.org/10.1016/j.proeng.2015.08.485

Saaksvuori, A. & Immonen, A. (2008). *Product Lifecycle Management*. Springer. https://doi.org/10.1007/978-3-540-78172-1

Secundo, G., Ndou, V., Del Vecchio, P., & De Pascale, G. (2020). Sustainable development, intellectual capital and technology policies: A structured literature review and future research agenda. *Technological Forecasting and Social Change*, 153, 119917.

Shamout, M., Ben-Abdallah, R., Alshurideh, M., Alzoubi, H., Kurdi, B. A., & Hamadneh, S. (2022). A conceptual model for the adoption of autonomous robots in supply chain and logistics industry. *Uncertain Supply Chain Management*, 10(2), 577–592.

Sirkin, T., & ten Houten, M. (1994). The cascade chain: A theory and tool for achieving resource sustainability with applications for product design. *Resources, Conservation and Recycling*, 10(3), 213–276.

Södersten, C. J., Wood, R., & Wiedmann, T. (2020). The capital load of global material footprints. *Resources, Conservation and Recycling*, 158, 104811.

Soh, S. L., Ong, S. K., & Nee, A. Y. C. (2014). Design for disassembly for remanufacturing: methodology and technology. *Procedia CIRP*, 15, 407–412.

St-Pierre, J., & Audet, J. (2011). Intangible assets and performance: Analysis on manufacturing SMEs. *Journal of Intellectual Capital*, 12(2), 202–223.

Taebi, B. (2021). *Ethics and Engineering: An Introduction*. Cambridge University Press.

Tao, F., Zuo, Y., Da Xu, L., & Zhang, L. (2014). IoT-based intelligent perception and access of manufacturing resource toward cloud manufacturing. *IEEE Transactions on Industrial Informatics*, 10(2), 1547–1557.

Tang, J., Fung, R. Y., Xu, B., & Wang, D. (2002). A new approach to quality function deployment planning with financial consideration. *Computers & Operations Research*, 29(11), 1447–1463.

Tassey, G. (2013). Beyond the business cycle: The need for a technology-based growth strategy. *Science and Public Policy*, 40(3), 293–315.

Tolio, T. (2008). *Design of Flexible Production Systems*. Springer.

Turino, J. (2012). *Design to Test: A Definitive Guide for Electronic Design, Manufacture, and Service*. Springer Science & Business Media.

Uckun, S., Mackey, R., Do, M., Zhou, R., Huang, E., & Shah, J. J. (2014). Measures of product design adaptability for changing requirements. *AI EDAM*, 28(4), 353–368.

Ulrich, K. (1994). Fundamentals of Product Modularity. In: Dasu, S., Eastman, C. (eds) *Management of Design*. Springer, pp. 219–231, https://doi.org/10.1007/978-94-011-1390-8_12

Ulrich, K. (1995). The role of product architecture in the manufacturing firm. *Research Policy*, 24(3), 419–440. https://doi.org/10.1016/0048-7333(94)00775-3

Ulrich, K. T., & Eppinger, S. D. (2016). *Product Design and Development*. McGraw-Hill.

Wang, Z., Zhang, M., Sun, H., & Zhu, G. (2016). Effects of standardization and innovation on mass customization: An empirical investigation. *Technovation*, 48, 79–86.

Weichel, C., Hardy, J., Alexander, J., & Gellersen, H. (2015, November). ReForm: integrating physical and digital design through bidirectional fabrication. In Proceedings of the 28th Annual ACM Symposium on User Interface Software & Technology (pp. 93–102).

Xue, Y., Tang, C., Wu, H., Liu, J., & Hao, Y. (2022). The emerging driving force of energy consumption in China: does digital economy development matter?. *Energy Policy*, 165, 112997.

Yassine, A. A., & Naoum-Sawaya, J. (2017). Architecture, performance, and investment in product development networks. *Journal of Mechanical Design*, 139(1), 011101.

Yin, Y., Kaku, I., & Liu, C. (2014). Product architecture, product development process, system integrator and product global performance. *Production Planning & Control*, 25(3), 203–219.

Yin, Y., Stecke, K. E., & Li, D. (2018). The evolution of production systems from Industry 2.0 through Industry 4.0. *International Journal of Production Research*, 56(1-2), 848–861.

Yong, J. Y., Yusliza, M. Y., Ramayah, T., Chiappetta Jabbour, C. J., Sehnem, S., & Mani, V. (2020). Pathways towards sustainability in manufacturing organizations: Empirical evidence on the role of green human resource management. *Business Strategy and the Environment*, 29(1), 212–228.

10 Design to Specification

> It is not enough that we build products that function, that are understandable and usable; we also need to build products that bring joy and excitement, pleasure and fun, and, yes, beauty to people's lives.
>
> —Don Norman

10.1 SPECIFICATION IN DESIGN

In product design, specifications bridge conceptual ideas with practical implementation. Imagine you are designing a backpack for mountain climbers. One climber tells you, "I need to carry gear safely on long hikes." That's a need: a broad, often fuzzy statement of what the user wants to achieve or solve. It is rooted in human experience, often emotional or functional, but not yet specific. Now, let us zoom in. To meet that need, you decide the backpack must have "a capacity of at least 40 liters, weather-resistant material, and straps that distribute weight evenly." These are requirements: clear, measurable, and actionable specifications that guide design and development. Think of needs as the "why" and requirements as the "how." In contrast, market requirements define the strategic rationale for developing the product, covering the broader market need, business objectives, target audience, and competitive positioning.

Product requirements are the features and functions of a product to meet the user's expectations. The product requirements are then translated into specifications, which are precise, measurable criteria that dictate design parameters, material choices, performance standards, and compliance considerations. There are four key areas that are essential for creating successful products in today's competitive markets.

DOI: 10.1201/9781003476177-10

(a) *Product specification*: These have been the backbone of design processes for decades. They explore foundational practices and their impact on the design outcome, drawing from understanding on integrating user needs into the product (Cooper, 2019) and other strategies in product design and development (Ulrich & Eppinger, 2015). Conventional specifications typically include clear definitions of product functions, performance metrics, material selection, dimensional tolerances, safety requirements, and environmental conditions for use. For example, a conventional specification for a consumer laptop may include requirements such as a 15.6-inch display, 1080p resolution, an Intel i7 processor, 16GB RAM, battery life of at least 10 hours, and compliance with Conformité Européenne (CE) and Federal Communications Commission (FCC) regulations. These measurable criteria help confirm that the final product meets customer expectations and industry standards.

(b) *Iterative specifications*: These are the dynamic specifications to emphasize flexibility and continuous improvement. This approach allows for ongoing refinements based on stakeholders' feedback and evolving project needs, highlighted through sketching user experiences (Buxton, 2007) and principles for making everyday things intuitive and user-friendly (Norman, 2013).

(c) *Regulatory compliance*: This highlights the importance of meeting relevant industry standards and regulatory requirements, which are essential for maintaining safety, quality, and marketability. While compliance itself is not a specification, it defines mandatory requirements that product specifications must fulfill. The discussion will include an overview of specific standards such as ISO 13485 (2016) for medical devices, EIA-632 (1999) for engineering systems, ISO 26262-1 (2018) for automotive functional safety, and DO-178C for aviation software, demonstrating how these regulations across diverse industries guide the formulation of specifications and are integral to successful product deployment.

(d) *Complex product and system specifications*: In contrast to standard product specifications that focus on individual components or straightforward assemblies, complex product and system specifications address the intricacies of integrated, multifaceted systems. These systems often comprise numerous interdependent subsystems, each with their own set of requirements, constraints, and interactions. Managing such complexity necessitates a holistic approach that considers the entire system's architecture, guaranteeing cohesive functionality across all components. For example, in pharmaceutical equipment, specifications must account for stringent regulatory compliance. Developing specifications for such complex systems involves coordinated efforts among multidisciplinary teams, rigorous interface management, and comprehensive documentation to capture evolving product specifications. These specifications must also accommodate scalability, adaptability, and compliance with stringent industry standards, all while maintaining system integrity and performance. In essence, complex product and system specifications extend beyond the scope of traditional product specifications by addressing the challenges of designing and integrating elaborate, interrelated systems.

Each of these four collectively builds a comprehensive view of the product specification landscape, emphasizing the need for a thorough understanding of both traditional practices and innovative approaches to meet the demands of modern product design.

10.1.1 PRODUCT SPECIFICATION

The roles of market requirements document (MRD) and product requirements documents (PRD) are pivotal in traditional product specification processes because they serve as *foundational blueprints* to align business strategy with product execution (Phaal et al., 2006; Cooper, 2019). Think of the MRD as "why should we build it." It uncovers market needs, while the PRD is the "what exactly are we e building," laying out the product blueprint with engineering precision. Together, they turn a good idea into a product that people want and know how to use.

Drawing from the foundational empathy work introduced in Chapter 4, the MRD effectively captures a comprehensive knowledge of market needs and customer expectations, serving as the initial blueprint for product development. The MRD outlines what the market requires and why a product is needed, providing a structured approach to gathering, prioritizing, and validating customer knowledge (Gaskin et al., 2015). By consolidating qualitative and quantitative market research, the MRD guarantees that the design process is user-centered and aligned with business goals.

The PRD translates these empathetically gathered and strategically framed market requirements into specific, technical specifications (Ulrich & Eppinger, 2015). This document outlines the engineering requirements, system architecture, materials, dimensions, compliance standards, and functional constraints (as discussed in Chapter 9) necessary to guide product development and maintain alignment with both user needs and business objectives (Otto & Wood, 2001). The PRD serves as a critical communication tool between cross-functional teams, helping designers, engineers, and stakeholders maintain a shared vision throughout development (Wiegers & Beatty, 2013; Pahl et al., 2007).

Imagine trying to build a house without knowing why it is needed (vacation home or family dwelling?) or what the owner expects (number of rooms, layout, budget). That is exactly what happens when you skip the MRD and PRD. The MRD sets the vision, and the PRD turns that vision into blueprints. Together, they ensure you are not just building something right but building the right thing (Figure 1.4). The specifications live primarily in the PRD, where the details are spelled out.

By effectively bridging user needs, strategic problem framing, and technical execution, the MRD and PRD together facilitate alignment across product vision, customer expectations, and engineering feasibility (Kumar, 2012).

Market Requirements Document (MRD)

Let's look at the example of a chair. The MRD serves as a foundational blueprint in the development of an ergonomic office chair, aimed at enhancing seating comfort in the workplace. Each section of the MRD (Table 10.1) is crafted to confirm that every aspect of the market requirements and the product's

TABLE 10.1 Market Requirements Document (MRD) for Office Chair.

Section Title	Description	Details
1. Executive summary	Provide a concise overview of the chair development aimed at enhancing seating comfort.	*Problem*: Prolonged sitting leads to discomfort among office users. *Product Solution*: Develop an ergonomic office chair with adjustable features to promote comfort and good posture. *Key Goals* —Achieve a 5% market share in the ergonomic chair segment within the first year. —Attain a customer satisfaction rating of 4.5/5.
2. Market opportunity	Analyze the market size, trends, for ergonomic seating solutions.	*Market Overview*: Increasing awareness of workplace ergonomics drives demand for adjustable office chairs. *Market Size and Growth*: The global ergonomic chair market is projected to grow from \$11.13 billion in 2025 to \$19.12 billion by 2033.
3. Customer and user profiles	Define the target customers and users, focusing on their specific needs and how the product will meet these needs.	*Target Market*: Office workers, remote employees *User Personas*: Remote software developer requiring adjustable lumbar support for long coding sessions. *Customer Needs*: Adjustable seat depth and height, lumbar support, breathable materials, and aesthetic design suitable for both offices and home offices.

(continued)

TABLE 10.1 (Continued)

Section Title	Description	Details
4. Market needs	Identify and prioritize the specific market needs the product will address	*Problem Statement*: How might we enhance sitting comfort in the office? *Needs* —Adjustable chair settings to fit different body dimensions. —Maintaining comfort during 8+ hour workdays. —Enhancing focus and productivity through ergonomic support.
5. Product scope	Provide a high-level outline of the product scope that will meet the market needs.	*Product Overview*: An ergonomic office chair featuring adjustable functions and breathable materials.
6. Competitive landscape	Analyze the competition, highlighting the gaps that your product will fill.	*Competitor Profiles* —Herman Miller Aeron: High-end ergonomic chair with extensive adjustability. —Steelcase Leap: Offers dynamic lumbar support and seat depth adjustment. *Competitive Differentiation* —Our chair offers comparable ergonomic features at a more accessible price point. —Targeting mid-range consumers seeking quality and affordability. *SWOT Analysis* —Strengths: Comprehensive adjustability; competitive pricing. —Weaknesses: Less brand recognition compared to established competitors. —Opportunities: Growing remote work trend increases demand for home office furniture. —Threats: Market saturation and aggressive pricing by competitors.
7. Go-to-market strategy	Outline the approach for bringing the product to market, including positioning and messaging.	*Product Positioning*: Marketed as a premium ergonomic chair offering high-end features at a mid-range price. *Value Proposition*: Combines comfort, adjustability, and design to enhance productivity and well-being. *Pricing Strategy*: Target retail price of \$350–\$400. *Distribution Channel*: Online platforms (company web site, Amazon), partnerships with office furniture retailers. Marketing Plan – Digital marketing campaigns focusing on ergonomics and health benefits; collaborations with workplace wellness influencers.

TABLE 10.1 (Continued)

Section Title	Description	Details
8. Success metrics	Define the key performance indicators (KPIs) that will determine the success of the product.	*Market Success Criteria*: Achieve $2 million in sales revenue within the first year; secure 5% market share in the ergonomic chair segment. *Customer Success Criteria*: Attain a Net Promoter Score (NPS) of 60+; maintain a return rate below 2 percent. *Product Success Criteria*: Ensure 90 percent of customers report improved comfort and posture; receive an average product rating of 4.5/5 across sales platforms.
9. Risks and Assumptions	Identify the key risks and assumptions underlying the product's market entry and success.	*Market Risks*: Potential market saturation; economic downturn affecting consumer spending. *Product Risks*: Supply chain disruptions; challenges in meeting ergonomic certification standards. *Assumptions*: Prolonged sitting leads to discomfort among users.

strategic direction is thoroughly explored and documented, setting the stage for the development of a product that is both technically robust and highly attuned to user needs. More information on creating an MRD can be found elsewhere (Gaskin et al., 2015; Phaal et al., 2006).

MRD and Business Plan

The MRD plays a crucial role in guiding product development by articulating market needs, customer expectations, and feature requirements. It is used early in the product development lifecycle to align the design and engineering teams with clearly defined product goals. The MRD grounds development efforts in validated market demand, helping teams prioritize user requirements, differentiate features, and create a product that solves real-world problems (Ulrich & Eppinger, 2015; Cooper, 2019). This document is especially important in the concept feasibility phases, where market opportunity and product direction are established.

In contrast, the business plan and strategy (discussed further in Chapter 12) serves a broader strategic function. It outlines the company's vision, financial goals, competitive strategy, and resource planning needed to support the product's successful introduction and long-term growth (Osterwalder & Pigneur, 2010). The business plan and strategy are essential when seeking investment, scaling operations, or launching a product into new markets.

While the MRD defines what the product should be and why, the business plan defines how the company will bring that product to market. Together, they form a cohesive framework that connects user needs with organizational execution, aligning technical innovation with operational readiness and market strategy.

Product Requirements Document (PRD)

The PRD translates the strategic direction and user requirements outlined in the MRD into detailed technical specifications essential for product realization. In the development of the office chair, the PRD (Table 10.2) serves as

TABLE 10.2 Product Requirements Document (PRD) for Office Chair.

Section Title	Description	Details
1. Product overview	Provide a detailed description of the product and its intended functionality.	*Product Overview*: A chair designed to wow users when seated. Core Capabilities: Engineered to provide the needed comfort and support, catering to various body sizes and preferences, aiding healthy sitting.
2. Product requirements	Specify the needs and expectations of the end-users that the product must satisfy.	*User Needs*: Based on user stories and ergonomic research, the chair is designed to provide a unique experience. Requirements gathered from user feedback and ergonomic studies guide the product specifications. *Emotional and Psychological Considerations*: Focus on the psychological impact of a comfortable solution on productivity and user satisfaction. Considerations include the aesthetic appeal of the chair and its effect on user morale and comfort during use. *Accessibility and Usability*: Design to accommodate various body types and disabilities, including features like auto-adjusting armrest height, seat, and lumbar support, easy-to-use controls, breathable fabric, and compact design.
3. Product specifications	Detail the aesthetic, functional, and ergonomic design specifications.	*Design Guidelines*: Guidelines focus on combining style with functionality, ensuring the chair not only fits into professional environments aesthetically but also supports good posture and optimum movements. *Aesthetic Details*: Selection of color and materials that are visually appealing and match contemporary office decor while being resistant to wear and easy to clean. *Design Illustrations*: Detailed diagrams and 3D mock-ups of the chair, illustrating ergonomic features, adjustment mechanisms, and overall design to facilitate understanding and further refinement.

TABLE 10.2 (Continued)

Section Title	Description	Details
4. Technical specifications	Outline the specific technical requirements necessary to develop the product.	*Engineering Requirements*: Materials chosen for durability and comfort, such as breathable mesh and memory foam. Dimensions, tolerances, material stiffness and resilience are specified to accommodate a wide range of users while ensuring stability and long-term use. *Hardware and Software Specifications*: Includes specifications for mechanical components like gas lift cylinders, various adjustments, and durable levers for stability on various floor types. Software might include systems for monitoring or updating for remote adjustments if applicable. *Compliance Standards*: Compliance with ergonomic certifications such as ANSI/BIFMA X5.1-2017 (2022) for office furniture to ensure safety and usability.
5. Performance criteria	Establish the performance benchmarks the product must achieve.	*Performance Targets*: Targets for durability under regular use (e.g., Support up to 150 kg, 10,000 cycles of adjustability), intuitive operation, and health benefits. *Safety and Security Protocols*: Incorporation of stability features to prevent tipping, use of non-toxic materials, and designs that minimize pinch points and sharp edges to enhance user safety. *Environmental Specifications*: Specifications for the chair to withstand various office environments, including resistance to temperature changes, sunlight exposure, and humidity. Recycling for a circular economy.
6. Regulatory compliance	Identify the legal and regulatory standards the product must comply with.	*Regulatory Compliance List*: Ensuring the chair meets all relevant local, national, and international standards for office furniture, focusing on ergonomic, safety, and environmental regulations. *Compliance Strategies*: Strategies to adhere to each regulatory requirement, including third-party testing and certification processes. *Certification Processes*: Outline steps to obtain necessary certifications such as GREENGUARD, BIFMA, and others that verify the chair's compliance with environmental and safety standards.

(*continued*)

TABLE 10.2 (Continued)

Section Title	Description	Details
7. Quality assurance	Define the quality standards and testing protocols for the product.	*Quality Standards*: Benchmarks for material quality, manufacturing consistency, and final product testing to ensure each chair meets predefined quality criteria before reaching the market. *Testing Protocols*: Protocols for testing at various stages of development, from initial prototypes to final products, focusing on durability, comfort, and ergonomics. *Review and Audit Schedule*: Planned reviews and audits throughout the design and production phases to ensure ongoing compliance with specifications and quality standards.
8. Risks and mitigations	Identify potential risks in the product development and market introduction phases.	*Risk Assessment*: Assessment of potential risks related to market entry, production, and user acceptance, focusing on areas like market demand, competition, and unforeseen production challenges. *Risk Management Strategies*: Strategies to address identified risks, ensuring proactive measures are in place to handle potential obstacles during product launch and lifecycle. *Contingency Planning*: Contingency plans for critical risks, such as supply chain delays or unexpected market shifts, to ensure the project remains on track and adaptable to changing conditions.

a technical blueprint, confirming that engineering teams, designers, and manufacturers work toward a unified vision that balances user requirements, market demands, and feasibility constraints. Best practices for defining product specifications and aligning them with both business objectives and engineering constraints can be found in Ulrich & Eppinger (2015) and Wiegers & Beatty (2013).

Beyond the PRD, the product feature specifications (PFS) document further refines the design by translating high-level product requirements into detailed functional and design attributes. This document specifies exact performance parameters, usability guidelines, component integration details, and material selections, confirming that engineering, manufacturing, and quality assurance teams have a clear, executable framework. The PFS is the final step before prototyping and production, providing the granular specifications needed to build a product that aligns precisely with customer needs and regulatory requirements.

TABLE 10.3 Component, purpose, process and output of MRD, PRD, and PFS.

Component	Purpose	Process	Output
Market requirements document (MRD)	Captures market needs and customer expectations to outline strategic goals for the product.	Facilitates communication among stakeholders (marketing teams, engineers, designers, and management) to align product objectives.	Results in a formal document defining market needs that guide product development.
Product requirements document (PRD)	Converts market requirements from the MRD into concrete, technical specifications for product development.	Specifies engineering requirements, materials, dimensions, and feasibility constraints so that the product meets expectations.	Produces a structured document detailing technical specifications, ensuring seamless transition from concept to development.
Product feature specifications (PFS)	Provides detailed feature definitions, material selections, and execution plans, ensuring precise alignment between design intent and final product attributes.	Translates PRD specifications into functional attributes, performance benchmarks, and component selection for engineering and manufacturing teams.	Produces a fully detailed specification document that engineers, manufacturers, and quality teams use to maintain compliance with functional, regulatory, and usability requirements.

This structured approach starts with broad market requirements in the MRD to detailed, actionable specifications in the PRD, culminating in the PFS (Table 10.3). It promotes a comprehensive, structured and user-focused product development process.

10.1.2 METHODS FOR TRANSLATING REQUIREMENTS TO PRODUCT SPECIFICATIONS

Various methods have been developed to support the translation from requirements into precise product specifications. These methods range from

TABLE 10.4 Strengths and Limitations of Product Specifications Methods.

	Methods	Strengths	Limitations
1	Functional analysis	Simplifies complex systems into manageable functions, providing clear and actionable specifications.	May oversimplify some aspects, potentially overlooking important interactions between functions.
2	Technical benchmarking	Provides a competitive edge by leveraging existing solutions, which can accelerate the design process.	Risks leading to incremental improvements rather than groundbreaking innovations.
3	Specification trees	Visual mapping aids in clear communication and helps address all user requirements in a systematic manner.	Can become cumbersome for extremely complex products with many layers of specifications.
4	Requirements engineering	Systematic approach for comprehensive coverage of all user requirements.	Primarily developed for software, it may require adaptations for physical product contexts.
5	Design metrics	Helps manage complex interdependencies in product design.	Can be complex to implement and maintain, requiring detailed knowledge of all product aspects.
6	Blueprinting	Excellent for service-oriented or experience-driven products, ensuring that all user touchpoints are considered.	Originally designed for services, it may require significant adaptation for physical products.

deconstructing complex systems to innovative problem-solving techniques, each tailored to bridge the gap between user expectations and technical requirements (Table 10.4).

(a) A functional analysis involves breaking down complex systems into simpler, manageable functions, so that each product function aligns with high-level user requirements, which are then detailed into technical specifications (Pahl et al., 2007).
(b) Technical benchmarking analyzes existing products to derive and enhance specifications based on user requirements, facilitating the translation of these needs into concrete product specifications (Porter, 1985).
(c) Specification trees visually organize product specifications from broad user requirements down to specific technical requirements, illustrating the detailed pathway from user expectations to concrete specifications (Ulrich & Eppinger, 2015).

(d) Requirements engineering, adapted from software development, systematically gathers, documents, and manages requirements, translating user requirements into detailed specifications through techniques like elicitation, analysis, and validation (Sommerville, 2011).
(e) Design matrices, such as the design structure matrix (DSM), manage complex relationships within product development, translating user requirements into specific design parameters and technical specifications (Eppinger & Browning, 2012).
(f) Blueprinting, traditionally used in service design. It maps service processes from the user's perspective and is adapted for physical products to define essential specifications that deliver the intended user experience (Shostack, 1984).

When selecting methods for translating requirements into product specifications, consider the context, complexity, and primary goals of your project.

- *Context*: Choose methodologies based on project context by assessing if the product is software-oriented, service-driven, or predominantly physical, as each method aligns differently with these domains.
- *Complexity*: Evaluate the complexity level by determining whether the product has simple, moderate, or intricate systems and select approaches like functional analysis or design matrices accordingly.
- *Primary goals*: Focus on the primary goals, such as incremental improvements, regulatory compliance, or user experience enhancement, to effectively match methods such as technical benchmarking, requirements engineering, or blueprinting to the objectives.

This strategic alignment helps translate user requirements into precise and actionable product specifications accurately and efficiently.

10.1.3 ITERATIVE SPECIFICATIONS

The adoption of iterative specifications represents a significant shift from traditional methods. This approach, central to agile and responsive design practices, embraces the principle that product specifications should evolve continuously as new information and user feedback become available. Iterative specifications allow for dynamic adjustments throughout the design and development process. By incorporating feedback loops into the design process, designers and engineers can identify and resolve issues more effectively, enhancing the product's relevance and user satisfaction.

The benefits of iterative specifications are many. They promote a more flexible approach to design that can adapt to changing market demands or user requirements. This agility is crucial in today's fast-paced market environments, where consumer preferences and technological capabilities evolve rapidly.

Iterative methods also reduce the risk of significant rework or redesign late in the development process, as continuous adjustments are made while the project progresses. The key to successful iterative specifications is the integration of user feedback into the design loop. Techniques such as prototyping, usability testing, and user interviews become invaluable tools. These methods provide direct insights into how real users interact with the product, highlighting areas for improvement that might not be evident from initial specifications alone.

The iterative approach aligns closely with the principles discussed in *Sketching User Experiences* (Buxton, 2007), where the design process is seen as an exploratory activity that helps to uncover the right design through repeated trials and revisions. Similarly, *The Design of Everyday Things* (Norman, 2013) emphasizes the need for designs to evolve in response to user needs and feedback, confirming that the final products are both intuitive and user-friendly. In summary, iterative specifications represent a modern approach to product design that emphasizes adaptability and user centricity. By allowing specifications to evolve based on continuous user input and testing, designers can verify that their products are not only technically sound but also deeply aligned with the users' needs and expectations, ultimately leading to more successful and innovative products in the marketplace.

10.1.4 REGULATORY COMPLIANCE

Industry standards and regulatory requirements play a critical role in shaping and validating product specifications. It is important to distinguish between compliance and specifications: compliance is not a specification itself but rather a determination of whether a product's specifications meet defined regulatory or industry standards. Specifications are technical and design parameters that a product must meet; compliance confirms that these parameters align with applicable external requirements.

For example, the ISO 13485 (2016) standard specifies requirements for a quality management system in the medical device industry. It helps organizations maintain consistent design, development, production, and delivery of safe and effective medical devices. Specifications developed during the product design phase must therefore be crafted to satisfy the applicable clauses in ISO 13485 (2016) to achieve compliance. Similarly, EIA-632 (1999), Processes for Engineering a System, provides guidance for system engineering practices, from conceptualization to implementation, so that specifications effectively support system reliability and functionality.

Industry standards vary widely across sectors. The automotive industry may follow ISO 26262-1 (2018) for functional safety, while consumer electronics might adhere to IEC 62368-1 (2018) for audio/video and ICT equipment safety. ISO 9241-210 (2019) emphasizes user-centered design by defining requirements for ergonomics and usability in interactive systems. Specifications that are aligned with these standards are more likely to be compliant and, therefore, acceptable for regulatory approval and market distribution.

Integrating compliance requirements into the specification phase involves understanding the standards relevant to a given product and translating those into measurable, testable criteria. This proactive integration improves documentation, supports quality assurance, and streamlines certification processes. In turn, it enhances product credibility, safety, and marketability.

10.1.5 COMPLEX PRODUCT AND SYSTEM SPECIFICATIONS

Designing complex systems isn't just about ticking boxes for compliance; it is like orchestrating a symphony where every instrument (or subsystem) must play in perfect harmony. Whether it is a medical device, a spacecraft, or a pharmaceutical production line, these products involve layers of advanced technology, tight tolerances, and high-stakes environments. That is why specifying them demands meticulous coordination, not just with external standards, but within and across teams, technologies, and countless tiny interfaces that must all click together flawlessly.

For instance, in pharmaceutical and biopharmaceutical industries, standards such as ASTM E2500-13 (2020) promote a risk-based approach that integrates design and development with comprehensive testing and validation strategies to uphold product quality and patient safety while meeting regulatory expectations and improving operational efficiency. Similarly, in aerospace and automotive sectors, specifications must align subsystems to operate cohesively. Standards like ISO/IEC/IEEE 15288 (2023), which outline lifecycle processes for complex systems, guide development from concept through production and operation, helping manage the intricate interdependencies among components.

However, integrating user-centered design principles into these specifications presents unique challenges. These include managing integration complexity, aligning stakeholder perspectives, and adhering to strict regulatory and safety standards, while keeping user considerations central to the design process. Strategies to address these challenges include early and continuous user involvement, interdisciplinary collaboration, and the use of iterative prototyping and testing to refine system design based on real-world user feedback. Moreover, the application of user-centered design tools, such as personas and task analysis (Chapter 5), is crucial in translating user requirements into specific, actionable product specifications.

By combining rigorous technical specifications with thoughtful integration of UCD strategies, complex products can achieve both operational excellence and enhanced user satisfaction. However, embedding user needs into these specifications demands more than technical precision, which requires a structured methodology to translate user needs into design criteria. This leads us to the next section to explore how to shape specifications that are truly centered on user experience.

10.2 INTEGRATING PEOPLE-CENTRIC APPROACHES

The critical transition from traditional product specifications to people-centric specifications is to align product design with the actual needs and user experience. This section focuses on how designers and product teams can integrate people-centric approaches into the specification process, thereby enhancing product relevance and user satisfaction.

People-centric specification focuses on the critical integration of user requirements into product specifications, aligning closely with design principles related to people's limits and capabilities (Chapter 2). This section highlights how user requirements and considerations are refined and incorporated throughout the product development process.

10.2.1 PRODUCT REQUIREMENTS

Effectively transforming refined design problems into concrete user requirements is essential for aligning the final product with user needs and preferences. This process begins with comprehensive user research methods such as interviews, surveys, usability testing, and contextual inquiries to gather an understanding of user behaviors, pain points, and expectations (more details in Chapter 4).

Once product requirements are identified (more details in Chapter 8), prioritization becomes key. The Kano model (Kano et al., 1984) offers a structured approach by categorizing the features of a product into three main types: (1) must-have attributes (fundamental features that users expect; their absence leads to dissatisfaction); (2) performance attributes (features where increased performance or quality leads to higher user satisfaction); 3) delighters (unexpected features that pleasantly surprise users and can significantly enhance satisfaction).

By applying the Kano model, design teams can prioritize features that will have the most significant impact on user satisfaction. This approach helps the product meet basic expectations while also including elements that elevate the overall user experience. Incorporating these prioritized product requirements into the design process keeps the product user-focused and functional. This approach enhances the product's relevance and appeal, ultimately leading to greater user satisfaction.

Building upon the product requirements established through prototype selection and testing (Chapter 8), the designers should then focus on translating those validated requirements into detailed product specifications.

10.2.2 PRODUCT SPECIFICATIONS

This section addresses the challenges of translating psychological, physiological, sensory, and physical product requirements into actionable product specifications. Designers must consider emotional responses, cognitive models, sensory interactions, and physical aspects to create designs that accommodate human limitations, diversity, and functional needs.

Psychological, sensory, and physiological aspects are often subjective and complex to quantify. Designers navigate areas such as user emotions, comfort, cognitive load, sensory perceptions (including visual, auditory, and tactile sensitivities), and other human factors (Norman, 2004; Hassenzahl, 2010). These factors often vary widely based on individual characteristics like body size, age, gender, sensory acuity, and other demographic characteristics, as discussed in Chapters 2 and 3, "Exploring Human Limits and Capabilities" and "Design for Diverse Populations." The physical aspect needs to add further consideration, including ergonomics, accessibility, and material properties, all of which directly impact usability, safety, and user satisfaction. Addressing these diverse requirements introduces products with more comfortable, inclusive, and practical for a broad range of users.

Translating this know-how into effective design strategies involves methodologies that integrate product requirements from psychology, ergonomics, sensory science, human factors engineering, and material science. Tools such as cognitive scales, mental models, sensory assessments, and ergonomic assessments, provide structured approaches to incorporate these factors into product design. Prototyping and user testing are instrumental in validating that these needs are effectively met. Detailed processes for validation will be covered in Chapter 11 ("Design Verification and Validation") to clarify their practical application.

Ultimately, this section aims to equip designers with the knowledge and necessary tools to develop products that not only meet functional demands, but also support and enhance the user's psychological, physiological, sensory, and physical well-being

10.2.3 PRINCIPLES FOR CREATING PRODUCT SPECIFICATIONS

This section outlines principles that guide the process to create the product specifications accurately reflect product requirements.

- *Subjective data handling*: It is difficult to quantify the subjective nature of psychological, physiological, sensory, and physical data, as emotional responses, cognitive states, comfort levels, and sensory interactions. Designers should use diverse data collection methods, including biometric tracking, cognitive workload assessments, sensory analysis (visual, auditory, tactile), and observational studies, to comprehensively understand user interactions. Engaging diverse user groups is essential for accurately capturing real-world needs, minimizing bias, and refining subjective experiences into concrete design considerations.

- *Diverse accommodation*: To create inclusive products, designers must accommodate a wide range of psychological (particularly cognitive), physiological, sensory, and physical differences among users (Chapter 3). Psychological diversity includes variations in cognitive abilities, emotional responses, and mental workload; physiological factors encompass age, body size, and fatigue levels; sensory diversity involves differences in sensory perception and processing; and physical considerations include mobility and physical capabilities. By integrating the design principles from ergonomics and cognitive psychology (Chapter 2), designers can create products that are functional, inclusive, and adaptable.
- *Soft metrics translation*: Designers must translate qualitative needs into concrete specifications. Key ergonomic design considerations include analyzing body pressure distribution, material interaction, sensory comfort, and adaptability to support both comfort and usability. Managing heat buildup, optimizing weight distribution, and providing postural support improve physical interaction with products. Ergonomic evaluations and structured user feedback are essential for refining these metrics.
- *Integration of multifaceted factors*: Balancing multiple design factors, such as intended usage, environmental conditions, and individual variations, which require careful coordination. Products must integrate usability, functionality, and aesthetics while maintaining technical feasibility. Since real-world conditions often differ from controlled design environments, specifications should incorporate adaptive ergonomic strategies allowing for situational flexibility and user-driven customization.
- *Complex configuration management*: Managing complex configurations in a design involves accommodating multiple physical constraints without compromising psychological and sensory usability. Ergonomic design principles must be tailored for specific contexts, ensuring modularity, adjustability, and appropriate dimensions for diverse users. Challenges include balancing factors like seat height, reachability, intuitive control placement, and maintaining accessibility for users with varying physical and sensory abilities.
- *Adaptability to dynamic factors*: Given the evolving nature of human factors, specifications must support long-term adaptability across different users and contexts. This includes integrating adjustable ergonomic features, modular configurations, and customizable elements enabling products to evolve with changing psychological, physiological, sensory, and physical needs. Planning for software and hardware updates, environmental adaptability, and personalization enhances the product's lifecycle usability.
- *Ethical data collection*: Ethical considerations in collecting psychological, physiological, sensory, and physical user data are paramount.

Designers must uphold privacy, transparency, and dignity, ensuring data collection practices prevent intrusive or exploitative methods. It is crucial to represent diverse populations in user testing to mitigate biases, fostering fair and inclusive ergonomic design decisions.

Table 10.5 outlines specifications for seat comfort, aligning each principle with specific applications and corresponding standards. These principles closely correspond to the ergonomic design principles discussed in Chapter 2, ensuring that seating solutions effectively address human limits and capabilities.

10.2.4 APPROACHES FOR EFFECTIVE PRODUCT SPECIFICATION

The integration of psychophysical studies with cognitive scales, mental models, and ergonomic principles (Chapter 2) can effectively translate psychological and physiological user needs into actionable product specifications. This approach bridges the gap between abstract user needs and concrete design elements; leverages the foundational knowledge on human limits and capabilities to foster designs that remain adaptable and responsive to the dynamic nature of human factors.

10.2.4.1 PSYCHOPHYSICAL STUDIES

Psychophysics, the scientific study of the relationship between physical stimuli and human perception, provides a critical framework for understanding how users experience product attributes under different conditions (Gescheider, 1997). It allows designers to quantify the sensitivity of users' perception to changes of product features, such as weight, texture, and sound. Identifying the relationship between the physical domain and psychological domain can support the development of specifications that align with perceptual thresholds and user preferences (Wolf & Drewing, 2020).

For example, in the development of handheld electronic devices, psychophysical studies can determine the optimal weight and texture to maximize comfort and reduce fatigue. Wolf & Drewing (2020) demonstrated that users perceive differences in weight relative to an internalized reference point, with a just-noticeable difference (JND) for heaviness of approximately 7–10 percent of an object's weight. If a standard smartphone weighs 200 g, users may only notice a weight change if the device becomes at least 14–20 g heavier or lighter. This reference is crucial for product designers, as a subtle 5 g weight reduction may go unnoticed, whereas a 15 g reduction could alter the user's perception of the device's portability and comfort.

Similarly, in the automotive industry, psychophysical methods play a key role in refining the acoustic properties of car interiors. By studying how humans perceive sound frequencies and volumes, designers can specify materials and

TABLE 10.5 Specifications for Seat Comfort by Principle.

Principle	Description	Application in Seating Comfort	Example of Product Specifications	Standard
Subjective data handling—sentiment assessment	Designers collected subjective data through user surveys and ergonomic assessments to tailor the chair design, enhancing comfort based on diverse user feedback.	Conduct ergonomic assessments to tailor the chair design, enhancing comfort based on diverse user feedback.	Adjustable lumbar support with three-level firmness control (soft: 20 N, medium: 35N, firm: 50 N resistance) based on user preference testing	ANSI/BIFMA X5.1-2017 (2022) adjustability requirements.
Diversity accommodation	Adjustable features like seat depth, seat height, and armrests were incorporated to accommodate varying body sizes and postures, ensuring the chair could be customized to suit different physical requirements.	Design adjustable features to fit 5th–95th percentile body sizes.	Seat height range adjustable from 380 mm to 520 mm. Assumption: seat height equals popliteal height plus standard shoe heel allowance (typically 25–30 mm), accommodating barefoot to standard shoe heights)	ISO 9241-5 (2024)—Ergonomic requirements for office work with visual display terminals—Part 5: Workstation layout and postural requirements.
Soft metrics translation—psychophysical mapping	Utilized pressure mapping technology to identify pressure points and selected materials that enhance airflow and adjust to body movements for optimal comfort.	Specify materials for pressure relief and air circulation.	Seat cushion made of pressure-relieving memory foam with breathable mesh fabric for airflow.	ASTM D3574-17 (2017)—Standard test methods for flexible cellular materials (foam seating).
Integration of multifaceted factors	Designed prototypes tested in different office settings to validate the chair's performance in various environments, reflecting a holistic approach.	Test prototypes in diverse environmental conditions.	Prototype testing conducted in open-plan offices, cubicles, and home office settings under controlled conditions.	ANSI/BIFMA X5.1-2017 (2022)—Office chairs testing and performance standard.

Complex configuration management	Features such as adjustable backrests, seat tilt, and armrests were designed to accommodate different activities, ensuring versatility across multiple usage contexts.	Provide mechanisms for easy adjustment to various user activities.	Backrest recline angle adjustable between 90° to 135° with locking mechanism; armrests pivot up to 15°.	ISO 9241-5 (2024)—Ergonomic requirements for office work with visual display terminals—Part 5: Workstation layout and postural requirements.
Adaptability to dynamic factors	Included durable materials and replaceable components to adapt to changing user needs and wear over time, enhancing the product's longevity and user satisfaction.	Use modular components for easy replacement and upgrades.	Modular seat cushion and armrests, designed for quick replacement with snap-fit connectors.	ISO 7173 (2023)—Furniture—Chairs and stools—Determination of strength and durability.
Ethical data collection	Followed ethical data collection practices by obtaining informed consent from all participants and anonymizing feedback to protect privacy and foster user trust.	Implement data protection measures and secure user consent.	Data storage secured with encryption; user consent required before data collection via a digital interface.	GDPR (General Data Protection Regulation) and ISO/IEC 27001 (2022)—Data security and privacy.

structural configurations that minimize undesirable noise while enhancing sound system performance (Kondo, Kato, & Tsuda, 2023).

For instance, Kondo et al. (2023) demonstrated that vehicle sound quality significantly influences user perception of overall attractiveness. Their study found that perceived comfort and luxury correlate with interior noise levels, particularly in the 20–200 Hz frequency range, which is associated with low-frequency engines and road noise. Consumers tend to rate cars with interior noise levels below 65 dB at highway speeds (120 km/h) as high quality, while noise levels exceeding 75 dB significantly reduce user satisfaction. To optimize acoustic comfort, many luxury vehicle manufacturers aim for interior noise levels in the range of 55–60 dB at 100 km/h, using acoustic laminated glass, active noise cancellation (ANC), and advanced soundproofing materials. Meanwhile, mass-market sedans typically range between 65 and 70 dB, whereas compact economy cars may reach 72–76 dB, making them perceptibly noisier.

Just as with weight perception and difference thresholds, auditory thresholds play a crucial role in shaping user expectations. If the noise level is reduced by 2–3 dB, many users will not consciously notice the change, but a reduction of 5 dB or more is typically perceived as a significant improvement in comfort. By applying psychophysical requirements, designers make modifications to sound insulation or speaker output perceptible and meaningful to users, directly enhancing the overall user experience and satisfaction.

10.2.4.2 COGNITIVE SCALES AND MENTAL MODELS

Cognitive scales, such as the NASA Task Load Index (NASA-TLX), quantify mental workload across dimensions such as mental demand, physical demand, temporal demand, effort, performance, and frustration. By evaluating these factors, designers can identify areas where a product may impose excessive cognitive load, informing specifications that aim to streamline user interaction and reduce mental strain. Mental models are used for understanding users' internal representations of system operations. This information allows designers to create the user experience that meets the user's expectations. When interfaces mirror these mental models, users can navigate systems more intuitively, reducing the need for extensive training and minimizing errors.

10.2.4.3 ITERATIONS FOR AN IMPROVED PRODUCT SPECIFICATION

People-centered design requires a collaborative approach where users are actively involved in the design process, so that product specifications reflect actual user needs and preferences. Through workshops, prototype testing, and continuous feedback, this method facilitates iterative refinements for fine-tuning the product specifications to better meet user needs.

For example, in the development of seats in an auditorium, a participatory design approach could involve a series of consultations with potential users from various professional backgrounds. During these sessions, participants could provide direct feedback on the different chair prototypes. By controlling the properties of the seat (input factors), such as the heights, widths, and depths, the support they provide, a potential user can report their seating experience (output effect) of sitting for short and long periods of time. The designers could use this feedback to make iterative adjustments to the seat design to accommodate a wider range of body sizes.

Another example could be seen in the development of a mobile health application for elderly users. Participatory design sessions could involve elderly participants using the app in a controlled setting, where they could provide feedback on the user interface, font sizes, navigation ease, and feature accessibility. The developers could then use this input to adjust the design, making it more intuitive and accessible for the elderly population, thereby enhancing user satisfaction and adoption rates.

10.3 A FRAMEWORK FOR DEVELOPING PRODUCT SPECIFICATIONS

A new framework for developing product specifications is proposed for a more dynamic and adaptive approach that integrates continuous user feedback and cross-disciplinary collaboration. The new framework enhances flexibility and allows product specifications to evolve alongside emerging user data and technological advancements. A practical case study on seating comfort is also presented to demonstrate how the use of the framework and these approaches can be applied in real-world product design scenarios.

10.3.1 PROPOSING A NEW FRAMEWORK FOR PRODUCT SPECIFICATIONS

Developing product specifications emphasizes the integration of product requirements into product specifications through various methodologies. A framework for product specifications (Figure 10.1) is presented to guide designers to develop a product based on the actual needs and preferences of its users, considering adaptability and compliance with ethical standards.

1. Simulation (more details in Chapter 8):
   - *Virtual and augmented reality (VR/AR)*: Use advanced simulation tools, including VR and AR, to prototype and test product specifications in virtual environments. This allows a more thorough understanding of how users interact with a product and lead to more accurate specifications before physical prototypes are developed.

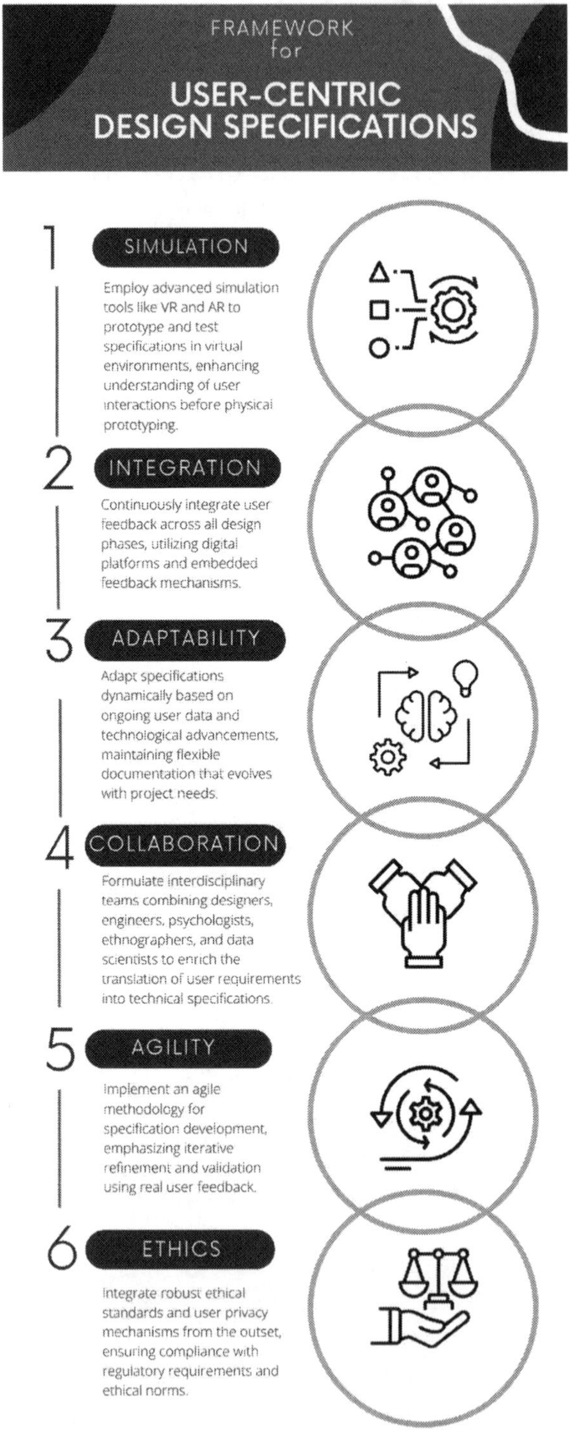

FIGURE 10.1 Framework for developing product specifications.

2. Integration
   - *Integrated user feedback*: Continuously integrate user feedback throughout the design process, not just during the initial requirements gathering or testing phases. This could involve more dynamic and continuous interactions with users, such as through digital platforms or embedded feedback mechanisms in products.
3. Adaptability
   - *Dynamic specification adjustment*: Develop a model where specifications can be adjusted dynamically based on ongoing user data and technological advancements. This involves creating a more flexible documentation process that can evolve as the project progresses, rather than static specifications set at the beginning of the project.
4. Collaboration
   - *Cross-disciplinary integration teams*: Formulate teams that include not only designers and engineers but also psychologists, ethnographers, and data scientists. This would enhance the ability to translate complex product requirements into technical specifications by incorporating diverse perspectives and expertise.
5. Agility
   - *Iterative refinement process*: Incorporate an agile methodology specifically adapted for specification development, emphasizing iterative cycles of refinement and validation with real user data.
6. Ethics
   - *Build in ethical standards*: Include robust mechanisms to address ethical considerations and user privacy from the outset, aligning with regulatory requirements and ethical standards.

10.3.2 CASE STUDY: APPLICATION OF FRAMEWORK FOR PRODUCT SPECIFICATIONS

This section applies the proposed six-step product specification framework to develop the solution of sitting experience, addressing the prevalent issue of discomfort during prolonged sitting. By systematically integrating product requirements into the design process, the case study demonstrates how iterative feedback, and interdisciplinary collaboration can lead to a product that enhances user comfort and well-being.

10.3.2.1 SIMULATION: ADVANCED SIMULATION AND PROTOTYPING

Virtual prototyping tools, including 3D modeling and pressure mapping simulations, were employed to visualize and assess design concepts (more details in Chapter 8) before physical prototypes were built. These simulations enabled the team to analyze how different seat configurations affected pressure distribution and posture, facilitating data-driven decisions that enhanced user comfort.

10.3.2.2 INTEGRATION: USER-DRIVEN REQUIREMENT ANALYSIS

Chair design has never truly been perfected for several reasons. First, the purposes and circumstances we use chairs vary greatly, leading to changing requirements for what makes a chair comfortable. Different criteria must be met for a person sitting on it to be comfortable. From the user's perspective, we can apply the basic principles of ergonomics outlined in Chapter 2: reducing forces, maintaining neutral body alignment, and avoiding static postures.

Studies have shown that forward-tilted seats reduce spinal deformation and help maintain spinal curvature, especially lumbar lordosis (Goonetilleke & Rao, 1999; Keegan, 1953). Looking into the specifics of seat design and its impact on the body is essential to explore the intricacies of seating further. As discussed in Chapter 5, consulting a subject matter expert can provide valuable insights. Knowing that seat inclination aids in maintaining spinal alignment, think of its effects on forces. Figure 10.2 illustrates a participant sitting at two different heights and three seat inclinations, showing the forces exerted by the feet on the floor and the interface pressure between the person and the seat. As the inclination increases, the load on the seat decreases, with the feet taking on more of the load. This means that seat inclination helps reduce spinal curvature and

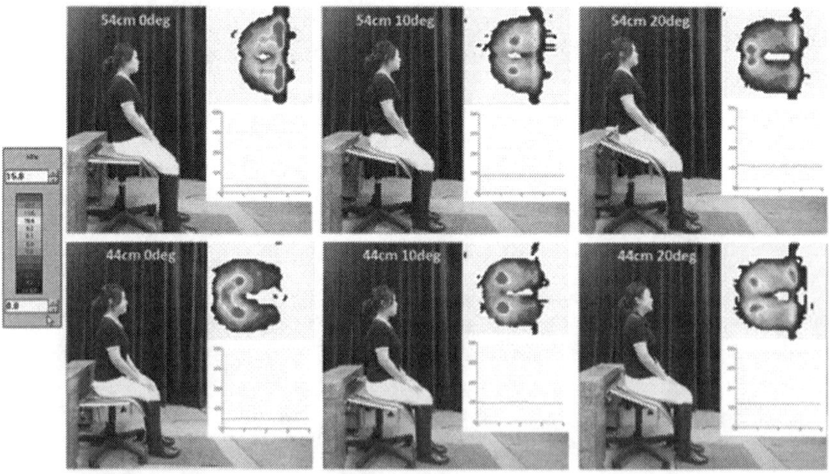

FIGURE 10.2 Top row shows the effect of 0 deg, 10 deg and 20 deg seat inclination at a height of 54 cm on seat interface pressure. The bottom row is at a seat height of 44 cm.

alleviates force and pressure on the buttocks. However, if the seat height exceeds the popliteal height (the height of the back of the knee), nerve compression on the thigh can occur, as indicated by the interface pressure plots in the top left figure. Increasing the seat pan angle can relieve this pressure, as seen in the top row. At lower seat heights, the seat pressure shifts more toward the rear.

Unlike in problem definition, we can take a more focused abstract laddering approach (Chapter 5) with the central theme of "Make sitting comfortable." This shift in focus highlights the versatility of the abstract laddering approach. Using the three principles (minimizing forces, maintaining a neutral posture, and avoiding static postures), we can ensure optimal comfort. The forces involved are related to spinal loading and the interface between the person sitting and the seat pan or seat back. The related abstract laddering chart is shown in Figure 10.3.

10.3.2.3 ADAPTABILITY: ADAPTIVE SPECIFICATION MODELING

The key variables controlling these aspects are the shape, size, height, inclination, and stiffness of the seat pan and seat back. On the flip side, the person's posture also significantly influences their comfort experience. Achieving the right posture for sitting requires awareness and practice. Therefore, we will focus on optimizing seat design from a product perspective. Table 10.6 shows the effects of various seat parameters on comfort, based on expert opinions. These ratings can be analyzed in two ways, offering product requirements into how each parameter impacts overall comfort.

1. Weighted objectives method for a combined rating (CR)

   $$\text{Combined rating (CR)} = \Sigma\, w_i\, c_j$$

   where c_j is the rating, and w_i is the weighting associated with the requirement. In this scenario, separating the three requirements is challenging, so we've given them equal weightings of 1. An alternative is to choose the weightings such that they total to 1. As they are equal, it makes no difference in what is chosen for the weights. The relevant values are highlighted in the table, with those scoring 10 or above in bold to emphasize their impact. A downward-sloping seat pan, a swiveling seat pan, or a soft front seat pan is particularly beneficial for a seat height above popliteal height. These features can significantly enhance comfort and support for the user.

2. Axiomatic design suggests that design parameters should be independent to meet functional requirements (Suh, 2001). This means each parameter should control only one requirement. However, since every design parameter typically influences multiple requirements, we propose a new metric: independent rating (IR). This metric uses

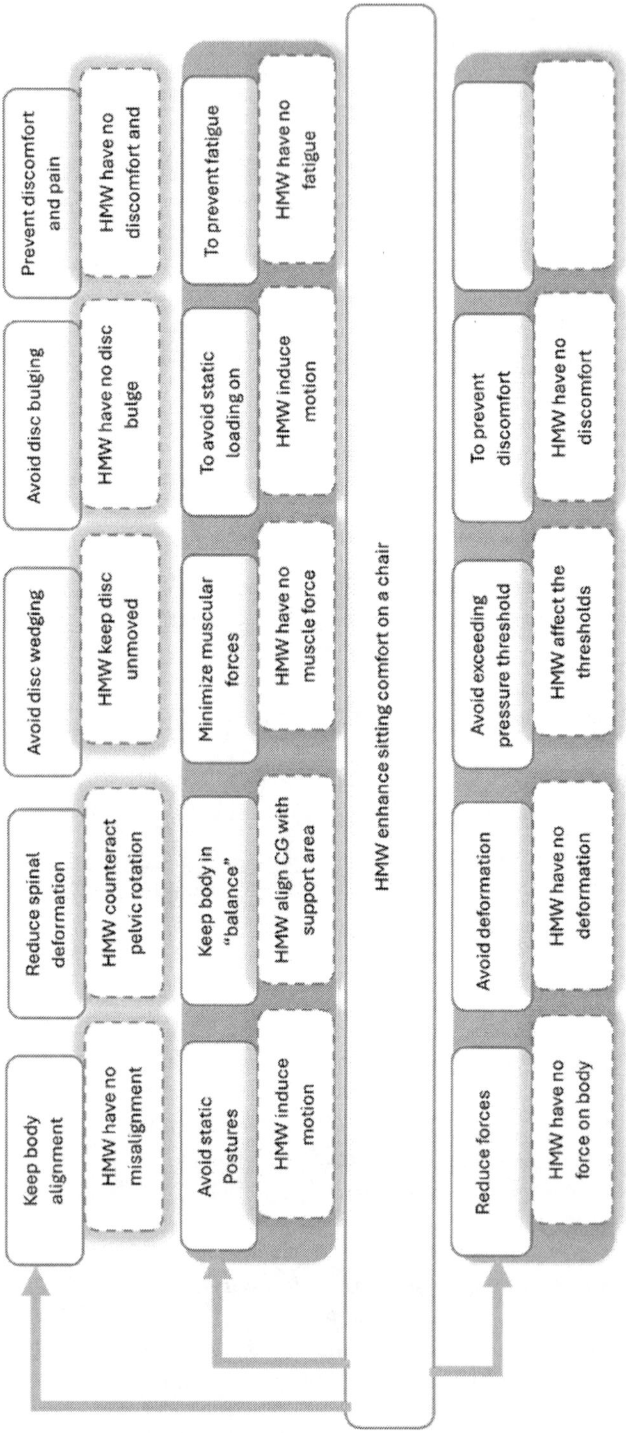

FIGURE 10.3 Abstract laddering chart. HMW stands for "how might we" as discussed in the chapter.

TABLE 10.6 Seat Characteristics to Achieve the Requirements for Seat Comfort.

How?	HMW minimize muscular forces and static postures to keep the body in balance	HMW minimize pelvic rotation and spinal deformations from a neutral posture	HMW Optimize the pressure distribution between the body and seat	Combined Rating	Independence Rating (ID)
Person					
Sit with an open hip angle	*	****	*	6	8
Sit "right" without pelvic rotation	*****	*****	*	**11**	**-5**
Strengthen muscles	****	***	*	8	0
Move/Get-up	*****	***	*	9	5
Exercise	****	**	*	7	4
Equipment					
Seat contouring	*	****	****	9	-4
Minimize seat depth	****	***	**	9	-4
Maximize seat depth	*	*	****	6	8
High seat height (above popliteal ht)—zero inclination	****	****	*	9	-4
High seat height—downward slope	*****	****	***	**12**	**-10**
Low seat height (below popliteal height)—no inclination	*	*	*	3	-3
Downward seat pan inclination	****	****	***	**11**	**-12**
Upward seat pan inclination	**	*	**	5	-2
Translating seat pan	***	*	*	5	3
Swiveling seat pan	****	****	**	**10**	**-8**

(continued)

TABLE 10.6 (Continued)

How?	HMW minimize muscular forces and static postures to keep the body in balance	HMW minimize pelvic rotation and spinal deformations from a neutral posture	HMW Optimize the pressure distribution between the body and seat	Combined Rating	Independence Rating (ID)
Variable stiffness—soft front, hard rear of seat pan	****	****	***	**11**	**-12**
Variable stiffness over the seat pan	***	**	*****	**10**	**0**
Limit posture	**	**	**	6	-4
Swiveling seat back	***	***	***	9	-9
Contoured seat back	***	*	**	6	0
High seat back height	*	*	**	4	0
Headrest	***	*	*	5	3
Armrests	**	*	***	6	0

Notes: A total rating of 10 or greater is shown in bold. The * represents the rating. The combined rating is the addition of these.
*Not effective; **slightly effective; ***moderately effective; ****very effective; *****extremely effective,

the highest rating multiplier to emphasize the importance of higher ratings, ensuring a balanced and effective design process. This is based on the premise that a feature influencing one aspect much more than others tends to be a more independent feature. Thus, a higher positive value is more desirable.

Independence Rating (IR) = Highest rating *(Highest rating – sum of all other ratings)

Evaluating the two ratings, CR and IR, it becomes clear that a combination of variables will work best, as no single factor can be independently controlled to achieve all three objectives in line with axiomatic design principles (Suh, 2001).

10.3.2.4 COLLABORATION: CROSS-DISCIPLINARY INTEGRATION TEAMS

The project assembled a diverse team comprising industrial designers, ergonomists, psychologists, and data analysts. This multidisciplinary approach facilitated a holistic understanding of user comfort, leading to innovations like variable stiffness in the seat pan, requiring soft at the front and firm at the rear. This configuration allows for optimal comfort and support, ensuring proper alignment and optimizing seat interface pressures based on pain and discomfort thresholds. The swiveling seat pan facilitates movement and adaptability, while the variable stiffness foam accommodates different body types and postures, contributing to an overall enhanced seating experience. Table 10.7 provides a description of the specifications for an ergonomic office chair.

10.3.2.5 ABILITY: SPECIFICATION AGILITY FRAMEWORK

An agile methodology was adopted, featuring iterative design cycles with regular user testing and feedback sessions. Each iteration refined the chair's features, such as adjusting the seat height range to 450–600 mm with a 3° slope, based on user comfort reports. With this consideration, the final product is closely aligned with user expectations and ergonomic standards.

10.3.2.6 ETHICS: ETHICAL COMPLIANCE AND USER PRIVACY ASSURANCE

Throughout the design process, ethical considerations were paramount. User data collected during testing phases were anonymized and securely stored, adhering to privacy regulations. Additionally, the design prioritized inclusivity, ensuring that features like adjustable armrests and seat depth accommodated

TABLE 10.7 Product Specification Details of an Ergonomic Office Chair.

Feature Category	Specification	Description	Value / Metric
Seating surface and ergonomics	Seat contouring	Moderate contouring to support pelvis and reduce spinal deformation	5–15 mm contour depth
	Adjustable seat depth	Adjustable depth to accommodate different user preferences	Depth adjustable from 420–450 mm
	Seat pan inclination	Downward inclination to promote open hip angles and spinal alignment	Inclination angle: 5°–10° downward
	Variable stiffness	Soft front, hard rear stiffness to optimize pressure distribution	Stiffness gradient: soft (2.5 N/mm) front, firm (6.0 N/mm) rear
	Swiveling and translating seat pan	Swiveling seat pan for mobility and reduced static postures	Seat pan swivel ±15°, translation range 30 mm
Seat height and adjustability	High seat height	Improves spinal posture	Seat height range: 380 mm to 520 mm, total range 140 mm
	Height adjustability	Accommodates a wide range of users for proper biomechanics	Adjustable in 25 mm increments, total range 150 mm
Seat back and lumbar support	Reclining seat back	Encourages movement and spinal flexibility	Backrest recline angle ±20°
	Contoured seat back	Provides moderate lumbar support to maintain a neutral spine posture	Lumbar curve depth 20-40 mm
	High seat back height	Offers moderate effectiveness but not a primary focus	Backrest height ≥ 500 mm
	Headrest	Supports upper spine and reduces neck strain	Height adjustable by 100 mm, tilt range: 15°
Armrests and additional support	Adjustable armrests	Essential for weight distribution and forearm support	Armrest pad width adjustable: 80–120 mm; Armrest height: 150–250 mm
	Exercise support	Built-in flexibility to allow shifts in posture	Flexible backrest zones support lateral movement up to 8° and forward/backward tilt up to 15°

a diverse user population, including individuals with varying body types and mobility needs.

The development of the ergonomic office chair exemplifies the effective application of the product specification framework. By integrating user feedback, ergonomic research, and iterative prototyping, the design process addressed the multifaceted challenges associated with prolonged sitting. Key features such as adjustable seat height above the popliteal region, a swiveling seat pan, and variable stiffness foam were incorporated to enhance comfort, support proper alignment, and accommodate diverse user needs. Consequently, the final product not only met technical specifications but also aligned with user expectations, resulting in a chair that promotes comfort, adaptability, and overall well-being.

10.4 CONCLUSION

The material presented reinforces the fact that user-centric design is not just a philosophy, but a practical approach for generating clear, actionable product specifications that align with user needs and business goals. This chapter has explored different methodologies from traditional to innovative, aimed at translating complex product requirements into concrete, actionable product specifications. By integrating approaches such as psychophysical studies, cognitive scales, ergonomic principles, and participatory design, we have addressed the multifaceted challenges of aligning product specifications with product requirements, enhancing both functionality and user satisfaction.

Despite these advancements, it is essential to acknowledge the ongoing challenges in achieving perfect specification alignment. The adaptive nature of product development necessitates assumptions and continuous iterations, underscoring the importance of flexible and dynamic specification processes. These challenges highlight the necessity for ongoing validation and refinement.

A significant addition to this discourse is the proposed new framework for product specifications, which integrates user feedback, advanced simulation tools, and cross-disciplinary collaboration. This framework aims to bridge the gap between traditional methods and the dynamic requirements of modern user-centered design.

As we move forward to the next chapter on "Design Verification and Validation," we will build upon the specifications developed here, emphasizing the importance of verifying that these specifications meet their intended functional and user satisfaction outcomes. Design verification and validation will involve rigorous testing and feedback mechanisms to refine the product continuously. The next chapter will further explore how to develop the products to adhere to technical specifications and deliver value and satisfaction to end-users, addressing the gaps and challenges outlined in this chapter.

> Excellence isn't about meeting the spec, it's about setting the spec.
>
> — Seth Godin

REFERENCES

ANSI/BIFMA X5.1-2017 (2022). General-Purpose Office Chairs—Tests. Business and Institutional Furniture Manufacturers Association (BIFMA).

ASTM D3574-17 (2017). Standard Test Methods for Flexible Cellular Materials -Slab, Bonded, and Molded Urethane Foams. ASTM International.

ASTM E2500-13 (2020). Standard Guide for Specification, Design, and Verification of Pharmaceutical and Biopharmaceutical Manufacturing Systems and Equipment. ASTM International. https://store.astm.org/e2500-13.html

Buxton, B. (2007). *Sketching User Experiences: Getting the Design Right and the Right Design*. Morgan Kaufmann.

Cooper, R. G. (2019). *Winning at New Products: Creating Value Through Innovation.* Basic Books.

EIA-632 (1999). Processes for Engineering a System. Electronic Industries Alliance (EIA). https://www.acqnotes.com/Attachments/EIA-632%20%E2%80%9CProcesses%20for%20Engineering%20a%20System%E2%80%9D%207%20Jan%2099.pdf

Eppinger, S. D., & Browning, T. R. (2012). *Design Structure Matrix Methods and Applications.* MIT Press.

Gaskin, J. E., Griffin, M., Burke, S., & Robertson, J. (2015). Customer insights and the evolution of market requirements documents (MRDs): A review of best practices. *Journal of Product Innovation Management*, 32(3), 293–309.

Gescheider, G. A. (1997). *Psychophysics: The Method, Theory, and Application* (3rd ed.). Lawrence Erlbaum Associates.

Goonetilleke, R. S., & Rao, B. G. (1999). Forward sloping chair effects on spinal shape in the Hong Kong Chinese and Indian populations. *International Journal of Industrial Ergonomics*, 23(1), 9–21.

Hassenzahl, M. (2010). *Experience Design: Technology for All the Right Reasons*. Morgan & Claypool.

IEC 62368-1 (2018). Audio/Video, Information and Communication Technology Equipment—Part 1: Safety Requirements. International Electrotechnical Commission (IEC). https://www.iecee.org/certification/iec-standards/iec-62368-12018

ISO 13485 (2016). Medical Devices—Quality Management Systems – Requirements for Regulatory Purposes. International Organization for Standardization (ISO).

ISO 26262-1 (2018). Road Vehicles—Functional Safety—Part 1: Vocabulary. International Organization for Standardization (ISO). https://www.iso.org/standard/68383.html?utm_source=chatgpt.com

ISO 7173 (2023). Furniture—Chairs and Stools—Determination of Strength and Durability. International Organization for Standardization (ISO). https://cdn.standards.iteh.ai/samples/73866/3a012a0646264ea48033c4444f5b4c95/ISO-7173-2023.pdf

ISO 9241-210 (2019). Ergonomics of Human–System Interaction—Part 210: Human-Centred Design for Interactive Systems. International Organization for Standardization (ISO). https://www.iso.org/standard/77520.html

ISO 9241-5 (2024). Ergonomic Requirements for Office Work with Visual Display Terminals (VDTs)—Part 5: Workstation Layout and Postural Requirements. International Organization for Standardization (ISO). https://www.iso.org/standard/86222.html

ISO/IEC 27001 (2022). Information security, cybersecurity and privacy protection—Information Security Management Systems – Requirements. International Organization for Standardization (ISO) and International Electrotechnical Commission (IEC). https://www.iso.org/standard/27001?utm_source=chatgpt.com

ISO/IEC/IEEE 15288 (2023). Systems and Software Engineering – System Life Cycle Processes. International Organization for Standardization (ISO), International Electrotechnical Commission (IEC), and Institute of Electrical and Electronics Engineers (IEEE). https://www.iso.org/standard/81702.html

Kano, N., Seraku, N., Takahashi, F., & Tsuji, S. (1984). Attractive quality and must-be quality. *Journal of the Japanese Society for Quality Control (in Japanese)*, 14(2), 39–48.

Keegan, J. J. (1953). Alterations of the lumbar curve related to posture and seating. *The Journal of Bone and Joint Surgery (Am)*, 35-A, 589–603.

Kondo, T., Kato, T., & Tsuda, K. (2023). The influence of sound on the attractiveness of automobile from a bird's eye view of perceived attractiveness. *Procedia Computer Science*, 217, 1806–1815. https://doi.org/10.1016/j.procs.2023.10.170

Kumar, V. (2012). *101 Design Methods: A Structured Approach for Driving Innovation in Your Organization*. Wiley.

Norman, D. A. (2004). *Emotional Design: Why We Love (or Hate) Everyday Things*. Basic Books.

Norman, D. A. (2013). *The Design of Everyday Things*. Revised and Expanded Edition. Basic Books.

Osterwalder, A., & Pigneur, Y. (2010). *Business Model Generation: A Handbook for Visionaries, Game Changers, and Challengers*. Wiley.

Otto, K., & Wood, K. (2001). *Product Design: Techniques in Reverse Engineering and New Product Development*. Prentice Hall.

Pahl, G., Beitz, W., Feldhusen, J., & Grote, K. H. (2007). *Engineering Design: A Systematic Approach*. Springer.

Phaal, R., Farrukh, C. J., & Probert, D. R. (2006). Technology management tools: concept, development, and application. *Technovation*, 26(3), 336–344.

Porter, M. E. (1985). *Competitive Advantage: Creating and Sustaining Superior Performance*. Free Press.

Shostack, G. L. (1984). Designing services that deliver. *Harvard Business Review*, 62(1), 133–139.

Sommerville, I. (2011). *Software Engineering*. Addison-Wesley.

Suh, N. P. (2001). *Axiomatic Design: Advances and Applications* (MIT-Pappalardo Series in Mechanical Engineering). Oxford University Press.

Ulrich, K. T., & Eppinger, S. D. (2015). *Product Design and Development*. McGraw-Hill Education.

Wiegers, K. E., & Beatty, J. (2013). *Software Requirements*. Microsoft Press.

Wolf, C., & Drewing, K. (2020). The size-weight illusion comes along with improved weight discrimination. *PLoS One*, 15(7), e0236141. https://doi.org/10.1371/journal.pone.0236141

11 Design Verification and Validation

Details matter. It's worth waiting to get it right.

—Steve Jobs

11.1 INTRODUCTION

With the specifications now clearly defined (as outlined in Chapter 10), the high-fidelity prototype is transformed into a pre-production prototype (see Chapter 7) that fully integrates and embodies these specifications. This pre-production prototype serves as a critical bridge between the high-fidelity functional prototype and the final product offering a platform for thorough evaluation of both technical compliance and production readiness.

Design verification answers the question, *"Did we build the product right?"* by checking if the design outputs like drawings and specifications, align accurately with the market, product, and safety requirements (Babuska & Oden, 2004; Maropoulos & Ceglarek, 2010). This step is crucial for confirming the design meets all technical and functional requirements before moving forward. It is accomplished through systematic inspections, tests, and demonstrations to ensure compliance with all specified product specifications (Deng et al., 2000).

Design validation answers the question, *"Did we build the right product?"* It involves testing the pre-production prototype in the actual or simulated use conditions and ensuring it meets the market's requirements and business needs (Babuska & Oden, 2004; Maropoulos & Ceglarek, 2010; AIAA 1998). The different testing methods presented in Chapter 8 continue to apply to the pre-production prototype during this validation phase.

DOI: 10.1201/9781003476177-11

Together, the verification and validation processes are important requirements in ISO 9001:2015 for controlling product design quality. These processes require meticulous execution through stakeholder feedback, systematic testing, and iterative refinement so that the final product meets quality standards, regulatory requirements, product requirements, and market requirements (Sargent, 2005). Ultimately, these processes mitigate risks, increase product reliability, and enhance customer satisfaction.

11.2 DESIGN VERIFICATION

Design verification is a systematic process that confirms whether design outputs accurately fulfill the specific design inputs and specifications. Design verification encompasses multiple methodologies including inspection protocols, performance testing and measurement, comparative analysis against previous designs, comprehensive design document reviews, and analytical assessments (Andreasen, 1988).

Effective design verification requires systematic planning and documentation (Guo et al., 2018). Organizations must establish clear verification protocols that demonstrate compliance with the specifications through comprehensive documented evidence, including detailed testing reports, standardized inspection checklists, formal design review records, and analytical summaries (U.S. Department of Defense, 2008). This documentation serves both as proof of compliance and as a foundation for continuous improvement processes.

For example, consider an air fryer manufacturer conducting design verification on their pre-production prototype. The verification process should confirm that the product meets predetermined specifications across multiple aspects: dimensional, safety, functional and regulatory specifications.

- *Dimensional verification*: Using precision measuring instruments to confirm that external dimensions, internal capacity, and component spacing align with engineering drawings.
- *Safety verification*: Conducting electrical safety tests, thermal performance assessments, and material compliance checks against relevant safety standards (such as Underwriters Laboratories (UL) or Conformite Europeenne (CE) specifications).
- *Functional verification*: Testing heating elements, fan performance, temperature controls, and timer mechanisms to ensure they operate within specified parameters.
- *Regulatory verification*: Confirming electromagnetic compatibility (EMC), energy efficiency ratings, and other regulatory specifications are met.

Each verification activity generates documented evidence that the design has been built according to specifications, instilling confidence before advancing to validation testing with end-users.

Research indicates that approximately 80 percent of product failures remain undetected until the final testing or actual use in the real-world environment (McMahon, 2010). Validation integrates quality assurance methods to identify potential risk and optimize reliability before market deployment. Key approaches include failure mode and effects analysis (FMEA) for proactive risk identification (Palady, 1997; Stamatis, 2003; Chen et al. (2009), Design for Six Sigma (DFSS) (Khalaf & Yang, 2006) for quality optimization (Creveling et al., 2002), axiomatic design (Suh, 2001), and robust design principles (Taguchi et al., 2005; Lopez et al., 2021) for performance consistency. By implementing structured validation protocols early in the development process, teams can identify potential pitfalls before finalizing detailed designs, reducing the likelihood of costly revisions or customer dissatisfaction (Jamnia, 2018).

The integration of quality assurance into validation processes enables a targeted testing sequence that focuses resources on the most critical failure modes and user experience factors (GHFT, 2004). This approach leads to more robust, people-centered designs that perform reliably. The following section gives additional methods for validating, ensuring comprehensive evaluation and refinement throughout the product development cycle.

11.3 DESIGN VALIDATION

While verification confirms technical compliance with specifications, validation addresses the fundamental question: "Does this product actually solve the user's problem?" This process evaluates whether the psychological, physiological, sensory, and physical design considerations create a positive user experience that aligns with market requirements.

Validation uses a structured approach to align product performance with user satisfaction. Unlike verification's focus on conformance to product specifications, validation assesses performance in actual use conditions through systematic evaluation of user experience and functionality. This requires comprehensive validation methodologies that capture the full spectrum of people–product interactions.

- *Psychological validation*: User perception, mental workload, and cognitive load assessments are conducted to verify if the product aligns with user expectations and mental models. Techniques such as NASA-TLX (Task Load Index) and eye-tracking analysis (Figure 8.3) help quantify cognitive effort and ease of use (Hart & Staveland, 1988).
- *Physiological validation*: Biomechanical studies, fatigue analysis, and sensor-based tracking measure the product's physical interaction with users. For example, wearable fitness trackers undergo heart rate

accuracy tests to ensure the reliability of physiological metrics (Shin et al., 2019).

- *Sensory validation*: Sensory validation focuses on how the product is perceived through sight, sound, touch, or even smell, especially in products where a multisensory experience significantly impacts usability, comfort, and user satisfaction. Sensory testing methods may include tactile surface assessments, visual contrast analysis for accessibility, auditory feedback tuning, or comparative texture evaluations. These methods are vital in inclusive or assistive design contexts, where sensory alignment with user expectations is critical (Marques & Correia, 2022; Spence & Gallace, 2011).
- *Physical validation*: Materials, durability, and accessibility are assessed to ensure long-term usability and compliance with industry standards. This involves stress testing, environmental simulations, and real-world trials to confirm that the design meets technical requirements (Maropoulos & Ceglarek, 2010). Quality assurance in design validation involves systematic approaches to identifying potential risks, assessing failure modes, and implementing strategies to improve reliability (Musale & Mankar,2023).

11.4 METHODOLOGIES FOR PEOPLE-CENTRIC VALIDATION

People-centric validation addresses critical factors, including usability, accessibility, human limits and capabilities, and overall user experience. Validation integrates qualitative and quantitative assessments, ensuring products align with user expectations and function reliably under real-world conditions.

Several validated assessment frameworks and methodologies have been developed to enhance product design and validation. The User-Centered Design (UCD) 11-item Measure (UCD-11) (Witteman et al., 2021) quantitatively assesses user involvement in the design and development process, ensuring that tools and systems are tailored to end-user needs rather than forcing users to adapt. This framework aids in evaluating design quality and improving workflows. Another methodology, Functionality, Usability, Look and Feel, and Evaluation (FULE) (Liberman-Pincu & Bitan, 2021), offers a structured approach to product design validation by focusing on four critical factors: functionality, usability, look and feel, and evaluation (see Section 11.8 for a related case study). Additionally, virtual reality (VR) testing can enhance interactivity and realism during validation (Joundi et al., 2024). VR-based provides an immersive evaluation of design elements, enabling designers to visualize procedures, simulate real-world use, and refine user interactions, though it may fall short in accurately mimicking real-world social interactions, necessitating a complementary approach with physical testing.

These methodologies complement traditional validation methods, such as usability testing, ergonomic assessments, and psychophysical studies, creating a holistic framework for evaluating product effectiveness before market deployment.

11.5 STRUCTURED APPROACH FOR VALIDATION

Validating relies on prioritizing user engagement and continuous feedback. Engaging directly with stakeholders is critical, as it allows for discovering their requirements, which can be translated into improved product offerings. Research shows that companies that work closely with stakeholders are more successful in product development (Bolton, 2013).

A structured approach to validation encompasses several key elements. First, it employs multi-method evaluation techniques, utilizing both qualitative and quantitative measures as discussed in Chapter 8, applicable to pre-production prototypes. Collaboration with stakeholders, including designers, engineers, human factors specialists, and end-users, fosters a comprehensive validation process (Persad et al., 2007). Additionally, ethical and inclusive testing methods ensure that diverse user demographics are considered, addressing important ethical concerns related to data collection and privacy (Tatlısu, 2023). Iterative validation cycles facilitate continuous design refinement through feedback loops, significantly reducing costly modifications after production (Beyer, 2010). Finally, once the pre-production prototype has been validated and design specifications finalized, the focus shifts to preparing tooling, manufacturing processes, and quality control metrics to ensure the product is ready for mass production.

By systematically evaluating each validation step, teams can confirm that the product performs as intended and aligns with the requirements defined in the design specification phase. This structured approach helps mitigate design risks, enhance product reliability, and ensure compliance with key performance criteria. Engaging directly with stakeholders is essential for collecting feedback and ensuring that market requirements are adequately fulfilled (Bolton,2013). Many successful companies achieve breakthrough product development by directly engaging customers to tap into what matters most to the people who will purchase and use their products and services. Such engagement enables the development of a pipeline of breakthrough products and services. The importance of connecting with customers and users is not a new phenomenon: building in the voice of the customer is a critical element of well-established tools such as quality function deployment (QFD) in large organizations (Ramanathan & Yunfeng, 2009). Validating human-centric design requires active user engagement, structured evaluation methods, integration of UX practices within Agile frameworks (Beyer, 2010), and a holistic approach to the user-product relationship. These practices collectively contribute to developing products that genuinely meet user needs.

11.6 ETHICAL IMPLICATIONS IN VALIDATION

As designers and developers, we have a moral responsibility to create products and services that benefit humanity. User testing during design validation raises several ethical implications that are critical to consider (McKenzie et al., 2018). Since design validation involves testing the product and getting feedback from stakeholders (users, customers, investors, or company executives), we must protect their anonymity and data privacy (Wexler & Largent, 2023). That is the case for many electronic wearable and bioengineering applications being developed, tested, and marketed to users (Tu & Gao, 2021). Many designers, engineers, and material experts may not know that they must obtain ethics approval from an independent review committee before testing products on people. This ensures that all procedures are ethical, protect participants' rights and welfare, and comply with established standards (McCradden, 2023).

Additionally, researchers must obtain informed consent from participants before conducting any tests. The independent review process involves researchers or developers submitting a detailed plan proposal to experts not affiliated with the designers or researchers, who can evaluate the protocols to ensure they comply with ethical standards and relevant research regulations. Independent review can be a legal requirement by government agencies, funding, device regulators, and institutions.

11.7 PEOPLE-CENTRIC VALIDATION IN ITERATIVE DESIGN

Design validation process is not a one-time event but an ongoing testing, refinement, and enhancement cycle. The iterative design process ensures that products evolve continuously based on user feedback, improving functionality, usability, and overall user experience. This approach enables teams to make incremental improvements while addressing emerging user needs and unforeseen challenges.

11.7.1 IMPROVING THE DESIGN

The iterative design process is based on field trials, systematic testing, data-informed decision-making, continuous stakeholder feedback loops, and ongoing refinement of the pre-production prototype. Field trials involve testing products in real-world conditions to evaluate their performance across diverse environments and user behaviors, yielding critical insights into reliability and adaptability. The testing methods outlined in Chapter 8 can be applied to pre-production prototypes, allowing for thorough evaluation. Data-driven decision-making plays a vital role by leveraging analytics from validation tests to refine product specifications; metrics such as error rates, task completion times, and usability scores inform iterative improvements. Continuous engagement with stakeholders through feedback loops ensures that market requirements remain at the forefront of product development. In addition, pre-production prototype refinement utilizes rapid prototyping and iterative adjustments to enhance design before final production, integrating new learnings from each iteration to minimize costly revisions later in the development process. By embedding iterative refinement into the validation process, teams can ensure continuous product improvement, resulting in higher user satisfaction and greater alignment with evolving market demands.

11.7.2 CHALLENGES IN IMPLEMENTING ITERATIVE DESIGN

Iterative design presents several challenges, including organizational, methodological, and operational hurdles that must be carefully managed to ensure successful implementation.

1. Organizational Challenges

In some organizations, iterative design operates independently of the main product development flow, creating potential disconnects that require strategic approaches to integrate iterative feedback effectively into the broader development lifecycle. This separation can hinder the seamless flow of insights and improvements throughout the development process.

Additionally, some companies demonstrate resistance to open innovation, which fundamentally relies on external feedback and ideas to drive improvement. This reluctance significantly limits the diversity and richness of input available for iterative refinement, potentially constraining the innovation potential of the design process. Organizations may also exhibit resistance to change when transitioning from traditional design methods to a more flexible, iterative approach. Stakeholders who are accustomed to linear, predictable processes often find the dynamic and evolving nature of iterative design to be unpredictable and cumbersome, creating barriers to adoption.

To address these organizational challenges, several solutions can be implemented. Fostering a culture of open communication and collaboration among the product development team is essential for ensuring a smooth product development process flow. This cultural shift helps break down silos and encourages cross-functional cooperation. Providing comprehensive training on open innovation principles and practices equips team members with the necessary knowledge and skills to participate effectively in open innovation initiatives, helping them understand the value and methodology of external collaboration. Finally, developing strong relationships with stakeholders and establishing clear communication channels ensures that all parties remain informed about the iterative design process and its goals, reducing resistance and building support for the approach.

2. Methodological Challenges

Iterative design depends fundamentally on consistent and measurable feedback mechanisms, presenting a significant methodological challenge for design teams. Drawing from established principles of user-centric approaches, this challenge involves adapting the specification process to be highly responsive to user feedback while addressing the inherent complexity of people's needs and the nuances of individual user experiences. Standardizing feedback mechanisms becomes particularly challenging when dealing with qualitative feedback that is difficult to quantify, requiring the development of structured methodologies that can transform subjective feedback into actionable data for iterative refinement. These methodologies are essential for ensuring that design specifications genuinely reflect user requirements rather than assumptions or incomplete interpretations.

To address these methodological challenges, several solutions can be implemented for qualitative data analysis. The process begins with organizing collected data, including notes and recordings, in a manner that facilitates easy analysis and retrieval. Once organized, the next step involves identifying themes, relationships, and insights within the data, looking for patterns that reveal user needs and preferences. Content analysis becomes crucial at this stage, requiring teams to recognize the frequency and patterns between words, phrases, or concepts that emerge from user feedback. Context consideration is equally important, as teams must account for factors such as user demographics, the specific tasks given during data collection, and environmental conditions to fully comprehend the meaning and implications of the data. Finally, the information gained from qualitative data analysis should be systematically used to refine and improve the design process, creating a feedback loop that continuously enhances the iterative design approach and ensures that user insights are effectively translated into design improvements.

3. Operational Challenges

Iterative design presents several significant operational challenges that organizations must carefully navigate to ensure successful implementation. Resource allocation and time management emerge as primary concerns, as iterative design can be resource-intensive, involving multiple design cycles, extensive testing phases, and continuous revision processes that potentially lead to increased costs and extended project timelines. This challenge is compounded by the difficulty of maintaining stakeholder engagement throughout the design process, particularly when projects span extended periods and stakeholders may lose interest or shift priorities over time. Additionally, navigating product complexity becomes increasingly challenging as products grow more sophisticated, requiring advanced project management skills to coordinate multiple iteration cycles for different features or components while maintaining overall project coherence and direction.

To address these operational challenges, several strategic solutions can be implemented to streamline the iterative design process. Planning the requirements for the iterative design process is crucial for avoiding surprises and preventing unnecessary costs, requiring teams to define overall requirements and allocate resources strategically to move through iteration cycles efficiently. Maintaining stakeholder engagement requires employing diverse feedback-gathering methods, including user surveys, interviews, usability testing, and analytics, to keep stakeholders actively involved and invested in the process. Demonstrating the value of stakeholder input becomes essential by meaningfully integrating their suggestions and addressing their concerns throughout the design evolution.

Continuous iterations based on stakeholder feedback create a responsive design environment that keeps participants engaged and invested in the outcome. Transparency plays a critical role in maintaining stakeholder confidence, requiring teams to consistently inform stakeholders about progress and clearly communicate the logic behind design decisions. Finally, helping stakeholders understand their vital role in creating something meaningful beyond just a product fosters deeper engagement and commitment to the iterative design process, transforming them from passive observers into active contributors to the project's success.

11.8 CASE STUDY

This case study was based on the research of Gutierrez et al., (2014) and presents how the process of validation for the respirators for the automotive painters was conducted. The most common respirators used by automotive painters in the Philippines are categorized as air-purifying respirators (APR). It has filters, cartridges, or canisters that eliminate the contaminants from the air to protect the user (Clayton et al., 2002). The safety function of the respirator and the protection it provides to the workers are influenced by factors such as fit of the respirator, the task, proper use, maintenance of the respirator, and most importantly, the capability to filter the contaminants (Clayton et al., 2002).

The respirators in the market were tested and design issues were identified (Gutierrez et al., (2014)). Based on these, a redesigned respirator prototype was created with a new facepiece having a cartridge or filter holder, an air-purifying element, and the inhalation and exhalation valve (Figure 11.1). The air-purifying element filters the paint fumes. As the cartridge weight is 0.68 kg (1.5 pounds), the inhalation valve that supports the air-purifying element is made of organic resin to prevent sagging. The flap folds inward and has a lining cushion made of silicone (Figure 11.1). The silicone lining rests along the user's nose, cheeks, and chin. The softness of the material assures comfort to the wearer.

The weight of the prototype is 1.04 kg and uses a single cartridge. The dimensions were 12 cm (length), 11.5 cm (width) and 7.5 cm (depth). The

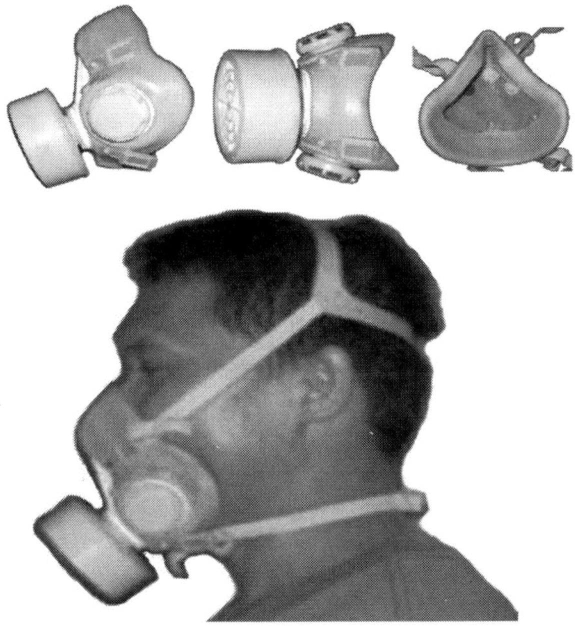

FIGURE 11.1 Redesigned respirator.

dimensions of the respirator were based on the landmarks proposed by Farkas (1981). Actual measurements of these facial dimensions were taken from the 85 employees of an automotive painting company (Zhuang & Bradtmiller, 2005).

11.8.1 VALIDATION: FIT REQUIREMENTS

The qualitative fit test, Occupational Safety and Health Administration Standards 29 (1910.134), was conducted by asking the painters to use the respirator in the working environment. First, they were asked if they could detect the paint odor in the surroundings. They all responded negatively and were asked to do a series of movements to check the fit (Table 11.1). After these activities, they were asked to rate on a scale of 1—lowest satisfaction and 5—highest satisfaction. Each exercise was performed for one minute, except for the grimace test, which was performed for 15 seconds. The respirator should not be adjusted once the test starts. Any adjustments voided the test, and the fit test was repeated.

Based on the fit test results (Table 11.1), the new respirator ratings are better than the preferred respirator available in the market in terms of normal breathing and talking. The result shows that the contour principle improved the respirator's fit to the employees' satisfaction.

TABLE 11.1 Mean and Standard Deviation of Satisfaction Rating of the Redesigned Respirator.

Fit Test Exercise	Redesigned Respirator Mean Rating	Redesigned Respirator (SD)
Normal breathing in a normal standing position without talking	3.92 (3.47)	0.40 (0.59)
Deep breathing in a normal standing position, breaking slowly and deeply, taking precaution not to hyperventilate	3.72 (3.49)	0.46 (0.59)
Turning the head slowly from side to side, while standing in place, with the employee holding his/her momentarily at each extreme so that the employee can inhale at each side	3.76 (3.81)	0.52 (0.82)
Moving the head up and down slowly while standing in place inhaling in the up position when looking toward the ceiling	3.82 (3.52)	0.55 (0.59)
Talking out loud slowly reading from prepared text such as the rainbow passage, counting backward from 100, or reciting a memorized poem or song	4.13 (3.81)	0.53 (0.50)
Bending at the waist as if to touch toes (jogging in place can be done when the fit test enclosure doesn't permit bending at the waist).	3.95 (3.72)	0.45 (0.54)
Normal breathing (as described above)	3.95 (3.34)	0.61 (0.48)

Note: The preferred respirator ratings are in parentheses.

11.8.2 VALIDATION: USING THE FULE METHOD

Table 11.2 presents the FULE method for evaluating the redesigned respirator. This structured approach makes sure that the respirator design is assessed objectively and subjectively through iterative evaluations. By focusing on critical factors such as functionality, usability, look and feel, and evaluation, the FULE method enhances the functionalities and user experience of the product.

11.8.3 PEOPLE-CENTRIC VALIDATION CHECKLIST

Table 11.3 summarizes the design validation for the redesigned respirator. The design validation checklist provides a holistic approach to evaluating the respirator. This method combines multiple evaluation methods and involves stakeholders considering the ethical testing and iterative validation to ensure that the final product will meet the specifications and ensure the product is ready for production.

TABLE 11.2 FULE Method.

FULE Method	Application on the Redesigned Respirator
Functionality	The respirator's facepiece has the following parts: a cartridge or filter holder, an air-purifying element, and the inhalation and exhalation valve. The air-purifying element filters the paint's fumes; since the cartridge weight is 0.68 kg (1.5 pounds), the inhalation valve that supports the air-purifying element is made of organic resin to prevent sagging. The headband strap design was used for better support.
Usability	Conducted usability testing in the actual environment of the automotive painters.
Look and feel	This flap design folds inward and has a lining cushion made of silicone. The silicone lining rests along the user's nose, cheeks, and chin. Actual measurements of the facial dimensions were taken from the 85 employees of the automotive painting company. The softness of the material assures comfort to the wearer. The headband strap design ensures the weight distribution of the respirator across the head.
Evaluation	Conducted the fit test following OSHA standards. Combined qualitative feedback (user surveys) and quantitative measures (satisfaction ratings).

TABLE 11.3 Design Validation Checklist for the Respirator.

Validation Principle	Checklist Items Met/Not Met	Application in the Respirator Case Study
Multi-method evaluation	☑ Used mixed methods ☑ Applied structured evaluation frameworks ☑ Matched evaluation to intended product use	Combined qualitative feedback (user surveys) and quantitative measures (satisfaction ratings, fit test).
Collaboration with stakeholders	☑ Engaged multidisciplinary teams ☑ Held design reviews ☑ Ensured seamless communication	Involved product designers, engineers, and ergonomics specialists in development.
Testing	☑ Conducted usability testing in varied conditions ☑ Used real-world environments ☑ Simulated testing scenarios	Iterated through multiple prototypes and conducted real-world environment testing. The fit test followed OSHA guidelines.
Ethical and inclusive testing	☑ Ensured diverse representation ☑ Obtained informed consent ☑ Secured user data ☑ Followed regulatory compliance	Included a diverse sample of workers, considering facial anthropometrics and ergonomic comfort. User consent was obtained before trials.
Iterative validation cycles	☑ Conducted multiple validation cycles ☑ Documented design changes ☑ Performed field trials	Conducted multiple rounds of feedback, improving fit, material choice, and usability. Future iterations will focus on weight reduction.

11.9 CONCLUSION

Ensuring that a product meets both its technical specifications and market requirements requires a structured verification and validation process that effectively bridges user needs with product performance. While Chapter 10 emphasized people-centric design specifications, this chapter focused on systematic methods for their verification and validation to ensure the product performs as intended across diverse user environments. Design verification confirms that the product satisfies technical and regulatory standards, whereas design validation ensures alignment with marketing requirements and functional user goals. Methodologies such as FMEA, FULE, and VR support the validation process. Best practices, including early user involvement, multi-method evaluation, iterative refinement, and ethical considerations, further strengthen the reliability and effectiveness of validation outcomes.

The respirator redesign illustrated how structured validation processes can significantly enhance the comfort, usability, and safety of a pre-production prototype. The findings highlight the critical role of iterative testing and continuous refinement in achieving optimal product performance. By integrating advanced validation techniques, designers can better align design intent with usability, ensuring that user-centered design principles remain central to product innovation.

Design verification and validation approaches provide a practical framework for confirming the quality and safety of pre-production prototypes. By adopting this structured methodology, entrepreneurs can identify potential issues early, streamline the development process, and improve product reliability, ultimately increasing their chances of success in a competitive marketplace.

> Here are three responses to a piece of design—yes, no, and WOW! Wow is the one to aim for.
>
> —Milton Glaser

REFERENCES

American Institute of Aeronautics and Astronautics (AIAA) (1998) Guide for the Verification and Validation of Computational Fluid Dynamics Simulations, AIAAG-077-1998.

Andreasen, M.M. (1988). *Design for Assembly*. Springer-Verlag, New York, USA.

Babuska, I., & Oden, J.T. (2004). Verification and validation in computational engineering and science: basic concepts, *Computer Methods in Applied Mechanics and Engineering* 193(36-38), 4057–4066.

Beyer, H. (2010). *User-Centered Agile Methods*. Morgan and Claypool Publishers.

Bolton, S. M. (2013). *Improving Product Design and Development Performances in SMEs with User-Centered Design Activities*. Manchester Metropolitan University.

Camburn, B., Viswanathan, V., Linsey, J., Anderson, D., Jensen, D., Crawford, R., Otto, K., & Wood, L. (2017). Design prototyping methods: state of the art in strategies, techniques, and guidelines. *Design Science*, 3, https://doi.org.10.1017/dsj.2017.10

Chen, L.H., & Ko, W.C. (2009). Fuzzy linear programming models for new product design using QFD with FMEA. *Applied Mathematical Modeling* 33(2), 633–647.

Clayton, M. P, Bancroft, B., & Rajan, B. (2002). A review of assigned protection factors of various types and classes of respiratory protective equipment with reference to their measured breathing resistances. *The Annals of Occupational Hygiene*, 46(6), 537–547. https://doi.org/10.1093/annhyg/mef071

Creveling, C. M., Slutsky, J. L., & Antis, D. (2002). *Design for Six Sigma in Technology and Product Development*. Prentice Hall.

De Beer, D., Campbell, R.I., Truscott, M., Barnard, L., & Booysen, G. (2009). Client-centered design evolution via functional prototyping. *International Journal of Product Development (IJPD)*, 8(1), 22–41.

Deng, Y.M., Britton, G., & Tor, S. (2000). Constraint-based functional design verification for conceptual design. *Computer-Aided Design* 32, 889–899.

Farkas, L.G., 1981. Anthropometry of the Head and Face in Medicine. Elsevier, New York.

Global Harmonization Task Force (GHFT) (2004) Quality Management System—Process Validation Guidance, second edition.

Guo, Z., Zhou, D., Chen, J., Geng, J., Lv, C., & Zheng, S. (2018). Using virtual reality to support the product's maintainability design: immersive maintainability verification and evaluation system. *Computers in Industry*, 202, 41–50.

Gutierrez, A. M. J. A., Galang, M. D., Seva, R. R., Lu, M. Cl., & Ty, D. S. (2014). Designing an improved respirator for automotive painters. international Journal of Industrial Ergonomics. 44, 131–139.

Hart, S. G., & Staveland, L. E. (1988). Development of NASA-TLX (Task Load Index): Results of empirical and theoretical research. In P. A. Hancock & N. Meshkati (Eds.), *Human Mental Workload* (pp. 139–183). North-Holland. https://doi.org/10.1016/S0166-4115(08)62386-9.

Jamnia, A. (2018). *Initial Design Failure Mode and Effects Analysis and Product Risk Assessment. Introduction to Product Design and Development for Engineers*. CRC Press, pp. 131–144.

Joundi, J., Baccarne, B., Robaeyst, B., Bombeke, K., De Marez, L., & Saldien, J. (2024). Comparing virtual and real-life rapid prototyping methods for user testing smart city interfaces: A case study. *Applied Sciences*, 14(21), 9918. https://doi.org/10.3390/app14219918

Khalaf, F., & Yang, K. (2006). Product development process—from deterministic to probabilistic; A design for 6-sigma approach to lean product validation, part II. *International Journal of Product Development*, 3(1), 18–36.

Liberman-Pincu, E., & Bitan, Y. (2021). FULE—Functionality, usability, look-and-feel, and evaluation of novel user-centered product design. Methodology—Illustrated in the case of an autonomous medical device. *Applied Sciences*, 11(3), 985. https://doi.org/10.3390/app11030985

Lopez, G., Jesus, A., Sanchez-Ramirez, C., & Garcia Alcaraz, J. (2021). *Techniques, Tools, and Methodologies Applied to Quality Assurance in Manufacturing*. Springer International Publishing.Maropoulos, P.G., & Ceglarek, D. (2010). Design verification and validation in the product lifecycle. *CIRP Annals—Manufacturing Technology*, 59, 740–759. https://doi.org/10.1016/j.cirp.2010.05.005

Marques, C., & Correia, E. (2022). An overview of sensory characterization techniques: from classical descriptive analysis to the emergence of novel profiling methods. *Foods*, 11(2), 301.

McCradden, M.D. (2023). Ethics, First. *American Journal of Bioethics: AJOB*, 23(9), 55–56. https://doi.org/10.1080/15265161.2023.2237459.

McKenzie, J., Carroll, K., & McIntyre, L. (2018). Stakeholder views regarding ethical issues in the design and conduct of pragmatic trials: study protocol. *BMC Medical Ethics*, 19, 90. https://doi.org/10.1186/212910-018-0332-z

McMahon, C.A., Liu, Q., & Zheng, L.Y. (2010). Integration of process FMEA with product and process design based on key characteristics. Proceedings of the 6th CIRP-Sponsored International Conference on Digital Enterprise Technology.

Musale, P., & Mankar, S. D. (2023). The quality-by-design approach is based on analytical method validation. *Asian Journal of Pharmaceutical Analysis*, 13(3) Retrieved from www.proquest.com/scholarly-journals/quality-design-approch-based-analytical-method/docview/2874765365/se-2

Palady, P. (1997). *Failure Modes and Effects Analysis*, 2nd ed. PT Publications, Inc..

Persad, U., Langdon, P., Clarkson, P. J., & Brown, D. (2007). Cognitive scales and mental models for inclusive design. universal access in human-computer interaction. *Coping with Diversity*, 4554, 703–712. https://doi.org/10.1007/978-3-540-73279-2_86

Ramanathan, R. & Yunfeng, J. (2009). Incorporating cost and environmental factors in quality function deployment using data envelopment analysis. *Omega* 37(3), 711–723.

Reik, M. P., Mcintosh, R. I., Culley, S. J., Mileham, A. R., & Owen, G. W. (2006). A Formal design for changeover methodology. part 1: theory and background. *Proceedings of the IMechE Part B Journal of Engineering Manufacture* 220(8), 1225–1235.

Sargent, R.G. (2005). Validation and verification of simulation models. Proceedings of the 2005 Winter Simulation Conference, pp. 130–143.

Shin, G., Jarrahi, M. H., Fei, Y., Karami, A., Gafinowitz, N., Byun, A., & Lu, X. (2019). Wearable activity trackers, accuracy, adoption, acceptance, and health impact: A systematic literature review. *Journal of Biomedical Informatics*, 93, 103153. https://doi.org/10.1016/j.jbi.2019.103153

Shoval, N., Auslander, G. K., Freytag, T., Landau, R., Oswald, F., Seidl, U., et al. (2008) The Use of Advanced Tracking Technologies for the Analysis of Mobility in Alzheimer's Disease and Related Cognitive Diseases. *BMC Geriatrics*, 8, 7.

Spence, C., & Gallace, A. (2011). Multisensory design: Reaching out to touch the consumer. *Psychology & Marketing*, 28(3), 267–308.

Stamatis, D. H. (2003). *Failure Mode and Effect Analysis: FMEA from Theory to Execution* (2nd ed.). ASQ Quality Press.

Stone, R.B., McAdams, D.A., & Kayyalethekkel, V. J., (2004). A product architecture-based conceptual DFA technique. *Design Studies*, 25(3), 301–325.

Suh, N. P. (2001). *Axiomatic Design: Advances and Applications*. Oxford University Press.

Taguchi, G., Chowdhury, S., & Wu, Y. (2005). *Taguchi's Quality Engineering Handbook*. John Wiley & Sons.

Tatlısu, E. (2023). *Bridging the Gap between Product and User: A Design Framework to Establish a Free Relationship with Technology*. Istanbul Technical University.

Tu, J., & Gao, W. (2021). Ethical considerations of wearable technologies in human research. *Advanced Healthcare Materials*, 10, 2100127.

U.S. Department of Defense (2008) Documentation of Verification, Validation & Accreditation (VV&A) for Models and Simulations.

Witteman, H., Vaisson, G., Provencher, T., Dansokho, S., Colquhoun, H., Dugas, M. Fagerlin, A., Giguere, A., Haslett, L., Hoffman, A., Ivers, N. Legare, F., Trottier, M.,

Stacey, D., Volk, R., Renaud, J. (2021). An 11-item measure of user-and human-centered design for personal health tools (UCD-11): development and validation. *Journal of Medical Internet Research,* 23(3), e15032. https://doi: 10.2196/15032

Wexler, A., & Largent, E. (2023). Ethical considerations for researchers developing and testing minimal-risk devices. *Nature Communications,* 14(1), 2325. https://doi.org/10.1038/s41467-023-38068-6

Zhuang, Z., & Bradtmiller, B., 2005. Head and face anthropometric survey of US respirator users. *Journal of Occupational and Environmental Hygiene,* 2 (11), 567e576.

12 Entrepreneurship and Intrapreneurship

> If something's important enough, you should try. Even if the probable outcome is failure.
>
> —Elon Musk

12.1 INTRODUCTION

For designers, engineers and "makers," startups shape the products that people use every day. When you blend design with business, you can ensure that creative ideas align with strategic goals. This means that products work and are produced efficiently and affordably. Embracing entrepreneurial principles allows them to tackle challenges head-on and adapt to the ever-changing market landscape. By doing this, you can make a real impact, enhance user experiences and drive sustainable growth in startups and established companies.

12.1.1 ENTREPRENEURSHIP AND INTRAPRENEURSHIP

Entrepreneurship and intrapreneurship are two concepts that revolve around creating and developing innovative ideas within the business world (Hisrich, 1990; Parker 2011). They are all about taking risks and seizing opportunities, but they differ in the context in which they occur. In some cases, splitting off departments to establish startups has become strategically important (Knossalla & Carbon, 2023).

Entrepreneurship is a journey of starting and managing a brand-new business venture (Hisrich et al., 2017). As an entrepreneur, you get to identify and bring those promising business opportunities to life. You are driven by passion, creativity, and determination to make things happen. You wear many hats, from securing funding to handling

DOI: 10.1201/9781003476177-12

operations, marketing, and sales. Embracing uncertainty is part of the game, but the potential rewards of success keep you motivated.

Intrapreneurship is where you showcase your entrepreneurial spirit within an organization (Bosma et al., 2010). You are like an entrepreneur working from within, coming up with fresh ideas and taking calculated risks to drive growth and create value for the company. Intrapreneurship fosters a culture of innovation, drives competitiveness, engages and retains employees, enables organizational adaptability, attracts top talent, facilitates corporate entrepreneurship, and contributes to social impact and sustainability. Intrapreneurship unlocks the full potential by empowering employees to think creatively, take risks, and make an impact within the organization. It promotes continuous learning, agility, and the ability to seize new opportunities, ultimately positioning the organization for long-term success and positive societal impact.

Entrepreneurship and intrapreneurship contribute to innovation, economic growth, and job creation (Figure 12.1). They are all about pushing boundaries, developing new products and services, and adapting to a changing market climate.

12.1.2 INNOVATION IN STARTUPS

Traditionally, the startup approach relies heavily on detailed business plans and forecasts, sequential phases of product planning, design, and development, and building comprehensive products upfront with the expectation that customers would naturally flock to them (Stevenson, 1983; Hills, 1985; Olson, 1985; Donckels, 1989). Recent studies have shown that as many as 90% of startups fail (Aminova & Marchi, 2021). This suggests that the traditional methods and thinking may need to change.

The high failure rates challenge assumptions, indicating a need for a more adaptable and customer-focused methodology. As the startup landscape

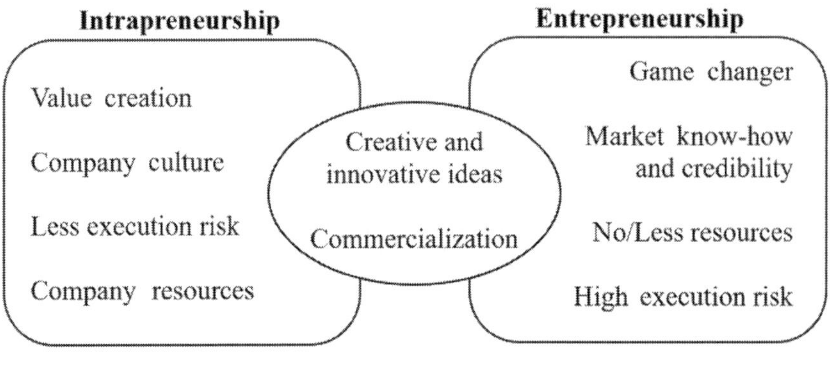

FIGURE 12.1 Characteristics of intrapreneurship and entrepreneurship.

evolves, researchers and entrepreneurs need to explore alternative frameworks that better address the realities and complexities of building a successful new venture (Rogers, 2003). As discussed throughout the book, any startup requires a deep understanding of people.

12.1.3 DESIGN AND BUSINESS MAPPING

Creating a prototype of an innovative product or service involves turning ideas into a tangible reality. The mapping from the design space to the business space affects the different entities (Figure 12.2). In a business setting, consumers are the individuals who purchase or use a startup product or service (Thulasipriya & Rodrigues, 2023), regardless of whether they are the primary users. Let's define users, customers, and consumers.

12.1.4 CUSTOMER STAKEHOLDERS

Customer stakeholders include users and consumers. Consumers are individuals who purchase products or use a startup's products or services, regardless of whether they are primary users (Thulasipriya & Rodrigues, 2023). Users are the individuals who actively engage with and use the product or service offered by the startup. They can have a significant impact on the success of a startup by providing feedback and insights, influence adoption and retention, and word-of-mouth marketing. For example, on YouTube, the users are the individuals who actively watch, upload, and engage with the content, while the consumers are advertisers on the platform.

When a company launches a new product or service, its impact can extend beyond target users. The offerings may directly or indirectly affect all people,

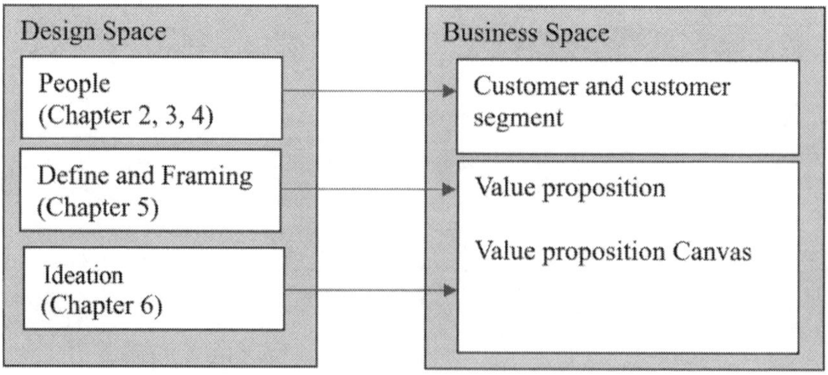

FIGURE 12.2 Mapping of terms in the design and business space.

FIGURE 12.3 Bicycles obstructing pathways, causing a hazard to pedestrians.

particularly when they disrupt or alter established norms and practices within a particular community or society (Ryan et al., 2023). Consider a bicycle-sharing business. While the users may appreciate the convenience and flexibility of the service, people in the community could experience inconveniences if the shared bicycles obstruct pedestrian pathways and pavements or pose environmental and safety hazards (Figure 12.3). Another example would be a social media platform that is perceived as detrimental to young users' mental health and well-being. Even though the platform's primary users may continue to utilize the service, concerns from parents, educators, and mental health professionals can significantly impact the company's public image and reputation. In these scenarios, the company's introduction of a new product or service has implications that extend beyond just the intended users and customers. The ripple effects on the general population can be substantial and may ultimately shape the overall reception and impact of the company's innovations within the broader community or society. In such cases, people can become vocal critics of the company, actively engaging in campaigns, boycotts, or legal actions that limit its growth or even force it to change its business model. This negative publicity and reputational damage can have far-reaching consequences, affecting the company's ability to attract new users, secure investments, or maintain its social license to operate.

Cultivating strong consumer loyalty can lead to repeat business, positive word-of-mouth, and increased brand recognition, all of which can contribute to the startup's long-term success. Commercialization is more than mass-producing a product or launching a service (Rogers et al., 2004). It requires careful planning, strategic thinking, and effective execution to ensure a successful market entry and sustainable growth. During the initial commercialization phase, individuals such as owners, co-founders, or entrepreneurs encounter many challenges. These hurdles encompass various aspects, including sharing ownership (Åstebro & Bernhardt, 2003), intellectual property (Baran & Zhumabaeva, 2018), partnership (Draper III, 2011), and financial and operational risks (Basu & Nair, 2015) associated with the venture. Additionally, they face resource constraints (Silva et al., 2021) and the need for funding (Bauer et al., 2023), navigate risks along the design and development lifecycle, and strive for growth and scalability to secure market share and long-term success.

12.1.5 BUSINESS STAKEHOLDERS

Business stakeholders (Westrenius & Barnes, 2015) are individuals or groups with an interest in a company's success, including internal stakeholders like employees and management, as well as external ones such as customers, suppliers, investors, trade associations, and regulatory bodies. Their influence can significantly shape business decisions, strategies, and reputation. Effective stakeholder engagement through transparent communication, feedback, and collaboration is essential for identifying risks, enhancing sustainability, and fostering loyalty, ultimately contributing to the long-term success of the organization.

12.2 STARTUP SEQUENCE

Let's look at the pre-seed stage of a startup journey (Figure 12.4). It is important to surround yourself with people who share your passion and are willing to roll up their sleeves and get to work. The founding team is typically small during the early days, usually just the founders and a few key members. Building the right team is extremely important at this stage, as you will need to rely heavily on their expertise and collaboration to bring your vision to life. The initial team should have complementary skills to reduce redundancy and optimize the number of members.

It all starts with an idea. In previous chapters, we have explored the ideation process, and the importance of prototyping. Not every creative idea will be commercially viable, making it essential to conduct due diligence to confirm that your product or service has real potential. Additionally, when pursuing projects focused on social innovation, nonprofit entrepreneurship must also consider viability and business development that are sustainable in the long run.

First, the startup team must work on the value of the product or service and consider total addressable market (TAM), serviceable available market

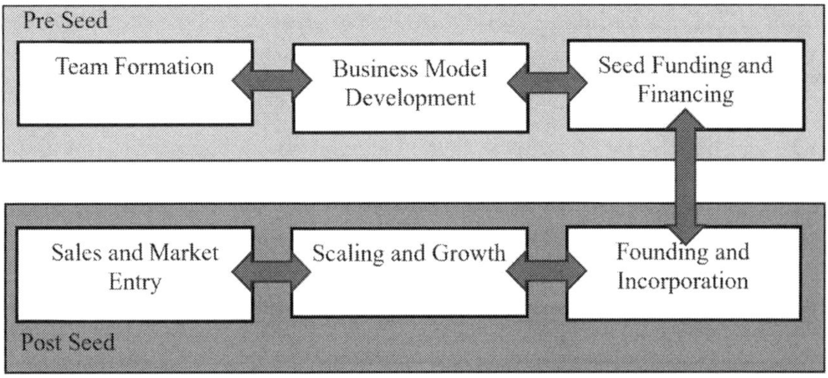

FIGURE 12.4 The startup stages.

(SAM), serviceable obtainable market (SOM), and minimum serviceable market (MSM) (Vlies, 2023; Kamps, 2020). TAM refers to the total revenue opportunity available for a product or service if it were to achieve 100 percent market share in a specific market. It represents the maximum potential sales volume and is typically calculated based on the number of potential customers and the average revenue per user (ARPU). SAM refers to the portion of the TAM that a company can target and serve based on its business model, resources, and strategic focus. It represents what is accessible through current distribution channels and aligned with the company's strategic focus and capabilities. SOM refers to the subset of SAM that a company can realistically capture within a specific timeframe, considering competition, market conditions, and internal resources. It represents the most achievable market share based on its current capabilities and strategies. MSM is the smallest segment of a market that a business can target and serve while still achieving sustainable operations and profitability. It represents the minimum level of market demand necessary for a product or service to be viable and for the business to function effectively.

Although designers may have completed several iterations on the suitability and relevance of the product, in the startup phase, the focus is on viability in terms of cost, pricing, sales and marketing, resources, and business activities. Let's not forget about funding. During the pre-seed stage, startups typically seek initial funding from various sources, including friends and family, government grants, angel investors, and small venture capital funds. The amount of seed funding can vary widely, depending on the needs and the investors' appetite. But, whatever the amount, this funding is key for further developing the product, validating the business model, and setting the stage for the next big steps. Funding can be pursued at any stage of the ideation, prototyping, design selection, design specification, and validation phases.

12.3 VALUE POSITIONING

Value positioning refers to the strategic process of creating a distinct and desirable image for a product or service in the minds of the target market (Makadon et al., 2010; Sethi, 2016). It involves defining and communicating the product's unique value proposition and competitive advantages to differentiate it from competitors and meeting the specific needs and preferences of the target customers. Effective value positioning helps establish a clear and compelling identity for the product, enabling it to occupy a distinct place in the market and match with the intended audience (Moon & Kamakura, 2017). Through thoughtful product positioning, organizations can shape customer perceptions, drive demand, and ultimately achieve a competitive advantage in the marketplace. When positioning any product, avoid the common mistakes (Table 12.1).

TABLE 12.1 Things that Can Go Wrong During Product Positioning.

Common Issues	What Happened
Ignoring emotional and psychological factors	Overlooked the emotional and psychological factors influencing customer purchasing decisions, such as brand perception, trust, and emotional appeals. Effective positioning goes beyond product features and benefits. Some psychological considerations are perception, cognitive biases, emotion and motivation, social influence, decision-making processes, branding and identity, pricing psychology (Chapter 2).
Lack of research	Not conducting thorough market research and understanding the target market (Chapters 3 and 4).
Vague or inconsistent messaging	Vague (or generic) language fails to differentiate the product from competitors. Effective product positioning requires clear and consistent messaging that communicates the product's unique value proposition (Chapters 8 and 11).
Neglecting competitive analysis	Lack of understanding of the competitive landscape, which is crucial for positioning a product effectively. Overlooked the importance of analyzing competitor offerings, strengths, and weaknesses (Chapter 11).
Overpromising or underdelivering	Overpromising the capabilities or benefits of the product without considering its actual capabilities or limitations. This can lead to customer dissatisfaction and damage the product's reputation (see Chapters 5, 9, and 10 for more discussion).
Failing to adapt to customer feedback	Resistant to adapting their positioning strategy based on customer input, resulting in missed opportunities for improvement. Product positioning is an iterative process and consumer feedback, and market dynamics must be continuously monitored (see Chapter 4).

12.3.1 CUSTOMER VALUE PROPOSITION

For a new startup, the final product is vague, and hence a customer value proposition (CVP) (Ries, 2011; Carvalho & Jonker, 2015; Frederiksen & Brem, 2017) is a good starting point to get initial investors and seed funding. CVP is a concise statement that summarizes the unique benefits and value that a product, service, or company plan to offer to its customers or target audience. It communicates the specific problem it solves, the benefits it provides, and why it is better than competing alternatives. A well-crafted CVP effectively conveys the unique value that a product or service offers to its customers, explaining why they should choose it over competing options (Table 12.2). CVP evolves from basic needs and functional benefits to psychological and physical well-being as integral components of customer value. The traditional value proposition canvas that focuses on pain points needs to evolve to address the changing expectations of people. A well-crafted CVP should include:

1. Clearly defined target audience: A specific group for whom the product or service is designed.
2. Compelling problem statement: A concise articulation of key challenges the target audience faces.
3. Innovative or distinct solution: A unique way the product or service addresses the identified problem.

TABLE 12.2 Examples of Customer Value Proposition.

"ZenSeat: Transform your work experience with our Ergonomic Wellness Chair—designed to enhance comfort, promote proper posture, and boost your productivity while prioritizing your well-being."	
Target audience	Office workers, remote professionals, and students need comfort and support during long hours of sitting.
Problem statement	Overcome discomfort, poor posture, and fatigue from prolonged sitting, leading to decreased productivity and overall well-being.
Unique solution	The chair is designed with adjustable lumbar support, breathable materials, and customizable seating positions to promote ergonomic posture and comfort.
Benefits and outcomes	Enhances physical well-being by reducing back pain and improving posture; boosts psychological well-being through increased comfort and reduced fatigue, leading to higher productivity and focus.
Differentiation	Unlike standard office chairs, the chair combines ergonomics with wellness features, such as autonomous resting and stress-relief adjustments.
Proof or evidence	Backed by research studies demonstrating reduced discomfort and improved posture; testimonials from users highlight enhanced productivity and comfort.

4. Tangible benefits and outcomes: The specific advantages customers will experience by using the product or service.
5. Competitive differentiation: A clear statement of why the offering is superior to existing alternatives.
6. Credibility and proof: Supporting evidence, such as testimonials, case studies, or data that validate the value proposition.
7. Clarity and simplicity: Concise, easy-to-understand language that effectively communicates the value.

By addressing these elements, businesses can craft a compelling value proposition that fits with modern consumers.

12.3.2 VALUE PROPOSITION CANVAS

The value proposition is represented using a value proposition canvas (VPC) (Figure 12.5). Osterwalder et al. (2015) conceptualized the value proposition canvas by focusing on pains and gains, where the goal of the product is to minimize customer pains and maximize the gains. This VPC is useful when we can see, observe, and discuss the pain points with the customers; however, with more emphasis on digital technologies and intangible products and services, the pain points are not always visible, and the users are not even aware of them. Peter Thomson (www.peterjthomson.com) proposed a VPC that centers around consumer needs, wants, and fears, which are addressed through a product or service's features, benefits, and experiences. This approach highlights the VPC as the critical intersection between business strategy and brand strategy, ensuring that offerings align with both customer expectations and the company's strategic vision (www.peterjthomson.com).

Chapters 2–4 explored aspects related to people. Today's products and services have shifted from physical products to digital services that seamlessly integrate into our lives. This transformation reflects the evolving needs outlined in

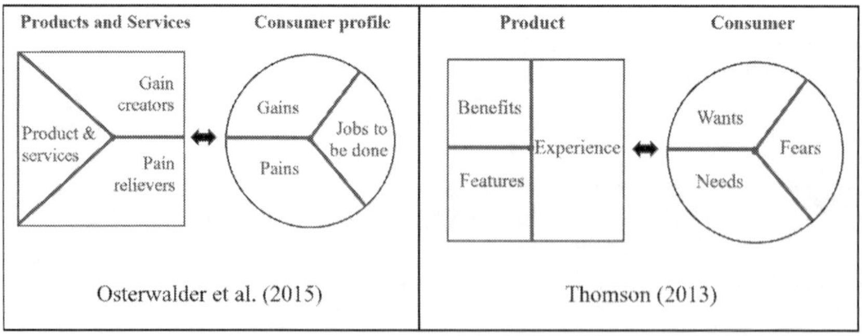

FIGURE 12.5 Two types of value proposition canvas.

Maslow's hierarchy (Figure 1.1), as people increasingly seek convenience, connectivity, and personalized experiences (McLeod, 2007).

Thus, to capture the changing challenges and aspirations, a more inclusive VPC is needed (Figure 12.6). Understanding people's needs and wants is necessary for creating products and services that can be functional, emotional and social. For functional, it involves the specific tasks customers want to accomplish, like completing projects or managing time effectively. But it doesn't stop there; emotional needs and wants are equally important. Customers seek comfort, security, and joy, often choosing brands that enhance their happiness or confidence. Social needs and wants reflect how customers want to be perceived, influencing their choices based on aspirations for approval or belonging. By addressing these functional, emotional, and social dimensions, companies can deliver a richer and more fulfilling customer experience.

Customers are driven not only by practical needs but also by deeper emotional and social motivations. They seek comfort, security, and joy, gravitating toward brands that elevate their confidence and overall well-being. Beyond functionality, products and services serve as expressions of identity. Customers often make choices based on how they wish to be perceived, influenced by their desire for social approval, connection, or belonging. By addressing this triad of functional, emotional, and social needs, companies can create more meaningful and rewarding customer experiences that resonate on multiple levels.

Chapters 6 to 11 of this book focused on how we can better serve people. Our offerings include a clear range of products and services designed to help users achieve their goals effectively while avoiding discontent for the people who do not directly use the product. We believe that fulfilling tasks goes beyond just functionality; it is about creating joyful experiences and fostering social innovation. Challenge alleviators tackle visible issues, like technical glitches and

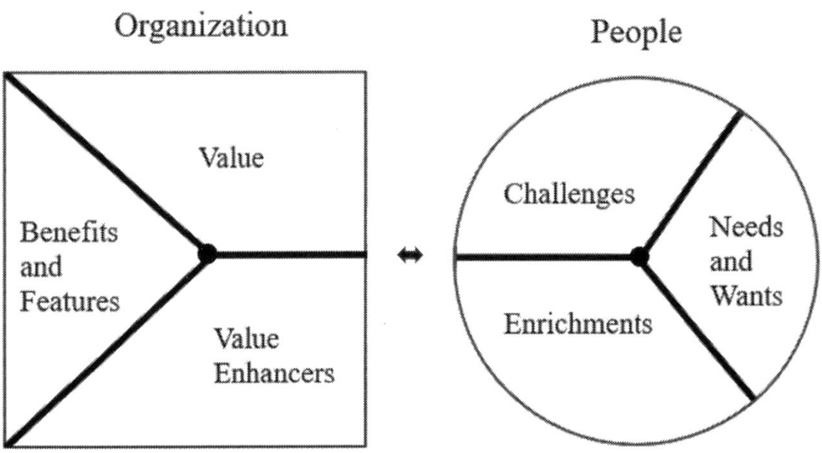

FIGURE 12.6 Inclusive value proposition canvas.

confusing interfaces, and hidden barriers that users may not recognize, such as fears, cognitive biases or emotional challenges. Contextual adaptors ensure that features adjust to users' environments, providing relevant support whether they are at home, at work, or on the go. To enhance value, we introduce experience amplifiers and aspirational catalysts. Experience amplifiers enrich user interactions with joy and ease, featuring intuitive designs and personalized touches. Meanwhile, aspirational catalysts help users pursue their long-term goals, guiding them toward personal growth, increased productivity, and overall well-being. By focusing on these dimensions, we create a holistic experience that meets needs and inspires users on their journey.

12.4 BUSINESS MODEL CANVAS (BMC)

The business model canvas is a strategic management tool that provides a visual representation of the key components and relationships within a business model (Osterwalder & Pigneur, 2010). It helps entrepreneurs and organizations understand, analyze, and communicate their business models effectively. The canvas consists of nine building blocks that describe various aspects of a business: value proposition, cost structure, revenue stream, customer relationship, customer channel, customer segments, key activities, key resources, and key partners (Figure 12.7).

To enhance alignment with the proposed inclusive value proposition canvas, we introduce an inclusive business model canvas (Figure 12.8). At the center of the canvas lies the inclusive value proposition canvas, which serves as the

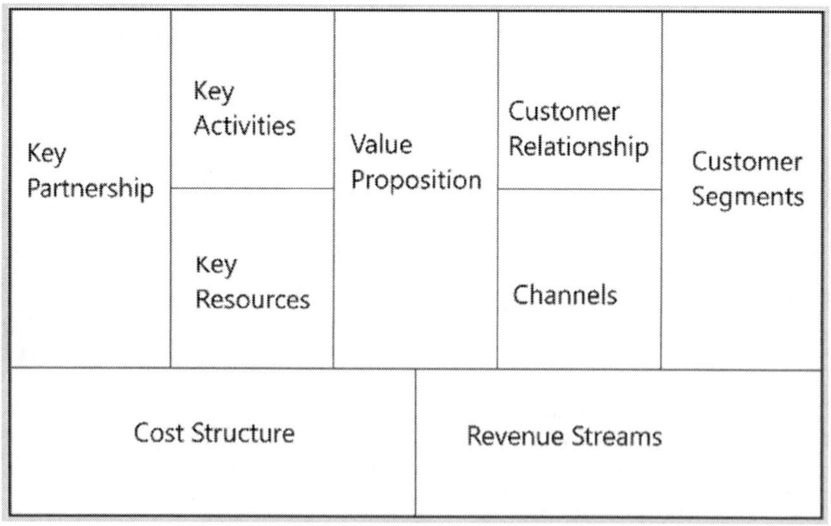

FIGURE 12.7 Traditional business model canvas.

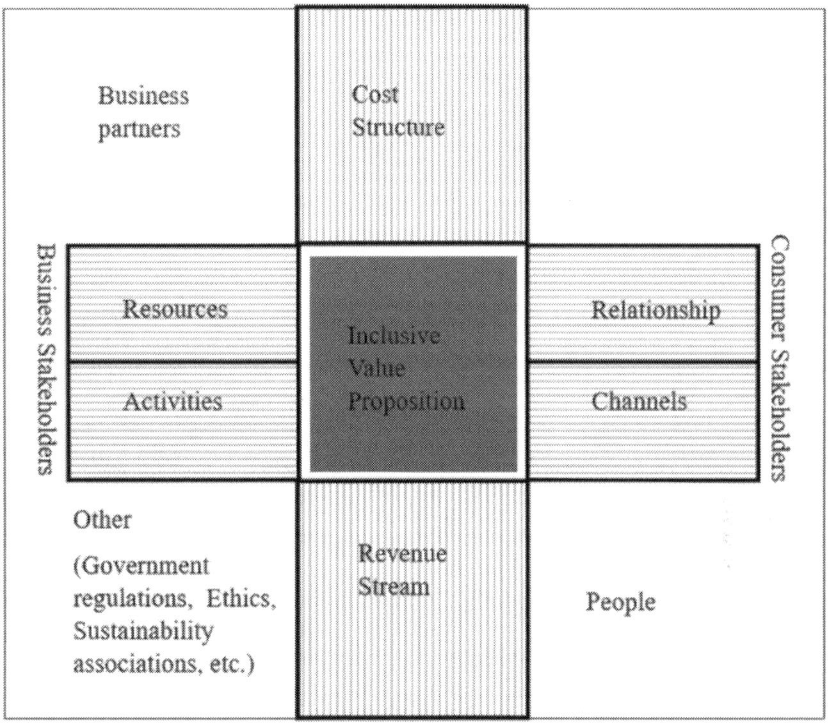

FIGURE 12.8 Inclusive business model canvas.

foundation for the entire model. The right-side focuses on the people involved, including consumers, and users. It outlines the necessary relationships the company must cultivate with these groups and the channels through which these interactions occur. The left side focuses on the company dynamics. This area encompasses the company itself, detailing aspects such as the startup team, available resources, key activities, and business partners. It also includes external influences from government entities, ethical organizations, and sustainability associations. The cost structure is positioned at the top, highlighting the financial aspects of the business. This section considers the consumers and the team but also acknowledges the impact of business partners and users on overall costs. The lower part of the canvas represents revenue streams. It considers the consumer and the team but can also involve various external agencies. In the initial stage, there is no clear sequence of which aspects must be considered first or last. A team of stakeholders can carry out "sticky note brainstorming." They write down ideas and stick on different boxes of the Inclusive business model canvas while considering the inclusive value proposition canvas. In later stages, each section can be consolidated and enhanced. The business model canvas is explained in detail in the following sections.

12.4.1 PARTNERSHIPS

Product and service companies often form strategic partnerships and collaborations to enhance their capabilities, access new markets, and leverage shared resources. These partnerships can take various forms and serve different purposes (Table 12.3). The key partnerships a company forms depend on its strategic objectives, market dynamics, and the specific needs of its business. Effective partnerships can provide companies with access to resources, expertise, market opportunities, and competitive advantages that they may not have been able to achieve independently.

12.4.2 ACTIVITIES

Product and service companies engage in various key activities to create, deliver, and support their offerings. These key activities can vary depending on

TABLE 12.3 Types of Partnership.

Supplier partnerships	Companies form partnerships with suppliers to ensure a reliable and efficient supply of raw materials, components, or specialized inputs.
Distribution partnerships	Companies collaborate with distributors, wholesalers, or retailers to expand their market reach and improve product or service distribution.
Strategic alliances	Strategic alliances involve two or more companies joining forces to pursue shared objectives, such as joint product development, market expansion, or technology sharing.
Joint ventures	Joint ventures are formal partnerships where two or more companies create a separate legal entity to pursue a specific business opportunity.
Co-marketing partnerships	Companies engage in co-marketing partnerships to jointly promote their products or services.
Technology and innovation partnerships	Companies collaborate with technology partners, startups, or research institutions to access new technologies, innovation, or research capabilities.
Licensing and franchising	Companies enter into licensing or franchise agreements to expand their brand presence and market penetration.
Reseller or value-added reseller (VAR) partnerships	Companies partner with resellers or VARs to sell their products or services.
Strategic supplier or customer relationships	Companies develop strategic partnerships with key suppliers or customers to foster long-term collaboration, preferential treatment, and mutual growth.
Industry partnerships and associations	Companies join industry associations, universities, consortia, or collaborative platforms to collaborate with peers, share knowledge, and address common industry challenges.

TABLE 12.4 Types of Activities.

Research and development (R&D)	This may involve conducting market research, developing prototypes, testing, and refining offerings based on customer feedback.
Production and manufacturing	This involves procuring raw materials, managing the production process, quality control, and ensuring efficient operations.
Supply chain management	This involves managing relationships with suppliers, optimizing the flow of materials, and ensuring timely delivery of inputs to support production and service delivery.
Marketing and sales	This includes market research, branding, advertising, pricing, distribution, sales channels, and customer acquisition strategies.
Customer service and support	This involves activities such as handling customer inquiries, resolving issues, providing technical support, and ensuring customer satisfaction throughout the customer journey.
Distribution and delivery	This includes warehousing, transportation, order fulfillment, and managing relationships with distributors or retailers.
Service delivery and implementation	This may involve project management, service implementation, customization, training, and ongoing support.
Quality assurance and control	This involves activities such as quality testing, quality control processes, adherence to regulatory requirements, and continuous improvement initiatives.
Technology and IT management	This includes activities such as managing IT infrastructure, developing and maintaining software applications, cybersecurity, and data analytics.
Financial management	Companies engage in financial activities such as budgeting, financial planning, accounting, billing, invoicing, and managing financial transactions.

the nature of the business and industry (Table 12.4). The specific key activities undertaken by a product or service company depend on factors such as industry, business model, value proposition, and customer needs. Effective execution of key activities is crucial for delivering customer value, maintaining operational efficiency, and achieving business objectives.

12.4.3 RESOURCES

Product and service companies rely on various key resources (Table 12.5) to operate and deliver value to their customers. The specific key resources required

TABLE 12.5 Types of Resources.

Physical resources	Physical resources may include manufacturing facilities, office spaces, equipment, machinery, vehicles, inventory, and distribution networks.
Human resources	Human resources refer to the people who work within the company, including employees, managers, executives, and specialized personnel.
Intellectual property	Intellectual property encompasses the company's intangible assets, such as patents, trademarks, copyrights, trade secrets, and proprietary technology.
Financial resources	Financial resources include the capital, funding, and financial assets that companies need to operate and invest in their growth. This can include cash reserves, lines of credit, loans, venture capital, or access to investment capital.
Information and data	Companies rely on data collection, storage, analysis tools, and information systems to gather and utilize valuable information.
Brand and reputation	Building brand equity, customer trust, and loyalty require investments in marketing, advertising, public relations, and customer relationship management.
Technology and infrastructure	Technology resources encompass the software, hardware, IT infrastructure, and digital tools that enable companies to streamline operations, enhance productivity, and deliver digital experiences.

by a product or service company may vary depending on the industry, business model, and strategic priorities. Companies must identify, acquire, and manage these key resources effectively to support their operations, growth, and competitive advantage.

12.4.4 RELATIONSHIPS

Product and service companies can build different types of consumer relationships to enhance customer satisfaction, loyalty, and engagement (Table 12.6).

12.4.5 MARKETING CHANNELS

Product and service companies have various channels available to reach users and buyers. The choice of channels depends on the target market, marketing goals, and the nature of the product or service (Table 12.7). Companies need to select channels that align with their target market, marketing strategy, and

TABLE 12.6 Types of Consumer Relationships.

Transactional relationship	The primary goal is to complete individual transactions and generate revenue. Limited emphasis is placed on long-term customer engagement or personalized interactions.
Customer service and support	This relationship model emphasizes providing excellent customer service and support throughout the customer journey.
Personalized relationship	Companies that prioritize personalized relationships seek to understand individual customer preferences, needs, and behaviors.
Community relationship	This can be achieved through online forums, social media groups, or dedicated platforms where customers can interact, share experiences, and provide support to one another.
Subscription or membership relationship	These relationships are built on recurring revenue streams, and the company focuses on delivering value, maintaining customer satisfaction, and encouraging long-term subscriptions or memberships.
Co-Creation relationship	Companies seek input, feedback, and ideas from customers, allowing them to contribute to the design, features, or improvements. This collaborative approach fosters a deeper connection and sense of ownership among customers.
Advocacy relationship	They encourage customers to share positive experiences, provide online reviews, refer to others, or actively promote the brand through word-of-mouth or social media. Companies may incentivize advocacy efforts through referral programs or exclusive rewards.
Loyalty programs	These programs often involve point systems, tiered rewards, exclusive offers, or VIP benefits. The focus is on cultivating customer loyalty and encouraging continued patronage.

budget. Often, a combination of multiple channels is used to maximize reach and effectiveness. Regular monitoring, evaluation, and adaptation of channel strategies are crucial to ensure optimal results.

12.4.6 REVENUE STREAMS

There are several ways for product and service companies to generate income (Table 12.8). Companies should assess their target market, competitive landscape, and business model to determine the most suitable income generation methods for their specific industry and offerings.

TABLE 12.7 Types of Channels.

Direct sales	Companies can engage in direct selling by establishing their own sales team or utilizing company representatives to reach out to potential users and buyers.
Retail stores	Selling products through physical retail stores is a traditional channel that allows companies to showcase their offerings to a wide range of customers.
E-commerce	This includes setting up their own online store or utilizing established e-commerce marketplaces like Amazon, eBay, or Alibaba.
Distributors and wholesalers	Companies can work with distributors or wholesalers who specialize in reaching specific markets or customer segments.
Business-to-business (B2B) sales	This can involve direct sales teams, industry-specific trade shows, business directories, or online platforms that connect businesses.
Online advertising and social media	Companies can utilize online advertising platforms, such as Google Ads or social media advertising, to reach a targeted audience.
Content marketing	Content marketing involves creating and distributing valuable and relevant content, such as blog posts, articles, videos, or podcasts, to attract and engage users. This can help build brand awareness, establish thought leadership, and drive traffic to a company's website or other online channels.
Influencer marketing	Influencers can promote products or services through sponsored content, reviews, or endorsements.
Events and trade shows	Participating in industry-specific events, trade shows, conferences, or exhibitions provides opportunities for companies to showcase their products or services, network with potential customers, and generate leads.
Referral programs	Companies can implement referral programs to encourage existing users or customers to refer to their friends, family, or colleagues.

12.4.7 COST STRUCTURE

Product and service companies incur various costs (Table 12.9). The costs incurred by a product or service can vary widely depending on the industry, business model, scale of operations, and other factors. Companies should carefully analyze and track their costs to ensure profitability and make informed business decisions.

TABLE 12.8 Types of Revenue Streams.

Sales of products or services	The primary source of income for product and service companies is selling their offerings to customers. This can be in the form of one-time purchases or recurring revenue models such as subscriptions or service contracts.
Licensing or royalties	Companies can generate income by licensing their intellectual property, such as patents, trademarks, copyrights, or trade secrets, to other businesses.
Franchising	Franchising allows companies to expand their business by granting the right to operate under their established brand and business model to independent franchisees.
Subscriptions or membership fees	This model is common in industries like software, media streaming, online memberships, and subscription boxes.
Advertising and sponsorships	Companies can generate income by selling advertising space on their platforms, such as websites, apps, or physical spaces.
Affiliate marketing	Companies can earn income through affiliate marketing by promoting other businesses' products or services and earning a commission on referred sales or leads.
Consulting or professional services	This can include services like strategic consulting, IT consulting, legal services, marketing services, or training.
Grants and funding	Some companies, particularly those in research, development, or social impact sectors, can obtain income through grants or funding from government agencies, foundations, or private investors.

TABLE 12.9 Types of Cost Structure.

Cost of goods sold (COGS)	For product companies, COGS typically includes costs such as raw materials, manufacturing or production expenses, packaging, and direct labor. For service companies, COGS may include costs related to delivering the services, such as labor, equipment, and materials specific to the service provided.
Operating expenses	Operating expenses are the ongoing costs incurred to support the day-to-day operations of the business. This category includes expenses such as rent or lease payments for facilities, utilities, insurance, office supplies, marketing and advertising expenses, salaries and benefits for employees, professional services fees (legal, accounting), software subscriptions, and general administrative costs.
Research and development (R&D) costs	These include expenses related to research activities, product design, prototyping, testing, and intellectual property protection.

TABLE 12.9 (Continued)

Marketing and sales expenses	This includes costs associated with advertising, promotional campaigns, market research, trade shows or exhibitions, sales team salaries and commissions, and marketing materials.
Distribution and logistics costs	These costs can include storage fees, transportation expenses, packaging materials, and logistics software or services.
Overhead costs	Overhead costs are indirect costs necessary to support the overall business operations, but not directly associated with producing or delivering specific products or services.
Depreciation and amortization	Companies that own physical assets, such as equipment, vehicles, or buildings, incur depreciation expenses as these assets lose value over time.
Financing costs	If a company has borrowed money or obtained financing, it may incur interest expenses on loans or credit lines.

12.5 MINIMUM VIABLE PRODUCT (MVP) TO MINIMUM LOVABLE PRODUCT (LP) AND MINIMUM WOW PRODUCT (MWP)

The concept of minimum viable product (MVP) was introduced by Frank Robinson in 2001 (Lenarduzzi & Taibi, 2016) but gained significant traction only after it was featured in *The Lean Startup* (Ries, 2011). Ries explains that a minimum viable product (MVP) helps entrepreneurs start the process of learning as quickly as possible. It is not necessarily the smallest product imaginable; rather, it is the fastest way to navigate the build–measure–learn feedback loop with minimal effort. An MVP is the version of a product with just enough features to satisfy early customers and provide feedback for commercial product development (Stevenson et al., 2024). With careful user feedback during the prototyping phase, prototypes can be used as MVP and tested for market readiness. An MVP includes only the essential features needed to address the primary user needs. The goal is to get the product to market quickly, so an MVP can be developed and launched in a relatively short timeframe. Releasing an MVP allows you to get real user feedback to inform future product iterations and development. Based on user feedback, the MVP can be improved and expanded over time through an iterative process. Developing an MVP requires less upfront investment compared to building a full-featured product from the start.

Creating an MVP emphasizes achieving validated learning with minimal effort, setting the stage for future product and business development. To stress the importance of making a positive impression and delighting users, the concept of "lovable" products has emerged as an evolution beyond the traditional MVP (de Haaf, 2015; de Haaff, 2017). A minimum lovable product (MLP) delivers just enough to delight your customers from day one. It is not about doing the least; it is about doing the least that truly matters. By deeply understanding your customers' problems, you create a focused solution they will love, then refine and expand it over time using your long-term vision and their feedback. Therefore, our goal should extend beyond merely developing viable products to crafting MLP. The entire focus shifts to creating products that influence the users emotionally (Hokkanen, 2017; Ghory et al. 2023). In addition to lovable products, the idea of a "minimum wow product" (MWP) offers potential advantages for startups and product development (James, 2002; Hashim et al., 2021). A Minimum WOW Product (MWP) goes beyond simply working; it is designed to surprise and delight. It delivers a standout experience from the start, setting a high bar for competitors and leaving users genuinely impressed, not just satisfied or loved. An MWP is specifically designed to maximize user delight, even if its initial functionality is somewhat limited.

As this book focuses on engaging the user right from the start, delight rather than just basic functionality is the emphasis. MWP teams can rapidly develop and launch products, gather user feedback, and iteratively enhance the overall experience. MLP or MWP can forge stronger emotional connections with users, ultimately fostering brand loyalty and advocacy more effectively than just an MVP. Overall, the MWP concept shifts the focus from the minimum set of features to the minimum set of experiences that can captivate and engage users. This approach may be particularly advantageous for startups aiming to make a strong initial impression, cultivate a devoted user base, and differentiate themselves in competitive markets.

Companies have used prototyping to gauge initial consumer feedback. Other methods include explainer videos, prototype apps, landing pages, the Wizard of Oz technique, preorder strategies, and social media platforms. The same can be used for lovable products and wow products, ensuring that user feedback remains central to the development process. By using these diverse methods, businesses can refine their offerings based on real user insights, ultimately leading to products that not only meet functional needs but also delight and engage users on a deeper level. Let us look at some real cases.

12.5.1 EXPLAINER VIDEO

One company that leveraged this approach is Dropbox. At first, MIT student Drew Houston (Shontell, 2013) developed the idea for Dropbox. In the early days of the company, Dropbox wanted to gauge interest and get early user feedback on the file hosting and sharing product before investing heavily in developing the full application. Instead of building out the entire Dropbox application from the start, Houston created a simple video demonstrating the core file syncing and sharing functionality of what they envisioned Dropbox to be (https://www.youtube.com/watch?v=7QmCUDHpNzE) in 2007. The video showed how Dropbox would work and the key problems it would solve for users. It highlighted features like automatic file syncing across devices, sharing files with others, and accessing files from anywhere. Dropbox then shared this video and used it to capture email addresses from interested users who wanted to be notified when the product launched. This allowed Dropbox to gauge real customer interest and demand before building the full product, get early user feedback to inform the product roadmap, and build a waitlist of potential customers eager to try the product. All of this was done without having to invest significant time and resources into developing the complete Dropbox application upfront. The video served as an effective MVP that helped validate the idea before the team began full-scale product development.

12.5.2 PROTOTYPE APP

One company that effectively leveraged a prototype app was Snapchat. In Snapchat's early days, Reggie Brown came up with the conceptual idea of creating an application for sending disappearing picture messages (Ritti, 2014). He discussed with Evan Spiegel, and later Spiegel used the idea and presented the prototype concept in one of his design classes (Colao, 2012). Later, they approached Bobby Murphy to write the code used to build the app, and eventually, Snapchat was incorporated.

12.5.3 LANDING PAGE

A landing page is a standalone web page designed for a specific marketing campaign or goal. It is where visitors "land" after clicking on an ad, email link, or social media post. One such example is Airbnb (or Airbedandbreakfast as it was known back in 2007). Brian Chesky and Joe Gebbia came up with the idea to convert a living room into a bed and breakfast place by putting an air mattress. Airbedandbreakfast.com officially launched on 11 August 2008, offered short-term living quarters, and breakfast. In the summer of 2008, they had their first three customers during the Industrial Design Conference because hotels were overbooked, and travelers had a hard time finding lodging in the city (Gallagher, 2017). In February 2008, Nathan Blecharczyk joined as the third co-founder of the new venture, and Airbnb was founded in 2008.

12.5.4 WIZARD OF Oz

The Wizard of Oz method is a research and prototyping technique where users interact with a system, believing the system is fully functional, but in reality, key components are operated manually by a human behind the scenes. This approach is often used to test ideas quickly and gather user feedback before investing in full development. Nick Swinmurn founded Zappos in 1999 (Göcke & Weninger, 2021) with the vision of creating an online shoe store offering a vast selection of footwear. To bring this idea to life, he launched the website Shoestore.com. However, purchasing and stocking thousands of shoes requires a significant investment, posing a financial risk if the idea failed. To validate his concept with minimal upfront costs, Swinmurn used the Wizard of Oz method (Ries, 2011). Instead of maintaining an inventory, he listed shoes online and, when a customer placed an order, he would personally purchase the shoes from local stores and ship them. This approach allowed him to test demand and refine the business model before fully committing to a large-scale operation.

12.5.5 PREORDER METHODS

The Tesla Cybertruck was unveiled in November 2019 as the first all-electric pickup truck, featuring a distinctive, angular, stainless steel exoskeleton design. Tesla opened the landing page test to preorder reservations for the vehicle (Stevenson et al., 2024). To place a preorder, customers were required to pay a $100 deposit (Reuters, 2019). It immediately generated a lot of interest, and 250,000 pre-orders were placed in five days. The preorder money was allocated toward various aspects of the manufacturing process, such as research and development, tooling, production line setup, materials procurement, and other associated costs.

12.5.6 CROWDFUNDING

Crowdfunding websites, such as Kickstarter (www.kickstarter.com), Indiegogo (www.indiegogo.com), GoFundMe (www.gofundme.com), startengine (www.startengine.com), Crowdfunder (www.crowdfunder.co.uk), Patreon (www.patreon.com), Fundable (www.fundable.com), and MightyCause (www.mightycause.com), provide platforms for inventor to launch their projects and connect with potential supporters and investors. Inventors use different methods to attract potential investors, including prototypes. For example, Oculus Rift presented a prototype for a virtual reality headset aimed at gamers. The campaign featured a basic working model demonstrating immersive gaming experiences, showcasing its potential without a fully refined product.

12.6 STARTUP EVALUATION

The value proposition and MVP/MLP/MWP are always evaluated by potential customers using surveys, interviews, A/B testing, and other techniques discussed in Chapter 8. In addition to assessing potential customers, the startup is evaluated by the startup team, investors and other key stakeholders to determine its viability.

Thoroughly evaluating startups across a comprehensive set of key criteria is essential for their successful commercialization and securing support from investors, accelerators, and other support programs (Table 12.10). This rigorous assessment helps identify a startup's market viability, product–market fit, and potential for scalable, sustainable growth. By assessing factors like the market opportunity, product or service, business model, founding team, traction and milestones, financial projections, competitive landscape, risk assessment, scalability, and investment potential, stakeholders can make more informed decisions that increase the chances of long-term success.

For startups, this evaluation process provides valuable insights that can guide their decision-making. If the assessment reveals significant challenges or weaknesses, it may signal the need for the startup to pivot its strategy, product, or business model. Conversely, a strong evaluation can bolster the founders' confidence and commitment to persevere through the inevitable obstacles of building a successful venture.

For investors, accelerators, and support programs, the comprehensive evaluation is crucial for making informed funding and investment decisions. Startups that score highly across the key criteria are more likely to secure the necessary capital, resources, and mentorship to accelerate their growth and reach commercialization. This support system provides vital funding, expertise, and industry connections that can propel the startup toward success.

Ultimately, a thorough startup evaluation is a critical step in the journey toward successful commercialization and sustainable growth, benefiting both the startup and the broader entrepreneurial ecosystem. By rigorously assessing startups across a comprehensive set of criteria, founders, investors, and support programs can make more informed decisions that increase the chances of long-term success.

To evaluate a startup, ten criteria can be used (Table 12.10). The percentage weight of each criterion may vary depending on the industry and expectations of stakeholders. Hence, the weights can be adjusted such that the sum of all weights is 100 percent. This is just one way to vet startup projects. Different methods should be used to better understand the startup potential (Zhang et al., 2023). The potential value is useful for evaluating either to persevere, pivot or have an exit strategy (Berends et al., 2021).

12.6.1 PERSEVERE

Perseverance is critical for startup founders, as navigating the challenges and uncertainties of building a successful business requires unwavering determination. However, knowing when to persevere and when to pivot is a delicate balance that startup founders must constantly assess. Startups should

TABLE 12.10 Ten Key Criteria to Consider When Evaluating a Startup.

Key Criteria	Description	Weight (Σweights=1)	Rating	Weighted value
Market opportunity	Size of the total addressable market (TAM). Growth potential and market trends. Identification of customer needs.	15%	88	13.2
Business model	Viability and sustainability of the revenue streams. Scalability and adaptability of the business model. Competitive advantages and barriers to entry.	15%	77	11.55
Founding team	Relevant experience and expertise of the founders. Complementary skill sets and ability to execute. Track record of success and problem-solving skills.	10%	70	7
Traction and milestones	Key performance indicators (KPIs) and growth metrics. Achievement of critical milestones (e.g., product launches, partnerships, funding). Evidence of customer adoption and engagement.	10%	81	8.1
Financial projections	Reasonableness of financial assumptions and forecasts. Funding requirements and capital efficiency. Potential for profitability and long-term sustainability.	15%	73	10.95
Competitive landscape	Identification of key competitors and their strengths/weaknesses. Startup's positioning and competitive advantage. Potential for market disruption and industry transformation.	15%	70	10.50
Risk assessment	Identification of potential risks (e.g., regulatory, technological, market). Mitigation strategies and contingency plans. Potential impact of risks on the startup's success.	5%	82	4.1
Scalability and expansion	Ability to scale the business model and operations. Plans for geographic expansion and product diversification. Potential for international growth and market penetration.	5%	70	3.5

(*continued*)

TABLE 12.10 (Continued)

Key Criteria	Description	Weight (Σweights=1)	Rating	Weighted value
Regulatory and compliance issues	Identify any regulatory challenges the startup may face. Ensure the business model complies with relevant laws and regulations.	5%	50	2.5
Exit and investment potential	Valuation and investment opportunities. Potential for successful exits (e.g., acquisition, IPO). Risk-reward profile and alignment with investor objectives.	5%	82	4.1
		Total weighted value		75.5
Start-up potential				
0–50 Not satisfactory	51–60 Satisfactory 61–70 Very satisfactory	71–80 Good	81–90 Very good	90–100 Excellent

Note: The weighted value = weight* rating.

persevere based on key factors such as high startup potential value, a validated market need, demonstrated traction and growth, positive unit economics, and a committed and capable team.

12.6.2 PIVOT

Pivoting or shutting down the startup may be the best course of action if there are considerations such as unsatisfactory startup potential value, lack of product–market fit, unsustainable economics, dwindling resources, and founder burnout. In these types of situations, pivoting the business model or shutting down operations entirely, as difficult as it may be, can be a wise strategic move. It allows the startup to avoid further losses, preserve whatever value remains, and prevents the founders from wasting more time and money on losing proposition.

The key for startup founders is to have a clear and objective view of their company's prospects and performance. They must have the courage to make tough decisions when the data indicates an unsustainable path. An honest assessment and a timely exit can be beneficial than pursuing a venture with little chance of meaningful success.

12.6.3 EXIT STRATEGY

An exit strategy is a planned approach for how and when a startup founder or investor will liquidate their ownership stake and transition out of the business. Some exit strategies are:

- Acquisition: selling to a larger business.
- Initial public offering (IPO): Going public by listing shares on the stock market.
- Merger: combining with another company to form a stronger entity.
- Management buyout (MBO): selling ownership to the existing management team.
- Shutdown or liquidation: closing the business and selling off its assets if necessary.

A startup's exit strategy will depend on factors like the industry, growth stage, competitive landscape, and financial performance. A well-defined exit strategy helps entrepreneurs and investors maximize value while minimizing risks. Savvy founders will consider potential exit scenarios from the very beginning and structure the business accordingly. Selecting the right exit path is critical, as it determines how and when founders and investors can realize a return on their investments of time and money. Ultimately, a well-planned exit strategy is a hallmark of a thoughtfully built startup that has its eyes on the endgame from the very start. It is a key piece of the overall business plan that startup founders need to carefully consider.

12.7 INTELLECTUAL PROPERTY

Intellectual property (IP) is a critical asset for startups, as it can provide a crucial competitive advantage and help secure their long-term success (Smiler & Erbisch, 2004; Schriek, 2022). As startups develop innovative products, technologies, or business models, protecting their IP through various means is essential.

One of the primary forms of IP protection for startups is patents. By obtaining patents, startups can safeguard their unique inventions and innovations, preventing competitors from replicating their core offerings. This exclusivity can be invaluable in a crowded marketplace, allowing startups to differentiate themselves, maintain pricing power, and attract investors.

In addition to patents, startups should also consider registering trademarks for their brand names, logos, and other distinctive identifiers. Trademarks help establish brand identity, prevent consumer confusion, and deter potential infringement. Building a strong trademark portfolio can be a powerful tool in building brand equity and customer loyalty.

Trade secrets can also be an effective IP strategy for startups, particularly for protecting proprietary information, processes, or formulas that are not easily replicated. By implementing robust security measures and nondisclosure agreements, startups can maintain the confidentiality of their competitive advantages.

Copyright protection is another important aspect of IP management for startups, especially for those working in creative industries or developing software and digital assets. Copyright safeguards the expression of ideas, providing legal recourse against unauthorized use or reproduction.

Startups should also consider the strategic use of licensing agreements, which can generate revenue streams and expand the reach of their IP. By licensing their patented technologies, trademarks, or copyrighted works to other entities, startups can generate additional income and potentially access new markets.

Effective IP management is crucial for startups, as it protects their innovations, enhances their value proposition, secures investment, and deters competitors. By developing a comprehensive IP strategy early on, startups can build a strong foundation for long-term growth and success. It should be noted that AI is redefining IP and how it will be protected (Lim, 2018).

12.8 CONCLUSION

This chapter explored entrepreneurship and innovation, highlighting the distinction between entrepreneurship and intrapreneurship and the process of building a business model and value proposition using the business model canvas. A key focus is on achieving the MWP rather than just a MVP. The goal is not merely to create a functional bare-bones product but to develop one that impresses and excites customers. Reaching this level of impact requires continuous iteration, customer feedback, and refinement to deliver a truly compelling offering.

The chapter also emphasized the importance of thoroughly evaluating a startup across key criteria before deciding to get involved, whether as a founder, investor, or partner. Factors such as market opportunity, the founding team, financial projections, and competitive landscape can make or break a startup's chances of success.

Additionally, the role of intellectual property (IP), including patents, trademarks, and trade secrets, cannot be overlooked. Protecting innovations is essential for securing a competitive edge and ensuring long-term viability. Overall, we have explored the multifaceted nature of entrepreneurship, addressed the ups and downs and balanced the analytical with the creative aspects of building a business.

Ultimately, this book has aimed to show that innovation is not the exclusive domain of geniuses or startups; it is a mindset and a method that anyone can adopt. Whether you are designing a product, shaping a service, launching a venture, or improving a system from within, the principles of human-centered design, critical thinking, creativity, and collaboration remain essential. We hope this journey has equipped you with the tools to ask better questions, engage diverse perspectives, and translate ideas into meaningful outcomes. As you move forward, remember: innovation is not a destination but a continuous process of learning, adapting, and co-creating a better future, with others, and for others, for the betterment of humanity.

> Entrepreneurs don't finish when we are tired. We finish when we are done.
>
> —Robert T. Kiyosaki

REFERENCES

Aminova, M., & Marchi, E. (2021). The role of innovation on start-up failure vs. its success. *International Journal of Business Ethics and Governance*, 41–72.

Åstebro, T., & Bernhardt, I. (2003). Start-up financing, owner characteristics, and survival. *Journal of Economics and Business*, 55(4), 303–319.

Baran, A., & Zhumabaeva, A. (2018). Intellectual property management in startups—problematic issues. *Engineering Management in Production and Services*, 10(2), 66–74.

Basu, P., & Nair, S. K. (2015). Analyzing operational risk-reward trade-offs for start-ups. *European Journal of Operational Research*, 247(2), 596–609.

Bauer, D., Junge, S., & Reif, T. (2023). May the resources be with you: a systematic review and framework of startup funding options. *Management Review Quarterly*, 1–32.

Berends, H., Van Burg, E., & Garud, R. (2021). Pivoting or persevering with venture ideas: Recalibrating temporal commitments. *Journal of Business Venturing*, 36(4), 106126.

Bosma, N. S., Stam, F. C., & Wennekers, A. R. M. (2010). *Intrapreneurship: An International Study*. EIM-Research reports (www.eim.nl), The Netherlands.

Carvalho, J., & Jonker, J. (2015). Creating a balanced value proposition—Exploring the advanced business creation model. *Journal of Applied Management and Entrepreneurship*, 20, 49–64.

Colao, J.J. (November 27, 2012). Snapchat: the biggest no-revenue mobile app since instagram. *Forbes*. Archived from the original on 28 November 2015. Retrieved 26 January 2025. https://www.forbes.com/sites/jjcolao/2012/11/27/snapchat-the-biggest-no-revenue-mobile-app-since-instagram/

de Haaff, B. (2015) How to create the minimum lovable product, *The Huffington Post*, Available (accessed 16 January 2016), www.huffpost.com/entry/how-to-create-the-minimum_b_7652844

de Haaff, B. (2017). *Lovability: How to Build a Business That People Love and Be Happy Doing It*. Greenleaf Book Group.

Donckels, R. (1989). New entrepreneurship: lessons from the past, perspectives for the future. *Entrepreneurship & Regional Development*, 1(1), 75–84.

Draper III, W. H. (2011). *The Startup Game: Inside the Partnership Between Venture Capitalists and Entrepreneurs*. St. Martin's Press.

Frederiksen, D. L., & Brem, A. (2017). How do entrepreneurs think they create value? A scientific reflection of Eric Ries' Lean Startup approach. *International Entrepreneurship and Management Journal*, 13, 169–189.

Gallagher, L. (2017). *The Airbnb Story: How Three Guys Disrupted an Industry, Made Billions of Dollars... and Plenty of Enemies*. Random House.

Ghory, S., Obeidat, B., & Masa'deh, R. E. (2023). Measuring café lovability using Google's HEART and understanding the roles of usability, sustainability innovation, and innovation cocreation in café lovability. *Sustainability*, 15(9), 7241.

Göcke, L., & Weninger, R. (2021). Business model development and validation in digital entrepreneurship. *Digital Entrepreneurship*, 71.

Hashim, W., Mkpojiogu, E. O., Hussain, A., & Abdul-Aziz, S. N. (2021). A product pain-pleasure framework for software product design in the usability and user experience domains. *Webology*, 18(SI01), 1–31.

Hills, G. E. (1985). Market analysis in the business plan: venture capitalists' perceptions. *Journal of Small Business Management* (pre-1986), 23(000001), 38.

Hisrich, R. D. (1990). Entrepreneurship/intrapreneurship. *American Psychologist*, 45(2), 209.

Hisrich, R. D., Peters, M. P., & Shepherd, D. A. (2017). *Entrepreneurship*. McGraw-Hill Education.

Hokkanen, L. (2017). From minimum viable to maximum lovable: developing a user experience strategy model for software startups, Doctor of Science in Technology, Tampere University of Technology.

James, K. (2002). *Improving Time to Profit: Customer Focused Strategies for Marketing and Sales*. iUniverse.

Kamps, H. J. (2020). *Pitch Perfect*. Apress.

Knossalla, C. E., & Carbon, C. C. (2023). Neither entrepreneurship nor intrapreneurship: a review of how to become an innovative split-off start-up. *Frontiers in Sociology*, 8, 1267706.

Lenarduzzi, V., & Taibi, D. (2016). MVP explained: A systematic mapping study on the definitions of minimal viable product. In 2016 42th Euromicro Conference on Software Engineering and Advanced Applications (SEAA) (pp. 112–119). IEEE.

Lim, D. (2018). AI & IP: innovation & creativity in an age of accelerated change. *Akron Law Review*, 52, 813.

Makadon, H. J., Bharucha, F., Gavin, M., Oliveira, J., & Wietecha, M. (2010). Value management: optimizing quality, service, and cost. *Journal for Healthcare Quality (JHQ)*, 32(1), 29–34.

McLeod, S. (2007). Maslow's hierarchy of needs. Simply Psychology, 1, 1–18.

Moon, S., & Kamakura, W. A. (2017). A picture is worth a thousand words: Translating product reviews into a product positioning map. *International Journal of Research in Marketing*, 34(1), 265–285.

Olson, P. D. (1985). Entrepreneurship: Process and abilities. *American Journal of Small Business*, 10(1), 25–31.

Osterwalder, A., Pigneur, Y., Bernarda, G., & Smith, A. (2015). *Value Proposition Design: How to Create Products and Services Customers Want*. Wiley.

Osterwalder, A., & Pigneur, Y. (2010). *Business Model Generation: A Handbook for Visionaries, Game Changers, and Challengers* (Vol. 1). Wiley.

Parker, S. C. (2011). Intrapreneurship or entrepreneurship? *Journal of Business Venturing*, 26(1), 19–34.

Ries, E. (2011). *The Lean Startup: How Today's Entrepreneurs Use Continuous Innovation to Create Radically Successful Businesses*. Crown Currency.

Rogers, E. M. (2003). Diffusion of Innovations (5th Ed). Free press.

Rogers, D. S., Lambert, D. M., & Knemeyer, A. M. (2004). The product development and commercialization process. *International Journal of Logistics Management*, 15(1), 43–56.

Reuters (2019). Musk suggests Tesla has 250,000 orders for Cybertruck. https://www.reuters.com/article/business/musk-suggests-tesla-has-250-000-orders-for-cybertruck-idUSL4N2871BI//. Retrieved 27 November 2024.

Ryan, M., Popa, E. O., Blok, V., Declich, A., Berliri, M., Alfonsi, A., & Veloudis, S. (2023). A model of social responsibility for start-ups: developing a cross-fertilisation of responsible innovation, the lean start-up approach, and the quadruple helix approach. *Journal of Responsible Innovation*, 10(1), 2264615.

Schriek, U. (2022). Basics and development of an IP strategy for start-ups. In *Intellectual Property Strategies for Start-ups* (pp. 33–52). Edward Elgar Publishing.

Sethi, A. (2016). *From Science to Startup*. Springer International Publishing.

Shontell, A. (2013) This is Drew Houston's 2007 Y Combinator application for A company that's now worth $4 billion, Dropbox, *Business Insider*, 13 September 2013, www.businessinsider.com/dropbox-y-combinator-application-from-2007-by-drew-houston-2013-9.

Shontell, A. (2014). Snapchat and Reggie Brown resolve dispute, *Business Wire*. (www.businessinsider.com/snapchat-settles-the-lawsuit-with-the-ousted-co-founder-and-fraternity-brother-reggie-brown-2014-9). Retrieved 9 June 2024.

Silva, D. S., Ghezzi, A., Aguiar, R. B. D., Cortimiglia, M. N., & ten Caten, C. S. (2021). Lean startup for opportunity exploitation: adoption constraints and strategies in technology new ventures. *International Journal of Entrepreneurial Behavior & Research*, 27(4), 944–969.

Smiler, B., & Erbisch, F. H. (2004). Introduction to intellectual property. *Intellectual Property Rights in Agricultural Biotechnology*, 1–21.

Stevenson, R., Burnell, D., & Fisher, G. (2024). The minimum viable product (MVP): theory and practice. *Journal of Management*, 01492063241227154.

Stevenson, H. H. (1983). A Perspective on Entrepreneurship (Vol. 13). Harvard Business School.

Thomson, P. (2013). *Tickle: Digital Marketing for Technology Companies: How to Land a Big Fish Using Social Media*. Peter Thomson.

Thulasipriya, B., & Rodrigues, L. J. (2023). Developing marketing strategies for startup entrepreneurs: A literature review. *IUP Journal of Entrepreneurship Development*, 20(3), 5–18.

Vlies, S. V. D. (2023). Assessing and advancing startup potential: A novel tool for the unlock_ startup incubator, Master's thesis, University of Utrecht.

Westrenius, A., & Barnes, L. (2015). Managing complex business relationships: Small business and stakeholder salience. *Journal of Developing Areas*, 49(6), 481–488.

Zhang, L., Yang, X., Zhu, S., & Xia, Z. (2023). business model innovation and performance of startups: the moderating role of external legitimacy. *Sustainability*, 15(6), 5351.

Index

D

G

H

T

Made in the USA
Las Vegas, NV
19 September 2026

ff4f6e93-bad4-4f4c-b32d-ca2d447d4257R01